D1787056

# STRUCTURE-BASED EDITORS AND ENVIRONMENTS

## Computers and People Series

*Edited by*

*BR GAINES and A MONK*

*Monographs*

Communicating with Microcomputers. An introduction to the technology of man-computer communication, *IH Witten* 1980
The Computer in Experimental Psychology, *R Bird* 1981
Principles of Computer Speech, *IH Witten* 1982
Cognitive Psychology of Planning, *J-M Hoc* 1988
Formal Methods for Interactive Systems, *A Dix* 1991
Human Reliability Analysis: Context and Control, *E Hollnagel* 1993

*Edited Works*

Computing Skills and the User Interface, *MJ Coombs and JL Alty (eds)* 1981
Fuzzy Reasoning and Its Applications, *EH Mamdani and BR Gaines (eds)* 1981
Intelligent Tutoring Systems, *D Sleeman and JS Brown (eds)* 1982 (1986 paperback)
Designing for Human-Computer Communication, *ME Sime and MJ Coombs (eds)* 1983
The Psychology of Computer Use, *TRG Green, SJ Payne and GC van der Veer (eds)* 1983
Fundamentals of Human–Computer Interaction, *A Monk (ed)* 1984, 1985
Working with Computers: Theory versus Outcome, *GC van der Veer, TRG Green, J-M Hoc and D Murray (eds)* 1988
Cognitive Engineering in Complex Dynamic Worlds, *E Hollnagel, G Mancini and DD Woods (eds)* 1988
Computers and Conversation, *P Luff, N. Gilbert and D Frohlich (eds)* 1990
Adaptive User Interfaces, *D Browne, P Totterdell and M Norman (eds)* 1990
Human–Computer Interaction and Complex Systems, *GRS Weir and JL Alty (eds)* 1991
Computer-supported Cooperative Work and Groupware, *Saul Greenberg (ed)* 1991
The Separable User Interface, *EA Edmonds (ed)* 1992
Requirements Engineering: Social and Technical Issues, *M Jirotka and JA Goguen (eds)* 1994
Perspectives on HCI: Diverse Approaches, *AF Monk and GN Gilbert (eds)* 1995
Information Superhighways: Multimedia Users and Futures, *SJ Emmott (ed)* 1995
Structure-based Editors and Environments, *G Szwillus and L Neal (eds)* 1996

*Practical Texts*

Effective Color Displays: Theory and Practice, *D Tavis* 1991
Understanding Interfaces: A Handbook of Human–Computer Dialogue, *MW Lansdale and TR Ormerod* 1994 (1995 paperback)

*EACE Publications*
(Consulting Editors: *Y WAERN and J-M HOC*)

Cognitive Ergonomics, *P Falzon (ed)* 1990
Psychology of Programming, *J-M Hoc, TRG Green R Samurcay and D Gilmore (eds)* 1990

# STRUCTURE-BASED EDITORS AND ENVIRONMENTS

Edited by

Gerd Szwillus

*Department of Computer Science,*
*University of Paderborn,*
*Paderborn,*
*Germany*

Lisa Neal

*EDS Technical Consulting Program*
*Lexington,*
*MA, USA*

ACADEMIC PRESS

*Harcourt Brace & Company, Publishers*
San Diego London New York Boston
Sydney Tokyo Toronto

This book is printed on acid-free paper



Academic Press, Inc.
525 B Street, Suite 1900, San Diego, California 92101–4495, USA

Academic Press Limited
24–28 Oval Road, London NW1 7DX, UK

ISBN 0–12–681890–8

Library of Congress Cataloging-in-Publication Data

A catalogue record for this book is available from the British Library

Typeset by J&L Composition Ltd, Filey, North Yorkshire
Printed in Great Britain by Cambridge University Press, Cambridge

96 97 98 99 00 01 EB 9 8 7 6 5 4 3 2 1

To Christine Szwillus and David Neal

# Contents

**Introduction** 1
*Gerd Szwillus* (Computer Science Department, University of Paderborn, Warburger Straße 100, D–33095 Paderborn, Germany)
*Lisa Neal* (EDS Technical Consulting Program, Lexington, MA 02137, USA)

PART I TRADITIONAL STRUCTURE EDITORS 17

1 **Coherent user interfaces for language-based editing systems** 19
*Michael L. Van De Vanter* (Sun Microsystems Laboratories, 2550 Garcia Avenue, Mountain View, CA 94043–1100, USA)
*Susan L. Graham* (Computer Science Division (EECS), University of California, Berkeley, CA 94720, USA)
*Robert A. Ballance* (Robert Ballance & Associates, Inc., 2400 Rio Grande Blvd. NW, Suite 1–247, Alburquerque, NM 87104, USA)

2 **Using Mjølner Orm as a structure-based meta environment** 71
*Sten Minör* (Q-Labs AB, Ideon Research Park, S-223 70 Lund, Sweden) and
*Boris Magnusson* (Department of Computer Science, Lund University, POB 117, S-221 00, Lund, Sweden)

3 **An OCCAM programming environment** 107
*M. Filali, Y. Jamoussi, A. Knani and J. C. Maurize* (IRIT, Université Paul Sabatier, 118 Route de Narbonne, F-31062 Toulouse Cedex, France)

PART II EVOLUTION OF STRUCTURE-BASED ENVIRONMENTS 133

4 **Automated customization of structure editors** 135
*Barbara Staudt Lerner* (Computer Science Department, University of Massachusetts, Amherst, MA 01003, USA)

5 **Support for software design, development and reuse through an example-based environment** 185
*Lisa Neal* (EDS Technical Consulting Program, Lexington, MA 02137, USA)

**6 Experimental data on the usefulness of a structured editor** 193
*Pierre-N. Robillard, Mario Simoneau, Jean Mayrand and Daniel Coupal* (Software Engineering Laboratory, École Polytechnique de Montréal, CP 6079, SUC centre-ville, Montreal, Qc., Canada, H3C 3A7)

PART III GRAPHICAL STRUCTURE EDITORS 205

**7 A uniform graphical view of the program construction process: GRIPSE** 207
*K. Halewood* (Department of Engineering, University of Aberdeen, Aberdeen, UK) and
*M. R. Woodward* (Department of Computer Science, University of Liverpool, PO Box 147, Liverpool, L69 3BX, UK)

**8 User interface definition based on graphical structure editors** 253
*Gerd Szwillus* (Computer Science Department, University of Paderborn, Warburger Straße 100, D-33095 Paderborn, Germany)

**9 A multimodal syntax-directed graph editor** 277
*Edwin Bos* (Nijmegen Institute for Cognition and Information, PO Box 9104, 6500 HE Nijmegen, The Netherlands)

PART IV EDITORS OF ALTERNATIVE STRUCTURES 299

**10 Formaliser – tool support for formal notations** 301
*Roy MacLean and David Brazier* (Logica UK Ltd., Betjeman House, 104 Hills Road, Cambridge, CB2 1LQ, UK)

**11 Environment for document structure recognition** 327
*Nenad Marovac* (Department of Mathematical Sciences, San Diego State University, San Diego, CA 92182, USA)

**12 Canae's structure-based editor components: rationale, description and field-tested applications** 339
*H. Tarumi, J. Rekimoto, M. Sugai, G. Yamazaki, T. Mori and C. Akiguchi* (NEC Corporation, 2–11–5 Shibaura, Minato-ku, Tokyo 108, Japan)

**Index** 369

# Introduction

*Gerd Szwillus and Lisa Neal*

Structure-based editors and environments have evolved from the earliest, highly restrictive syntax-directed editors to graphical structure editors to environments supporting new languages, methodologies, representations and views; this evolution has taken place primarily in the research community over the past 25 years. These systems have varied considerably in mechanisms for entry, modification and display, and the treatment of errors.

Structure-based editors were initially intended as teaching vehicles for novice programmers, but as the systems increased their sophistication and capabilities, their intended audiences moved to skilled programmers involved in the design, development and maintenance of complex systems. In order to provide greater and more sophisticated levels of support, the functionality and interfaces of structure-based editors and environments have changed far more significantly than have the internal representations. Hence, our focus in this book is on issues relating to human–computer interaction including design and evaluation.

## The focus on structures

Structures exist in a variety of forms; examples of structure types are trees or hierarchies, graphs, lists and tables. Even spoken and written natural language has a structure, which is parsable and can therefore be tested or decomposed into structural units. Structures show up as dynamically changing hierarchical structures within text processors, spreadsheets or programming tools – in these tools semantic structure is imposed on the documents under development. We find fixed hierarchies in special applications such as the structure of electronic mail documents within mail software; graphs are explicitly used within hypertext systems through the link concept, and more implicitly within text processors with the power of annotations, and references to other text parts (e.g. figures, tables or entries within an index) can be included. These structures are used to give the user access to other semantic structures within a textual, graphical or symbolic document by providing special structure-based functionality, views or transformation operations. Examples are outlining functions and automatic generation of table of contents and indices. Frequently structure information is used for providing more flexible navigation through a structure, as opposed to movement along the canonical, trivial structure of a document. This is especially evident within hypertext systems, which consider structure-based navigation as a central concept.

Structure-based editors or structure-based systems have therefore found their way into a large number and a great variety of application systems without their inherent basis on structure being explicitly mentioned. In particular, outside of the research community there has been an enormous influence by structure-based editors and environments in commercial products. Their impact is obvious in mechanisms for program text entry and retrieval and in editing tools of more general structures that support different phases of the software life-cycle. The latter includes CASE tools and document preparation systems and, more recently, editors and environments for HTML, VRML and Java for use on the World Wide Web.

Current structure-based editors and environments evolved from the syntax-directed editors developed over 25 years ago. The original syntax-directed editors were primarily for the entry of programs and were oriented toward novice programmers. The interfaces were non-graphical and the functionality limited. Some well-known examples include EMILY, the Cornell Program Synthesizer and POE. Later editors varied in a number of ways: entry, modification and display of text, and the internal representation, traversal and navigation, treatment of semantics, treatment of errors and graphical capabilities. They were oriented toward a wide range of users and operated on a number of languages, especially as editor and environment generators were developed. Examples of these editors include GENIE, Pecan and numerous commercial environments; examples of generators include Gandalf and the Cornell Synthesizer Generator.

## How the traditional approach to programming has evolved

The traditional approach to program development is to use a non-integrated set of tools including a compiler, a text editor and a debugger. Neither tools to aid in the design, specification and development progesses nor tools to aid learning to program are available in most traditional programming environments. Such environments do not encourage a structured approach to program development, and the coding process is difficult to learn, time-consuming and error prone. Programming and software development environments can provide an integrated and structured approach to program development and improve all phases of the software lifecycle. This is especially important with languages of greater complexity than Basic and Pascal; in particular, with the object-oriented and visual programming languages where graphical representations are more necessary for comprehension or, in some cases, essential because of the graphical nature of the language.

Text editors are one of the most important components of any software development environment as they are used to enter and modify all text. Syntax-directed editors, which are also known as language-based editors or structured editors, are more sophisticated tools which are based on an underlying knowledge of a programming language. While all text is viewed as equivalent in a text editor (i.e. there is no meaning attached to specific tokens), a syntax-directed editor understands the tokens in a program and provides capabilities such as automatic pretty-printing and the prevention or detection of errors. Syntax-directed editors therefore view programs in a more meaningful way than do text editors and have the capability to provide a more visual or graphical representation since the meaning and underlying structure are recognized.

The concepts of syntax-directed editors can be generalized to editors for tree-structured texts, extending the use of these systems to many roles within the software production process. Since many documents are tree-structured, an editor supporting these structures can ease development. The use of directed graphs as an internal representation structure allows a more powerful treatment of semantics. The inclusion of graphics produces a graphical depiction of the internal structure, which can be very informative as a supplement to a formatted textual representation. An advantage of a graphical approach is that while programming languages are one-dimensional, humans are good at processing two and three dimensions (Myers, 1986). Recent universal or generative graphical editor systems allow the production of graphical structure editors based on a language specification. The support of graphical languages addresses multiple needs within the software engineering lifecycle, such as specification and documentation.

In this introduction, we discuss the different approaches taken by syntax-directed and graphical structure editors, and examine the current state of the technology. The lack of success of syntax-directed editors is also discussed, along with the directions that syntax-directed and graphical structure editors are taking. The chapters in this book further discuss the recent approaches taken. While many of these tools are primarily research prototypes, their influence in the commercial world has already been felt and is outlined here.

## Before structure-based editors: text editors

Text editors are multipurpose tools, which a user interacts with in terms of characters, words, lines or regions of text. Traditional line editors allow the user to operate on a single line of text at a time, while full-screen editors allow the user to view a screen of text and move and edit within this window. Editors vary enormously in their complexity and in their design; for example, they can be moded or modeless.

Text editors have no knowledge of the format or the meaning of the text they manipulate. They neither provide help on a conceptual level for the programmer nor aid in learning good programming methodologies or a programming language. Other tools are necessary for the development of software because a text editor has no knowledge of a language; for instance, compilers are often used merely to obtain error diagnostics, rather than to generate machine code, and pretty-printers are used to make the syntactic structure and flow of control clear in a program, aiding visualization.

## The first structure-based editors: syntax-directed editors

Syntax-directed editors are editors that are based on an underlying knowledge of a programming language; this knowledge translates into a structural focus which is intended to aid novice programmers in learning, as they are freed from remembering the details of a language. Programmers at all levels of expertise can benefit from the typing time saved, the formatting, the error detection and, most significantly, the ability to conceptualize at a higher level of abstraction rather than focusing on the language details. Since the user interacts with the editor in terms of language constructs, the means of expression is, in theory, closer to the programmer's understanding of the task. An environment should conform to the programmer, rather than force a programmer to make his or her concepts of how a system should work conform to an environment (Reiss, 1986). Syntax-directed editing can, therefore, potentially improve the software development process. However, their success has been limited for many reasons, the most widely accepted reason being that the advantages gained through use of a syntax-directed editor are not sufficient to outweigh the cost of the technology from the increased processor complexity and the added difficulty for users of learning a more complex tool (Lang, 1986).

Syntax-directed editors have varied enormously, with their only common feature being knowledge of a programming language and use of this knowledge to produce formatted and correct code. The approaches taken by syntax-directed editors range from the purely textual to the purely syntactic. In the former case, text entry is the same as with a text editor except that incremental parsing of the code provides automatic pretty-printing and error detection. In the latter case, programs are built by the repeated expansion of templates. Templates are predefined, pretty-printed patterns of code which consist of keywords, punctuation and non-terminals. The non-terminals are placeholders that the user fills in with either templates or typed code. The most restrictive editors force the programmer to use templates for all language constructs, directly typing only identifiers and constants. Templates can be inserted only at syntactically valid points in the program, or they can also be deleted or transformed, making syntax analysis by the editor unnecessary. This approach to editing is often used in a teaching environment, where its restrictiveness is arguably advantageous in teaching a programming language and a structured approach to program development.

Most syntax-directed editors fall in between the purely textual or purely structural approaches to editing. Some editors merge the two approaches by classifying language constructs based on how they are edited, i.e. by template insertion or typing. Each class is then treated consistently by the editor, so that templates can never be edited as text, and constructs for which templates exist can never be typed. The awkwardness of template use for some language constructs is avoided; for example, while "i" or "if" identifies the template for an "if statement," many expressions can only be identified by an operator that appears in the middle of the construct. This is also a problem for assignment statements.

Some syntax-directed editors allow both templates and typed code to be used for all language constructs by allowing the programmer to use special function keys to expand non-terminals into their right-hand-side productions, or to type in text directly. This allows the user to switch between the two methods of entering program text, using whichever style is more natural for the programmer and the type of text being added.

Templates are the most common means of entering language constructs in a syntax-directed editor and are invoked by special function keys, by pointing to a palette with a mouse or by typing a unique abbreviation of the first token in the construct. While new programmers have little familiarity with or loyalty to an editor, more experienced programmers are usually very comfortable with a text editor. Just as people rarely switch editors, there is little incentive for a programmer who is comfortable with program entry using a text editor to adjust to a new text editor and a new, often rigidly structured, means of entering a program (which is certainly what the template-based approach is).

Since the coding phase is a small fraction of software development, it is important that an editor facilitates the modification of text, especially outside of a teaching environment. Many structural editors allow code transformations where, for example, the expression and statements from an "if statement" become the expression and statements in a "while statement." It is often simpler to transform code by editing the textual representation, rather than the underlying structure.

With all syntax-directed editors, the user operates on and navigates through the internal representation of the code, while seeing the textual representation. The internal representation is usually an abstract syntax tree, which is unparsed to present a textual display. Some syntax-directed editors allow cursor movement based on the textual display or the language syntax; however, purely structural editors only allow cursor movement based on a traversal of the tree.

Because programs are heterogeneous, they need a display more complex than that of text (which is generally homogeneous), especially when a program is too large to fit on a screen. Many editors, therefore, use techniques such as ellipsis and holophrasting to display the frame of a program with the details abbreviated, to display the text to a depth that is either a function of the program size and the available space or is user-specified, or to hide comments. The user has a larger window onto a program than with normal display techniques, and these methods preserve the structured approach of the syntax-directed editors by displaying the higher level structure of a program, essentially allowing the program to be viewed at a different level of granularity.

A generalized fish-eye view is a viewing strategy based on an analogy to a wide angle or a fish-eye lens (Furnas, 1986), which has the potential of providing a better interface between an internal representation (a rooted tree structure) and an external display. The fish-eye view trades superfluous detail around the focus of attention for more remote, but higher level, contextual information; screen displays are calculated using a degree of interest function that computes how interested the user will be in seeing a structure, where interest increases with importance and decreases with distance from the current focus of attention. This approach is applicable for displaying large programs or any hierarchically organized text but has not been used by syntax-directed editors (which have not typically taken "a priori" importance into consideration). Multifocus fish-eye views could also be useful for the editing process, during which the focus of attention frequently shifts, for instance, between the declaration of an object and its use.

Various screen layouts are used by different syntax-directed editors. Many editors use the top or bottom few lines of the screen for commands or the display of syntactically valid template names. Sometimes multiple windows are used to display different parts of views of a large program, and the programmer can move between the windows, having, for example, a window for declarations and windows for different procedures.

Another use of windows is to provide diagnostic information to the user. Syntax-directed editors that allow incorrect programs to be entered display error messages to the user or use highlighting or reverse-video to mark the erroneous code. Temporary nodes are used in the internal tree, so that structural correctness is maintained, and they are also used to allow the user to delay filling in a non-terminal in a template. Syntax-directed editors provide varying degrees of semantic checking. Errors such as undeclared variables or incompatible type assignments can be detected.

## The usefulness of the first generation of structure-based editors

The basic concept of syntax-directed editors is applicable for a wide range of tasks, in fact, for any editing which emphasizes the underlying structure of the text. Syntax-directed editors also provide unifying interfaces to programming environments. The integration of the tools within a programming environment and the fluid interactions between them provide an integrated environment for software development. There exists systems for building programming environments from attributed language descriptions; other syntax-directed editors and programming environments are language-independent or are created from generators. They are generated from specifications that are extended grammars for a language.

While a programming environment integrates a set of tools, software development environments integrate the software development process. They are concerned with program development through its entire lifecycle, supporting conceptualization, specification, implementation, validation, verification, maintenance and management of software. A syntax-directed editor provides the basis for such a system, with code specifications edited through stepwise refinements into programs. Some syntax-directed editors and programming environments provide a program design language, so that an initial design is entered and each statement of the design is expanded either into a more detailed design or into source code. Syntax-directed editors encourage a structured approach to program development. The user conceptualizes a problem at a higher level than with traditional editors and development environments. Some editors focus on a class of users, while others encompass a wide range of users. Many syntax-directed editors have features that aid in the acquisition of knowledge and expertise about a language.

The programming language that a system is for is important in terms of the likely users and the applications, and in terms of the type of editing that makes the language easier to learn and use and takes advantage of its capabilities. For example, different editors would be needed for Ada and Lisp if the editors are to fully exploit their knowledge of the language and maximize their capabilities. An editor for Ada, for instance, would benefit from a template capability, because of Ada's emphasis on readability over writability, and from a multiple window capability, because of Ada's emphasis on modularity. It would also have a greater concern with providing teaching tools and help capabilities, because the language is new. A Lisp editor, however, would be dealing with a different set of issues, users and applications. These distinctions become even more apparent with visual languages and object-oriented languages because of the complexity and the even clearer advantages of more graphical representations and displays.

While there seems to be a current trend toward syntax-directed editors (which take a textual approach) supplying the user with little more than formatting, additional approaches are possible. Example-based programming (Neal, 1989) augments a syntax-directed editor with a window for displaying examples of programs. It is oriented towards novice programmers who often do not understand the terminology used in templates, and it allows them to program using examples as visual aids or to fully or partially copy into programs. The example programs can be used as examples of language constructs, thus providing syntactic information through instantiations of templates, or as examples of algorithms or programs. This concept has been carried further in some of the commercially available PC programming environments.

## The generalization of structure-based editors

The basic concept of syntax-directed editing has been generalized in two directions. The first drops the binding of the editor to a fixed context-free grammar. This binding can be seen as a restriction of the editor's generality, which makes sense when a specific language has to be supported. However, the concepts of syntax-directed editing can be exploited without making use of a fixed underlying grammar, leading to editors for tree-structured text. The second generalization uses directed graphs instead of trees as the internal representation structure. This gives the editor more power for treating the semantics of a language.

Tree structures exist in all text, not just in programs. An illustration of this is the documentation for a modular program, in which the tree structure of the modules implies an equivalent structure of the documentation; the single module description again may be sub-structured into EXPORT-part, IMPORT-part, RESTRICTION-part, etc. Another example is books, which can be seen as being structured into head and text, the head into title, author and publisher, the text into chapters, the chapters into paragraphs, and so on to the word level.

As the advantages of tree-based editing operations became accepted, editors based on tree structures were developed that incorporated ideas from syntax-directed editing without explicitly or implicitly feeding a grammar into the editor. This concept was shown to work well on such different things as Lisp programs, SNOBOL debugging structures and representations of simple graphic diagrams.

Other tools have been based on or influenced by tree structures (e.g. utilities such as outline programs and word processing systems). These tools do not usually force the user to use the structuring mechanism; for instance, headers may be entered as plain text. However, if the system is told explicitly where headers are placed and what their relative levels are, the system allows tree-based operations. Features like these aid in the structuring of text and support top-down document production.

The descriptive power of syntax-directed structure editors is restricted to what is expressible by context-free syntax. The addition of attribution to the internal tree representation is one way of overcoming this problem. Other systems use directed graphs as the internal representation, which produces more problems for the system implementation but allows very elegant and natural specifications of semantic relations within the edited structure.

For specifying graphs, graph grammars are frequently used. Graph grammars are a natural generalization of the string grammar concept; the symbols handled by the grammar are not character strings but are directed graphs that are linked to higher structures by the use of graph production rules. The resulting complete directed graph is used as the internal representation of the editing object throughout the environment. Different tools can operate on this representation using their own special-purpose edges and attributes. Hence, the structure editor integrated into such a system serves as a uniform, general-purpose interface to the entire system. The use of graphs instead of trees allows all semantic relations to be stored and handled as edges in the internally represented graph.

## Graphical structure editors

Graphical languages that are based on directed graphs are used to describe discrete structures, expressing concepts like connectivity, hierarchy and "is-part-of" relations geometrically. Graphics can more clearly express objects and their relationships, and graphical structure editors used to specify designs can aid greatly in the design and implementation processes. Flow charts, for instance, have long been used to represent any hierarchically organized set of instructions, such as the control flow in a program (Miller, 1962). Other examples include:

- Nassi-Schneiderman diagrams (which impose structure on program control) breaking down the components of a program
- state diagrams and Petri-nets (which are used to describe hardware)
- SADT-diagrams
- data flow nets
- VLSI schematics
- PERT/CPM charts.

In software engineering, diagrams like these are widely used for documentation and specification, especially in the early design phases. Today, they have found their way into various CASE tools as parts of more general methodologies such as SA/SD (Structured Analysis, Structured Design), SADT or, more recently, for the specification of object-oriented systems, such as Booch diagrams.

Graphical structure editors support these languages and allow operations that are expressed in terms of the objects relevant to the language. For example, a graphical structure editor for Petri-nets is able to handle a "transition" object, as opposed to the single picture elements by which a transition may be represented. Just as systems such as the Synthesizer Generator are used to generate new syntax-directed editors for new or modified languages, universal or generative editor systems can readily deal with new and modified graphical specification languages.

Graphical structure editors are very useful, especially within the field of software engineering. The large number and increasing success of industrially available CASE tools supporting visual notations for software design and specification proves this convincingly. Many graphical languages exist, and a flexible, easy to use tool for treating most of them under a uniform interface is very useful. Since there is much research on specification languages and little potential for standardization, there is a need for systems to experiment on design problems using specification languages. In general, it is useful to create and modify editors quickly and easily according to the user's needs, or to customize editors to users or application.

## Different approaches taken by graphical structure-based editors

Graphical structure editors, just like syntax-directed editors, are dedicated to a language. For a graphical structure editor, the graphical language is entered using one of two techniques: the editor may be a universal one that is driven by language-dependent tables containing the language-specification, or it may be generated for a graphical language from an appropriate language specification.

When the universal editor approach is used, one powerful editor is implemented that is able to treat different types of diagrams whose structure is based on directed graphs. The graphical language is described in a special specification language. Such an editor is usually customized using additional "style" information for the depiction of nodes and edges. The specification language and its expressive power are the crucial points of this type of system. It can be based on two separate parts, one specifying hierarchical substructures with traditional grammar techniques and the other adding non-context-free edges; alternatively it can be based on graph grammars that are on a more general level from the start. When the generative approach is taken, the program representing the graphical structure editor is created from a language specification. The main advantage is higher efficiency of the executing editor paid for through the loss of flexibility and the possibility to modify the system dynamically at runtime.

In both cases, the technique for specification of graphical languages is central for the systems. Whereas for textual languages established concepts exist and are in widespread use, this is not the case for graphical languages. There is a general agreement that directed graphs are the appropriate concept for representing the structure of graphics. Usually these graphs are typed, giving node types and edge types. Also, all systems use attribution, especially for attaching information about the graphical representations of nodes and edges.

A very basic difference between systems can be seen in the explicit distinction between a graphical language's objects and the links between them. Simpler systems map objects on the graph's nodes, and links to its edges. In more general systems connections are objects that may be depicted as lines or in a different way. This approach allows treatment of a larger class of graphical languages, as links may be expressed graphically very differently than by lines (such as by nesting, touching or covering).

More differences show up when dealing with the specification of the class of "valid" graphs underlying the language. Some systems do not include any structural semantic check; using a graph grammar allows all rules that can be expressed by graph grammar rules to be checked. Other systems allow the specification of assertions written in a mathematical, first order logic notation, or use path expressions, i.e. properties of graph traversals. Some systems make explicit use of a hierarchical sub-structure within the graph structure underlying a diagram. Hence, the directed graph contains an explicitly defined spanning tree. The context-free sub-structure eases the writing of a structure specification as it expresses a "natural" hierarchy that is found in many graphical languages.

Concerning graphical representation, systems differ in the degree to which they deal with automatic layout. Some systems allow simple rules about the elements' placement relative to each other; others use sophisticated automatic layout algorithms. Systems may or may not be able to deal with different layers of a graphical representation (allowing "opening" and "closing" of objects) or with different views of a given semantic structure.

There are different approaches to treating errors, which are dependent on the possibilities to express semantic conditions and rules on the graphical outlook. In most systems there are errors that are intolerable, usually concerning connectivity rules. Usually there is no problem handling structures that are only partially completed. Some conditions are tested whenever the user edits the graph and others are tested only on explicit request. Some systems allow the specification of rules about how to change the picture without changing the underlying graph. Typical examples include information about stretching or shifting of picture elements. Certain element types may be specified as fixed in size or form, or their position may only vary within given limits. Accordingly, the system does not allow operations violating these rules, or it executes them but signals the problem.

## Directions for the future of structure-based editors and environments

Increased programmer productivity and higher quality software can result from improved tools and environments. Syntax-directed editors have been found useful for learning a programming language and as an aid in code entry and manipulation. They have not been widely accepted, especially by experienced programmers (Neal, 1987). Concepts from syntax-directed editing have been integrated into text editors, word processors and document preparation systems where there is minimal underlying knowledge of the language, particularly semantically. Commercially available, integrated language-based programming environments provide automatic pretty-printing and provide some syntactic error detection as a result of the formatting: essentially a non-tree-based, purely textual syntax-directed editor with no template capability. These systems, many of which have achieved commercial success, are viewed as far more useful than traditional syntax-directed editors.

Empirical studies tried to gauge the usefulness of structure-oriented editors in software production, examining program text entry systems for novice programmers as well as sophisticated editing tools of more general structures that support different phases of the software life-cycle. After a period of wide enthusiasm – often most strongly expressed by the implementors of such systems – the general opinion seemed to be that syntax-based editors do not really provide any advantage to the user, or that the advantages were outweighed by the disadvantages. The advantages included that novice programmers learned and understood the underlying structure of the programming language and that they were freed from focusing on details, such as the placement of colons and semi-colons, so that they could focus their attention on the algorithm. The perceived disadvantages were primarily that tools were highly restrictive in their input and that there was a fear that programmers were not sufficiently learning the syntax and semantics of languages when the tool took over much of the work for them. There was also a perception of syntax-directed editors as crutches that a programmer could not program without, once trained with one initially.

The concept of structure-based tools, despite the availability of commercial systems, had not managed to show its value by a really "successful" system used by a large number of people. The perception of available tools was typically that these tools were of limited usefulness or that the difficulties with their use or the restrictiveness of the interface overshadowed any benefits. The most frequent criticism was that structure editors restrict or even hinder the user, especially the experienced user, with their structure operations. Furthermore, it was speculated that even for novices (with respect to the structure under consideration, such as program syntax) the potential advantage was very small – basically restricted to helping the user with basic syntactical details. Many of the studies examining the use of structure-based editors looked solely at student environments. However, structure-based editors have evolved from their early stages to very powerful systems. The notions arisen from structure editors have also been incorporated into widely available commercial systems, moving them out of the domain of experimental systems or research vehicles.

## Structure of the book

The book is organized into four parts, each with three chapters from authors working in the field. The first two parts cover the traditional, text-based structure editor and its modern incarnations, reflecting the state of the art reached in this area: Part I is dedicated to the discussion of three different program editors; Part II deals with the evaluation of, and extensions to, program editing systems. Part III deals with the generalizations that the principles have undergone when being applied to computer graphics, leading to what we call graphical structure editors. Part IV presents three chapters which exploit structure editing concepts either in non-standard fields, such as document structure recognition, or in an innovative way, such as for constructing user interfaces by combining base editors.

Our goal in the book is to clearly show the concepts, methods and techniques of structure-based editing. We find that, in many applications, structure editing methods are used without being made explicit. For instance, many commercial applications are in fact structure-based tools operating on abstract, textual, or graphical structures. Commercially successful development environments rely heavily on the use of concepts from the structure-based environments presented in this book. User interface development and the development of interactive software in general can benefit from applying structure editing techniques, as there is a rich set of principles and experiences available to be built upon.

## References

Furnas, G. W. (1986) Generalized fisheye views. In *Proceedings of CHT'86*, Boston, USA

Lang, B. (1986) On the usefulness of syntax directed editors. In *Proceedings of IFIP WG2.4 International Workshop on Advanced Programming Environments*, Trondheim, Norway, June

Miller, G. (1962) The study of intelligent behavior. In *Proceedings of a Harvard Symposium on Digital Computers and their Applications*, Cambridge, MA, April 1961. Harvard University Press, Cambridge MA, USA.

Myers, B. A. (1986) Visual programming by example, and program visualization: a taxonomy. In *Proceedings of CHI'86*, Boston, MA, USA.

Neal, L. R. (1987) Cognition-sensitive design and user modeling for syntax-directed editors. In *Proceedings of CHI'87 + GI'87*, Toronto, April.

Neal, L. R. (1989) A system for example-based programming, In *Proceedings of CHI'89*, Austin, April.

Reiss, S. P. (1986) GARDEN tools: support for graphical programming. In *Proceedings of IFIP WG2.4 International Workshop on Advanced Programming Environments*, Trondheim, Norway, June.

## Acknowledgements

We are indebted to our editors, Kate Brewin and Andrew Monk, and to Thomas Green, who is a source of inspiration and suggested that we produce a special issue of the International Journal of Human Computer Studies on this topic. In addition, we would like to thank Ugo Gagliardi and Tom Cheatham, both of Harvard University, who provided guidance and support in the pursuit of this area of research.

# Part I Traditional structure editors

1 **Coherent user interfaces for language-based editing systems**
*Michael Van De Vanter, Susan Graham and Robert Ballance*
(reprint from the Special Issue of *Int. J. Man–Machine Studies* on Structure-Based Editors and Environments Vol. 37, 4

2 **Using Mjølner Orm as a structure-based meta environment**
*Sten Minör and Boris Magnusson*

3 **An OCCAM programming environment**
*M. Filali, Y. Jamoussi, A. Knani and J. C. Maurize*

The traditional view of structure editors is that of an editor that aids in the construction and modification of a program. Typically this is for a structured programming language such as Pascal. These environments have been prevalent as instructional tools. Some have a greater complexity than others in terms of features or capabilities. These traditional structure editors are important to examine, not just for their capabilities, but for their influence on environments in more complex domains. The underlying principles behind tools supporting programming tasks have been generalized to other fields, such as graphical editing. Also, ideas from these tools have been utilized in commercially available software products in widespread use.

The predecessors of the tools described in Part I were syntax-based structure editors (often referred to as syntax-directed editors) for simple structured languages. Later generations of editors not only support more complex languages but also different paradigms, such as object-orientation and parallelism. Further, editors support structures beyond the scope of programming, including specifications and documents.

The three chapters which form Part I describe structure-based programming environments. The striking difference between the chapters is in the nature of the language supported: a procedural programming language, an object-oriented language and a parallel language. The first two systems are universal systems in the sense that they are instantiated for a language from a description of the language, whereas the third system operates on only one language.

Chapter 1, by Van De Vanter, Graham and Ballance, describes *Pan*, a highly functional structure-based editor which supports multiple structured languages within a session. A user-centered view was taken in the design in order to provide a better understanding of program structure to the intended user population. (A discussion of evaluation issues will appear in Part II.) *Pan's* capabilities include unrestricted text editing. This is important for experienced programmers since other environments, which did not offer this capability, were rejected by users as too inflexible and restrictive. *Pan* has been widely used within educational environments.

Chapter 2, by Minör and Magnusson, presents Mjølner Orm, a universal system for object-oriented programming which supports SIMULA. Not only is an object-oriented language supported but the environment itself is object-oriented which is depicted visually in that windows are hierarchically nested. Orm can be viewed as a language laboratory; it has been used practically for grammars for "little" languages such as Unix "make". Language-specific grammar formalisms can be developed for syntax, semantics, and appearance of window structure. This extends beyond the scope of most programming environments. These principles were applied to SIMULA, as illustrated in the chapter.

Chapter 3, by Filali, Jamoussi, Knani and Maurize, presents a programming environment for the concurrent language OCCAM. The environment is generated using the Cornell Synthesizer Generator. The main goal is to support program transformations which maintain the semantics. The aim is to allow the user to switch from a client–server architecture to distributed memory models. This illustrates a very different benefit of a structure-based environment. In fact, the three systems discussed in this section represent the development from syntax-based program editors to systems supporting semantic aspects of program editing.

# CHAPTER 1

# Coherent user interfaces for language-based editing systems

*Michael L. Van De Vanter, Susan L. Graham and Robert A. Ballance*

## Abstract

Many kinds of complex documents, including programs, are based on underlying formal languages. Language-based editing systems exploit knowledge of these languages to provide services beyond the scope of traditional text editors. To be effective, these services must use the power of language-based information to broaden the options available to the user, but without revealing complex linguistic and implementation models. Users understand complex documents in terms of many overlapping structures, only some of which are related to linguistic structure. Communications with the user concerning document structures must be based on models of document structure that are natural, convenient and coherent to the user. *Pan* is a language-based editing and browsing system designed to support development and maintenance of complex software documents. *Pan*'s implementation combines several approaches: unrestricted text editing, language-based browsing and editing, description-driven language definition for incremental analysis and support for multiple languages per session. *Pan* uses a variety of mechanisms to help users understand and manipulate complex documents effectively, in terms of underlying language when necessary, but always in the framework of a coherent, user-oriented interface. This chapter describes that interface, the mechanisms needed to support it, and the complex relationships between interface design and implementation techniques demanded by the goals of the system.

Structure-based Editors and Environments
ISBN 0-12-681890-8

## 1.1 Introduction

Interactive language-based editing and browsing systems will be a central and crucial component in software development environments of the future, serving as the primary interface among expert practitioners, expert systems and the complex software being managed. They will provide access to a wealth of information created both by human developers and by increasingly powerful and sophisticated tools. A successful system of this kind must exploit the power of language-based technology in ways that enhance its users' abilities to carry out their tasks, but do not hinder their flexibility of action. In particular, the multiplicity of structures inherent in a document (or collection of documents) must be made available and comprehensible to the user. Since the various structures in a document overlap meaningfully, it is essential that the system provide a *coherent* user interface – one that clarifies and augments understanding without hindering the user.

*Pan*[1] (Ballance *et al.*, 1990) is an interactive editing and browsing system that addresses this challenge. *Pan* operates on *documents* – traditional natural language texts as well as objects with formalized structure, such as formal designs and program components. Such documents have both textual and structural aspects. Every document is encoded with one or more languages, simple text being the universal base language. *Pan* is truly a hybrid (Bahlke and Snelting, 1992) text and structure editor, providing unrestricted textual editing anywhere within a document together with operations requiring language-based analysis.

*Pan* grew out of our research into how language-based technology can be made useful in interactive computing environments. This investigation led us to the design and construction of a prototype editing/browsing system, now in use as a platform for that research. The fundamental characteristics of the prototype system demanded a number of advances in language-based technology, reported elsewhere, and some new approaches to designing the user interface, reported here.

[1] Why "Pan"? In the Greek pantheon, Pan is the god of trees and forests. Also, the prefix "pan-" connotes "applying to all" – in this instance referring to the multilingual text- and structure-oriented approach adopted by this project. Finally, since an editor is one of the most frequently used tools in the software professional's tool box, the allusion to the lowly, ubiquitous kitchen untensil is apt.

### 1.1.1 The user-centered view

We view the design of *Pan* as a user interface design problem in a deeper sense than is usually construed for this kind of system. Although it is important for the system itself to have a well-designed user interface that makes its behavior easily understood and controlled, we feel it is just as important to think of the entire system as an interface between people and documents – an interface whose purpose is to help make documents more easily understood and modified.

The design began with consideration of intended users: their tasks, expectations, knowledge, expertise and working environments. We developed a vision of the role to be played by language-based systems in future software development environments. The primary users are seen to be software professionals fluent with their languages and tools who manage documents written in many different languages and who spend more time reading and trying to understand documents than they do authoring. The general design for *Pan* describes the kind of support we feel can be provided using language-based technology. This design involves rethinking many of the functional aspects usually associated with browsers and editors.

### 1.1.2 The prototype

*Pan I* (Ballance and Van De Vanter, 1988; Ballance *et al.*, 1988b) is now in use for ongoing research in language description and processing techniques, user-interface design, advanced document viewing methods and related areas. *Pan I* is a full-featured, multi-window, multiple-font, mouse-based editing system that is highly customizable and extensible. It incorporates two new language description and analysis techniques: *grammatical abstraction* (Ballance *et al.*, 1988a; Butcher, 1989) and *logical constraint grammars* (Ballance, 1989).

*Pan I*'s design addresses the integration of four fundamental characteristics: unrestricted text editing, multiple languages per editing session, an accessible language description mechanism, and a description-driven incremental analyser that gathers and maintains information derived from each document, even in the presence of editing changes. We believe that this combination provides considerable leverage for developing more advanced editing and viewing capabilities that can be used to enhance a user's understanding. The system was constructed both to be usable and to provide a flexible, extensible platform for ongoing research.

With respect to our long-range vision, this prototype has some limitations: current description techniques are aimed at a particular class of formal languages; the implementation supports only one language per document; a single analysis may span multiple documents, but only within one language; the system provides only part of the desired flexibility in generating visual presentations. Related issues, including support for novice programmers and learning environments, support for program execution, graphical display and editing and the actual design and implementation of a persistent software development database, were deferred. The *Ensemble* project, which will create a successor to *Pan I*, is addressing some of them.

Within this framework, aside from addressing the technical challenges of making such a system work at all, we have been developing a number of services for users based on the kind of information the prototype can derive and maintain. These begin with our own version of services traditionally provided by language-based editors, but without many of the restrictions they impose. These services have been produced using a number of general mechanisms, whose potential for extension to new services we are now exploring.

### 1.1.3 Major accomplishments

The prototype has enabled us to realize many of the features important to achieving our goals. Those characteristics are summarized here and discussed in more detail in subsequent sections.

*Pan I* provides the user with unrestricted text editing, requiring none of the limitations or compromises traditionally imposed by systems that perform language-based analysis. The user need not switch editors when moving from formally specified documents to simple text, nor follow a system-prescribed discipline for authoring or modifying structured documents such as program components.

When structural information exists the system provides structure-oriented operations, but without interfering with text editing. For example, the user can smoothly mix structural browsing with textual editing. The user can also move from one language to another during a session.

Description-driven document analyzers maintain incrementally for each document a repository of information that is used to provide language-oriented editing operations and other language-based services. Information derived from other analysis tools can be incorporated as well.

Document analysis includes, but is not limited to, checking for well-formedness with respect to the underlying language. For example, one might extend a language description to check for violations of project-specific naming or stylistic conventions.

The system can be extended by the addition of language descriptions. These can have varying degrees of completeness, permitting quick prototyping and gradual enhancement.

In *Pan I* what have traditionally been called "language errors" are simply another kind of analysis-derived information, available to the user upon request. We attempt to place no more restrictions on the user in their presence than a standard text editor does in the presence of spelling errors. Analysis never "fails", and documents are never "illegal" (although they may be incomplete or ill-formed in some way).

A guiding principle has been to design and present services based on a conceptual model of document structure that is appropriate to the user. Our intent is to hide the internal complexity that supports those services. The user should be unaware of the underlying representations, such as the form of a syntax tree, and of implementation-oriented distinctions, for example between syntax and static semantics.

The author of a *Pan I* language description can use a variety of mechanisms to specify and tune aspects of the user interface for the language, such as appearance attributes and conceptual components. There may be more than one description for a given language, each suited for particular tasks or classes of users.

### 1.1.4 Retrospective

A major problem with the design and implementation of systems like *Pan* is the degree to which different aspects interact. For instance, the goal of presenting conceptual models of documents to users while hiding internal complexity has tangibly affected our language description techniques, document analysis algorithms and the design of internal data structures. One theme of this chapter concerns these interactions and their effect on interface design.

We are pleased with our success so far in presenting users with a document model that is not visibly rooted in underlying language technology and is not subject to the same technical constraints. More and more, we have been able to hide language technology, providing *services* instead. We believe this will continue to be a workable approach.

We have been struck by the importance of adjusting the user interface for each language. Careful choices of appearance make a difference, as does consideration of the major conceptual notions appropriate to each language.

### 1.1.5 Overview

This chapter describes our approach to the design of *Pan* and reports on our experience with the current prototype. Section 1.2 reviews the analysis that drove the design and shows how it moves beyond previous systems. Section 1.3 introduces the prototype, its organization and fundamental mechanisms. Sections 1.4–1.9 introduce various services for users, in each case discussing historical precedent, special design problems and the actual implementation adopted in the prototype. Finally, Section 1.10 briefly presents our ongoing and future work. Throughout, we attempt to show how the requirements of a coherent user interface model are woven into the design and implementation of *Pan*'s mechanisms and services.

## 1.2 Goals for the *Pan* system

The *Pan* project was motivated by a particular vision of the role to be played by language-based browsing and editing systems. This section describes that vision and shows how it defines the fundamental requirements for *Pan*'s design.

The term *language-based* indicates that one or more of the facilities provided by the system makes use of language-specific information derived from the documents known to the system. In the context of this chapter, the term *system* (or *editing/browsing system*) encompasses the entire collection of services that are used to browse, manipulate and modify one or more documents interactively. The term *editing interface* refers to the fact that those services are provided to the user through a generalization of the services of a traditional interactive editor. Fraser and Lopez (1981) point out that the user interface to many interactive services can be modelled as a form of editing activity;

examples include editors for simple tables, file system directories and general data structures. In that sense, editing is a fundamental part of an integrated software development environment.

### 1.2.1 The working environment

*Pan* is intended to support experienced professionals who manage large collections of interrelated documents. Many common assumptions about traditional language-based editors do not hold in this domain. In particular, the kind of support most helpful to experienced users is different from the support needed for beginners.

Yet even experienced users are confronted with unfamiliar languages or novel situations when they need the support normally provided for novices. We believe that the kinds of analysis and support necessary to support experienced users can be adapted to support novices. The converse – supporting the experienced practitioner using tools tuned to the novice – is far more difficult.

Neal (1987) distinguishes three dimensions of expertise in users of program editors: computer, language and programming expertise, in addition to expertise with a particular tool. When designing *Pan*, we assumed a high level of expertise with computers (and eventually with *Pan*) and a relatively high (but not uniform) level of expertise with languages and with programming.

*Understanding is the primary activity*

Editors tuned for authoring fail to address today's problems. Software systems have become so large and complex that developers spend far more time trying to read, understand, modify and adapt documents than they do creating them in the first place (Goldberg, 1987; Winograd, 1979). A successful interactive development environment must support understanding by recognizing, exploiting and making visible complex relationships within and among related documents. At the same time, the system should provide related, but different support tuned for authoring and modifying documents.

*There are many languages*

Software developers typically use several formal languages daily: design languages, specification languages, structured-documentation languages, programming languages and small languages for scripts, schemas, mail messages and the like. Even a program written in a single programming language may contain embedded "little languages" that impose their own conventions. For example, many subroutine packages such as libraries for developing window-based applications effectively define mini-languages that determine how subroutine calls, and in particular long sequences of arguments, must be used. A language-based system must support all these different languages as smoothly and uniformly as possible.

Further, it must be convenient to add support for new languages using natural, declarative, language-description mechanisms. New languages arise often, as do enhancements to existing ones. Declarative descriptive mechanisms allow a description writer to focus on what is being described rather than on how document analysis proceeds.

Finally, since the services that the system can provide are based on the data it derives from the document, a language *description* may include elements beyond the scope of the basic *definition* of a language. Thus, someone adding a new service to the system may need to extend some existing language descriptions to derive new data or maintain new annotations.

*Users are fluent*

Users know their primary languages and tools well. An editing interface must augment productivity, without any sacrifice of flexibility and power in the name of safety or learnability. For example, experienced developers will not trade away the flexibility of unlimited text editing for the safety of enforced syntactic correctness. Finer grained support should be available in situations where the user is dealing with an unfamiliar language.

### 1.2.2 The role of the system

A language-based editing/browsing system provides the primary interface between people and integrated environments containing the documents they manage. Positioned this way, between user and documents, the system is uniquely situated to share information *about* documents that may be provided by both user and tools, as shown in Figure 1.1. In this model, users interact with documents through the editing interface;

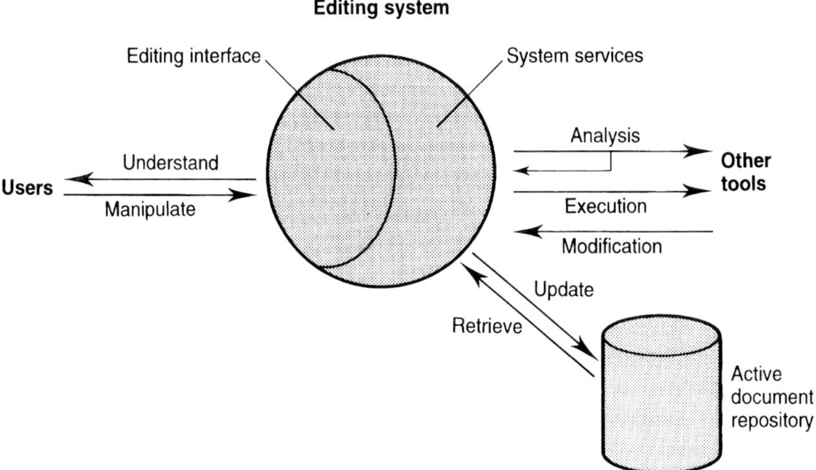


**Figure 1.1** Editing interface and system services in relation to the environment.

tools interact with documents through the system services; they communicate with one another via an active data repository. The editing interface and the system services provide alternate projections (views) of the document as well as analysis for the user.

## *Gathering and presenting information*

Users opportunistically exploit many forms of information to help them understand and modify complex documents (Letovsky, 1986). "Information gathering" is the primary task associated with important activities like program maintenance (Holt *et al.*, 1987). The system can support these activities by gathering and presenting many kinds of information.

For example, many language-based systems check that a document is well-formed. The same analysis can enable a user both to edit the document in terms of its underlying language and to locate document components that violate restrictions in the formal language definition. Other kinds of interaction may require more elaborate analysis. For example, language-based formatting (sometimes called pretty-printing), traditionally based only on surface syntax, should exploit information about scopes, types, local usage or even distinctions such as "main-line" vs. "error handling" code. These kinds of analysis move far beyond simple error-checking: they involve knowledge of particular organizations, techniques and systems, not just languages. Although this kind of information must be broad in subject domain, it need not be deep (in the sense of program *plans* (Letovsky and Soloway, 1986; Soloway and Ehrlich, 1984) or *clichés* (Rich and Waters, 1988) to be useful.

*Maintaining and sharing information*

Complex, expensive analyses to support an editing interface make sense only in an environment in which many tools share the information maintained by the system. The same checking that is used to tell the user that a document is type-correct can provide type information to a compiler, an interface consistency service or an auditing tool. In some cases information produced by other tools should be made available to the editing interface. For example, the results of performance analysis or information derived from version control can be used to produce helpful views of programs or prototypes.

### 1.2.3 The structure of documents

Documents represent richly connected, overlapping webs of information having many structural aspects. Each aspect is more relevant for some kinds of users than for others and for some tasks more than for others. An editing interface in the role we envisage must support many kinds of users, many tasks, many structural aspects.

The multiplicity of structural aspects has the potential to confuse the user, but it need not if the system supports structures already understood by users. People routinely think about complex objects from different perspectives and are remarkably adept at shifting perspective. An effective editing interface need only support these shifts without imposing any extra overhead on the user.

This discussion reviews the document aspects most important to users who are the experienced professionals in *Pan*'s intended audience.

*Textual display*

Text remains the display medium of choice for documents in most languages, since most languages are by definition textual. Graphical presentations of such documents have some potential value for overviews and summaries, but it is very unusual for a graphical display to carry *all* the information present in a textual rendering and it is usually not a comfortable visual field in which users can specify editing operations.[2]

An effective editing interface should exploit the full power of the textual medium. The value of high-quality typography for natural language documents is well established. Recent studies suggest the same potential benefits for programs (Baecker and Marcus, 1990; Oman and Cook, 1990); these studies involved printed versions of programs typeset from source code by language-specific formatters.

### *Text-oriented editing*

An effective editing interface must support text-oriented editing with as few restrictions as possible.

This history of language-based editors is marked with controversy over text-oriented editing. In practice, people will not do without it. It fits naturally with the textual display medium, and people are accustomed to it. Furthermore, most kinds of documents contain textual chunks that have no structural properties beyond the textual. Examples include sentences or paragraphs in most natural language documents, labels in spreadsheets and comments in programs.

Despite arguments to the contrary (Waters, 1982; Wood, 1981) many language-based editors impose varying degrees of restrictions. Some of these projects have since concluded that text-oriented editing should not be limited (Bahlke and Snelting, 1992; Chandhok *et al.*, 1990).

### *Language-specific structures*

Language-specific structures represent relationships among the parts of a document, or even among documents, that are defined by the underlying formal languages in which the documents are encoded. For example, consider the relationships in a natural language document defined by the connection between a figure and references to it, the relationships between declarations (definitions) and uses of variables in

[2] This remark does not apply, of course, to those languages which are by definition graphical. For those languages, graphical display is the medium of choice.

a computer program, the structures represented by a call graph in a program, or the relationships among grammatical units as specified by a formal syntax for the language being edited. Although the syntax of a language has often been taken to be the primary (and most interesting) decomposition, other language-oriented structures are at least as important to the user.

The formal definition of the syntax of a language is an artifact of language-description techniques and may not always produce structural components that are consistent with the way people think about document structure. In particular, there are aspects of structure that the user may wish to ignore. An effective language-based system must present and allow manipulations of syntactic components when appropriate, but must do so in terms of a conceptual model of documents that is natural and convenient for users.

For the implementor of a language-based system, the distinction between the formally defined syntax and other derived structures is quite convenient. In contrast, it is useful to blur that distinction for users since their understanding of "the syntax" of a language may be far removed from the formal definition used by the system for internal processing.

### *Other structures*

A variety of other decompositions can induce structures within or among documents. For instance, editors like ED3 (Strömfors, 1986) and the Cedar editor Tioga (Teitelman, 1985) allow users to specify explicitly a hierarchical decomposition for each document, where the structure may or may not be correlated with structures in the underlying formal language. Outline processors are editors for precisely this kind of structure where the underlying language is simple text.

Another interesting structure arises when a single document is encoded in more than one language. In this case, the relationships across languages and their appearances within the document become pertinent. The programming environment Mentor (Donzeau-Gouge *et al.*, 1984a) supports such nesting of languages (Donzeau-Gouge *et al.*, 1984b).

Finally, relationships within and among documents such as those generated by hyperlinks provide a third example of document structures that are neither textual nor language induced.

### 1.2.4 Pragmatics

Other issues must be addressed for an editing interface to be practical and usable in *Pan*'s intended role. Some of these issues have been complicated in *Pan* by the necessity for an early implementation to serve as a research platform:

- *Open architecture*: An effective system, especially one used for research, must be built on a flexible framework designed to accommodate many kinds of variation and evolution (Halme and Heinnan, 1988; Lang, 1986).
- *Customization and extension*: A usable editing interface must be customizable and extensible to accommodate the enormous variations among individual users, among projects (group behavior) and among sites (Lang, 1986; Stallman, 1981).
- *Performance*: Experience has shown repeatedly that an editing interface must be acceptably fast. Users are seldom willing to compromise, even in the name of additional or improved functionality.
- *Familiarity*: In the modern workplace potential users often view a new tool from the perspective of a rich, well-established working environment. An editing/browsing system whose user interface departs dramatically from those to which users are accustomed suffers the handicap of perceived isolation.

### 1.2.5 Precursors

Development of *Pan* began in 1984. At that time, a number of language-based editing systems had already been implemented or designed. The systems that influenced *Pan* can be roughly classified into three categories based on underlying models of editing: display-oriented text editors, syntax-directed structure editors and syntax-recognizing editors.

As important as the underlying model is the choice of how to delineate and maintain structural information. It can be done *explicitly*, by interpreting the sequence of user interactions as a document is constructed, or the structures can be *derived* by a component of the editing system that examines some representation of the document after it has been constructed. Both approaches can coexist in the same editing system. For instance, Mentor can derive structure initially from a textually represented document, but structure may only be edited using explicit structure-oriented operations.

The two choices, explicit maintenance of structure vs. derivation from some other representation, are matters of implementation but they affect the design of the user interface. For example, *syntax-directed editors* rely on the user to construct structures explicitly. *Syntax-recognizing editors* attempt to derive a structural representation for the document. A purely *text-oriented* editor makes no real use of linguistic structures at all; its "language-based" operations examine only the surface representation of a document, perhaps inferring a very simple syntactic decomposition.

### *Text-oriented editors*

A *text-oriented editor* operates on a document modelled both as a stream of characters and as a two-dimensional plane of characters. (The coordinates in this plane are commonly called "lines" and "characters within a line".) Users are free to operate on any character at any time.

While groups of characters such as words or lines might be accorded special status and operations, no special structural constraints are imposed on the document and few well-formedness constraints are enforced. Bravo (Lampson, 1978), EMACS (Stallman, 1981), MacWrite (Apple, 1984) and Z (Wood, 1981), are examples of text-oriented editors. Both EMACS and Z provide some language-specific editing operations.

Language-based text editors provide commands that operate on easily recognizable elements of a language, but they do not provide syntax-oriented commands or deeper analysis of linguistic structures, thus limiting the kinds of services provided.

### *Structure and syntax-directed editors*

*Structure editors* present a document as having a definite internal structure, with editing operations modelled as operations upon that structure. Most often, the document is tree-structured, with operations defined on subtrees. Outline processors are structure editors for tree structures; spreadsheet editors are structure editors for tables. Structure editors may or may not interpret the structures that they edit. When they do attribute language-specific meanings to the structures, they are often called syntax-directed editors.

A *syntax-directed editor*[3] is a structure editor that requires that the document be syntactically correct at all times: editing operations must "follow" the syntax of the language. As in a pure structure editor, the user can add new material to a document only at those points where it can be successfully grafted onto the existing structure. ALOE (Medina-Mora and Feiler, 1981) and the Gandalf editors (Notkin, 1985), Centaur (Borras *et al.*, 1988), the Cornell Program Synthesizer (Teitelbaum and Reps, 1981) and the Synthesizer Generator (Reps and Teitelbaum, 1984), Emily (Hansen, 1971; Hansen, 1984), Mentor (Donzeau-Gouge *et al.*, 1984a), PSG (Bahlke and Snelting, 1986) and the SbyS editor of the Mjølner project (Minör, 1990) are all syntax-directed editors.

The syntax-directed approach greatly simplifies editor design by restricting the changes users can make. However, some syntax-directed editors, including the Gandalf editors, the Cornell Program Synthesizer and PSG, provided limited forms of text-editing beyond that strictly required for program entry. While the syntax-directed development style is useful in some cases, (e.g. in environments for novice programmers or in environments for constructing mathematical proofs (Constable *et al.*, 1986)) the practising programmer has more general needs.

### *Syntax-recognizing editors*

A *syntax-recognizing editor* (Budinsky *et al.*, 1985) derives structural information from text, in order to check for correctness without necessarily demanding the consistency constraints of a syntax-directed editor. Syntax-recognizing editors can provide structural operations as well as text-oriented editing. Babel (Horton, 1981), the Saga editor (Kirslis, 1986), SRE (Budinsky *et al.*, 1985), Syned (Gansner *et al.*, 1983) and the UQ editors (Welsh and Toleman, 1992) are all syntax-recognizing editors, as is *Pan*.

The syntax-recognizing approach[4] permits text-oriented editing by

[3] The term "structure-oriented" (Notkin, 1984) is sometimes used in place of "syntax-directed".

[4] The syntax-recognizing approach does not preclude a user interface that simulates syntax-directed editing. In fact syntax-directed editing can be provided easily in a syntax-recognizing editor.

users at any time in any context, but at the cost of considerable complication in document representation, incremental analysis algorithms and user-interface design. A common problem in early syntax-recognizing editors was a failure to hide this complexity from the user.

## 1.3 The *Pan* system

From our conception of the role to be played by *Pan* we created a vision of what *Pan* should be. To its users it should appear to be a fast, convenient and powerful text editor that happens to be extremely knowledgeable about the local working environment: the many languages in use, local conventions and perhaps the user's own personal working habits. One might use *Pan* for editing text all day, without ever giving a thought to its other capabilities. But at any time one might choose to broaden the dialogue, to draw on many kinds of information *about* the document, information being gathered and maintained by *Pan*. *Pan* could then be directed to use this information to guide editing actions, to configure and highlight selectively the textual display, to present answers to queries and more.

### 1.3.1 Technical challenges

Implementing this vision required solving three significant technical problems in combination:

- supporting unrestricted text- and structure-oriented editing;
- performing incremental analysis to maintain a database of information about each document;
- supporting multiple languages, preferably using declarative language descriptions.

To preserve the appearance of being a "smart text editor" while providing the benefits of language-based editing, *Pan I* is a *syntax-recognizing* editing system. All language-oriented structures, including the primary syntax used to drive other analyses, are derived from analysis (parsing) of a textual representation. Structure-oriented (and even syntax-directed) editing is implemented on this base.

Adoption of syntax recognition did not solve all of the problems. One major problem in syntax-recognizing editors was their need to keep large and unwieldy internal representations of the syntax in order to provide incremental (local) re-analysis. No such problem occurs in purely syntax-directed editors. The theory of *grammatical abstraction* (Ballance *et al.*, 1988a; Butcher, 1989) was developed to resolve this problem in *Pan*.

Maintaining derived information in the presence of change was a second technical hurdle. Structure-oriented and syntax-directed editors localize the extent and impact of changes by localizing the user's editing actions, an approach that is antithetical to *Pan*'s goals. To solve this problem, the notion of *logical constraint grammars* (Ballance, 1989) was developed.

A logical constraint grammar associates goals expressed in a logic programming language with structures in the formal abstract syntax. Goals can express context-sensitive well-formedness requirements of a language, but may also describe extra-lingual constraints, such as site- or project-specific naming conventions. A document is considered "semantically" well-formed whenever all goals are satisfiable at each instance of a document's internal structure. A logical constraint grammar may additionally specify that information produced during goal evaluation be recorded in a document's logical database and made available for use by other language-based services in *Pan*.

### 1.3.2 User-interface goals

As important as the implementation challenge, however, was the design and delivery of user-oriented services of the kind suggested by the discussion in Section 1.2. This presented a double-edged problem in user-interface design.

First, the design of an effective user interface for any interactive system is challenging, especially for one with conflicting goals. For example, *Pan*'s design called for a model of text editing that is familiar to a wide variety of potential users, but it also demanded the integration of rich new functionality at all levels of the system.[5]

Second, we view the entire system as an interface between user and

[5] For example, complex low-level data representations that could efficiently support incremental document analysis had to be implemented and refined. Support for document analysis was one major reason why we were unable to build directly on top of an existing editing system.

document, subject to some of the same considerations but at a different level. *Pan*'s ultimate goals depend on the quality of its user interface at both levels.

As *Pan*'s design proceeded, common themes emerged that shaped the user interface.

- Provide text editing based on familiar models.
- Don't try to "do too much" (Neal, 1987). Derive information from document text, but present it to the user only upon request.
- Use *Pan*'s rich repository of derived information to build specific user-oriented services, but don't expect users to know much about the repository or analysis methods.
- Communicate with the user in terms of a useful conceptual model of documents rather than *Pan*'s representation model, particularly when communicating about structures other than visible text.
- Provide generic services and a uniform interface for all languages, but make it possible to adjust the interface for each particular language and for each class of users.
- In the spirit of *user-centered system design*, treat ill-formed documents not in terms of user "errors" but in terms of successive approximations of what the user intends (Lewis and Norman, 1986). Maintain full service in the presence of "errors", including, but not limited to, the presentation of diagnostics.
- Allow *Pan*'s users to customize the user interface and control important policies. Provide for experimentation with new services defined upon existing infrastructure.

Many of these embody diSessa's (1986b) principle of *continuous incremental advantage*: do not confront the user with steep learning thresholds, but provide a steady learning progression in which each quantum of learning offers immediate benefits.

### 1.3.3 System organization

Figure 1.2 shows how the *Pan* system is organized from the perspective of its users. This organization supports three different kinds of users: clients, customizers and language-description authors.

For *clients*, *Pan* is the interface for browsing and editing a database of documents. Clients may use libraries and language descriptions but

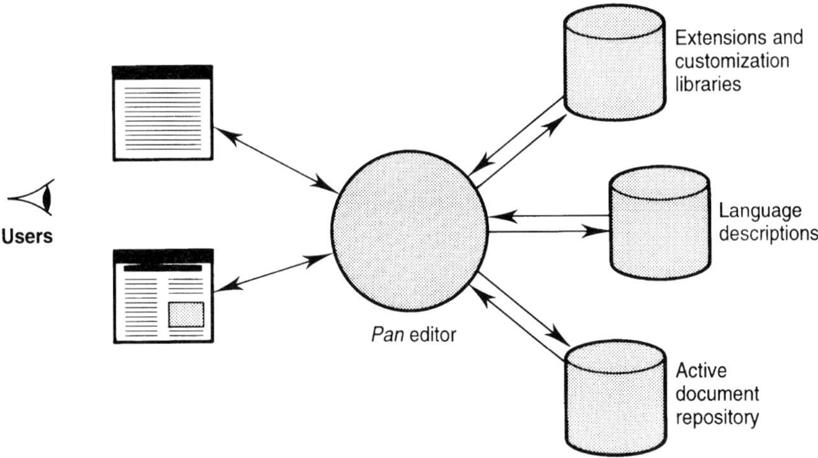


**Figure 1.2** The *Pan* system.

need to know little about them.[6] *Customizers* provide new services by adding to extension and customization libraries; this requires expertise ranging from the shallow (adding new groups of key bindings, for example) to the deep (adding a new kind of directory editor, for example). *Language description authors* add languages to *Pan*'s collection, requiring some expertise in language processing and with the specific techniques used by *Pan*'s document analyzer.

### 1.3.4 Customization and extension

Although language definition is *Pan*'s most technically intricate form of extension, other facilities support user customization, simple extension and rapid prototyping of editing environments for experimentation with user-interface designs.

- *Bindings* for keystrokes, mouse buttons and menus[7] as well as generalized *option variables* permit extensive personal customization using simple declarations. The standard behavior of the editor

[6] All but the most naive clients customize their environments in some way, blurring the boundary between clients and customizers.
[7] Keyboard bindings are crucial, since expert users seldom use the inherently slower menu systems (Lang, 1986).

is determined entirely by a configuration file using the same declarative mechanisms.

- *Pan*'s configuration variables and bindings are *scoped*. Values may be assigned at one of three nested levels (document instance, document type and global) and are dereferenced dynamically to support runtime reconfiguration.
- Simplified *command definition* faciities in *Pan*'s extension language (a superset of the underlying Common Lisp language) make it possible to add new layers of functionality in straightforward ways. These facilities automatically integrate extension programs with a generalized undo facility, with generic exception detection and recovery mechanisms (Van De Vanter, 1989), and with *Pan*'s elaborate help system.
- A *run-time library* permits special editor environments and extensions to be loaded dynamically.

### 1.3.5 Language description

Adding new languages is the most important extension mechanism in *Pan*, supported throughout by description-driven system components. All aspects of language description are declarative for readability and convenient modification.

A *Pan language description* supplies information for several aspects of document analysis and user-interface configuration: abstract syntax for internal structural representation, concrete syntax for parsing, contextual constraints[8] and user-interface definitions. One important kind of language-specific user-interface information categorizes structures appearing in the abstract syntax into *operand classes* (corresponding roughly to the phyla of operator-phylum trees (Kahn *et al.*, 1983), but potentially overlapping) as well as templates for structural elaboration.

The language description facilities developed for *Pan* are presented in detail elsewhere (Ballance, 1989; Ballance *et al.*, 1988a; Butcher, 1989). For this discussion, the following features are noteworthy:

[8] Almost all language description techniques similarly distinguish syntax and contextual constraints (also known as static semantics). Section 1.10 mentions some of the shortcomings resulting from this distinction.

- Different formalisms, each suited for different aspects of language descriptions, are used.
- Multiple language descriptions for a single underlying language may be written, possibly by "layering" of the formalisms. Alternate descriptions may vary the language in some ways, as seen by users, and they may provide different services for different classes of users.
- Language description techniques and associated analysis algorithms permit more than one language to be used for a single document, but the prototype implementation does not.
- Default definition and handling of "errors" in ill-formed documents is automatic. However, the error-handling mechanisms can be refined and optimized for each specific language description.
- Language descriptions can be loaded dynamically by *Pan* as needed, although standard language descriptions are often preloaded.
- Language descriptions are written in two formal languages: *Ladle* and *Colander*. *Pan*'s document analyzer is used as the preprocessor for one of these, and it may be run as a batch program.

To date, syntactic descriptions have been written for Modula-2, Pascal, Ada, *Ladle* and *Colander*. Descriptions are being developed for a variety of other languages, including C, C++ and FIDIL (Hilfinger and Colella, 1989). A complete language description for Modula-2 (including the language's contextual constraints) is available; other complete descriptions are being developed. Finally, like other aspects of the prototype, improvements to our language description mechanisms are underway. These are discussed in Section 1.10.

## 1.4 Text editing

One of the first commitments in *Pan*'s design was to provide powerful text-editing services that people would be willing to use. A closely associated commitment was to prevent *Pan*'s rich language-based features from ever interfering.

Figure 1.3 shows how *Pan I* appears during ordinary text editing. The same text-oriented services are available for every document, whether or not language-based analysis is taking place. As a matter of policy, derived information is never used to hinder text-oriented

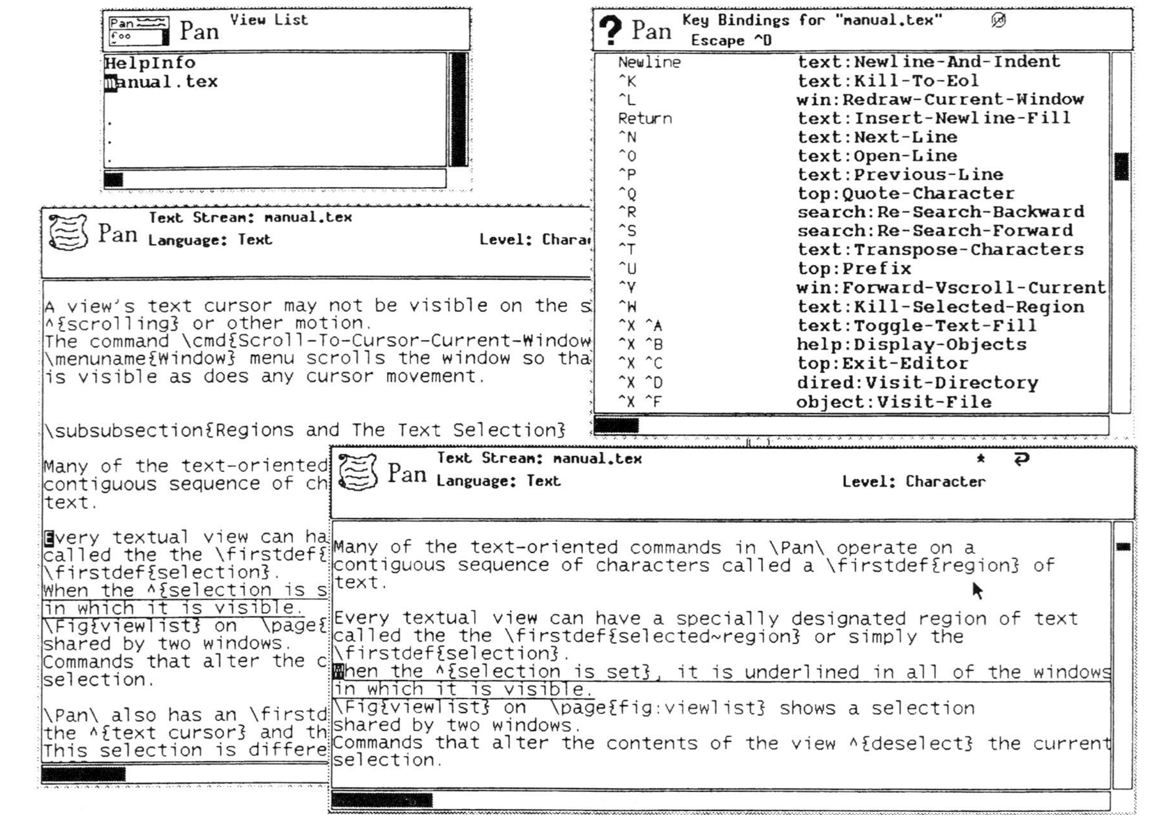


**Figure 1.3** Editing text using *Pan I.*

services, so it is possible to consider the *Pan* text-editing interface independently of any language-based features.

### 1.4.1 Models of text editing

Designing a usable text editor is by itself a significant undertaking. Among the many challenges is the need to exhibit a coherent conceptual model of what is being edited, a model reflected by both visual presentation and editing operations. *Pan* must also present an effective model of language-oriented editing, subject to the requirement that it does not clash with the textual model (see Section 1.6). Models clash when they suggest conflicting interpretations of how the editor is behaving or should be expected to behave.

Although seldom explicit, the design of effective conceptual models for text editors has always been a problem. For example, some of the earliest interactive editors operated on a virtual deck of punch cards; users could think of these editors as card punches with an erasing backspace key. Two important historical improvements added support for lines of varying length and eventually replaced the line-oriented model with a stream model. Meyrowitz and van Dam (1982) argue that the long term significance of these changes was that,

*displayed text was no longer considered to be a one-to-one mapping of the internal representation, but rather a tailored, more abstract view of the editable elements.*

The conceptual distinction between a one-dimensional text stream (the "editable elements") and a virtual two-dimensional page (the "displayed text") usually makes itself evident in the mysterious and idiosyncratic behavior associated with whitespace characters (spaces, newlines and especially tabs). For example, many text editors require that space characters be inserted explicitly at the end of a line before the cursor can be positioned in the right margin. This kind of behavior contributes greatly to the difficulty novices encounter when learning text editors, especially when misled by the *typewriter* and *blank page* metaphors (Douglas and Moran, 1983; Mack *et al.*, 1983).

This bit of historical reflection offers two lessons for the design of *Pan*. The first is that a well-designed user interface can compensate for an apparently inconsistent underlying model. People become quite proficient with the text editors, presumably building something like diSessa's (1986a) *distributed models* of the editing domain. The effect is so powerful that experienced users of text editors are often tempted to instruct novices with obviously inaccurate metaphors like *blank page*. The second lesson is that learning to use any text editor is a difficult task, one that should be expected of users as seldom as possible.

### 1.4.2 The text-editing interface

The *Pan* text-editing interface appears to the user as a bit-mapped, multiple-font, mouse-based text editor with multiple windows, in the spirit of Bravo (Lampson, 1978) and its many successors. A *Pan* user may open any number of *windows* onto each document's virtual two-dimensional text display. All windows on a document share a single, visible *selection* that appears as underscored text in any window in which it happens to be visible. Each window has its own scroll position and text *cursor*, both of which persist when the window is made invisible. Some text-oriented commands operate exclusively on the selection; others, including ordinary character insertion, operate at the text cursor. All editing commands are undoable.

### 1.4.3 The text model

*Pan*'s text-editing model is a hybrid based on two familiar models: the Macintosh (Rose, 1985) and EMACS (Stallman, 1981). Like both of those and many other editors, *Pan* treats text as both a stream and a two-dimensional page. Like Macintosh editors, *Pan* distinguishes the insertion point from the selection,[9] has a global *clipboard*, and supports the menu- and key-driven commands `Cut`, `Copy`, and `Paste`. Like EMACS, *Pan* offers a rich set of text-oriented editing commands and EMACS-compatible key bindings (Ballance and Van De Vanter, 1988).

The hybrid text model was developed to ease the cognitive burden on users, the intention being that *Pan* would behave like a familiar text editor until the user began to request additional language-based services. The combination has succeeded in some ways, but has revealed model clashes in others. For example, the model of `Undo` supported by

[9] This conflicts with the EMACS model where the insertion point *defines* one boundary of the selection.

*Pan I* and by the Macintosh editors differs from the EMACS `Undo`. Both *Pan* and EMACS have text cursors, but their behaviors differ in minor ways, as do the semantics of the *kill ring* and the correspondence between documents and windows. A user can ameliorate somewhat the effects of these clashes through customization, in particular by hiding features that cause confusion.

### 1.4.4 Displaying other information

Each *Pan* window may display optional *panel flags*, modelled on physical control panels. In Figure 1.3 the flag "*"[10] appears, indicating that the document has been changed since last saved. The flag "←⤸" also appears, indicating that automatic text-filling (line-wrapping) is in effect.

*Pan*'s text display, in addition to its conventional role for text-oriented editing, plays an important role as a medium for displaying derived information about documents. *Pan* associates with each character a font code, mapped indirectly to fonts via a user-configurable *font map*. Any mixture of fixed and proportionally spaced fonts of varying sizes may be combined in a font map. Like other options in the *Pan* editor, font maps are scoped by document instance, by document type, or globally. Additional facilities are available for superimposing more information upon the textual display: underlining, stipple patterns, colored inks and colored background shading.

## 1.5 Structure vs. presentation

A central problem in the design of any interactive editor is that the structure of an object being edited seldom maps nicely to the object's *presentation*, the visual display of the object created by the system. The previous section mentioned how this kind of discrepancy makes ordinary text editors difficult to learn.

For documents with the kind of rich structure we described in Section 1.2.3 the problem is potentially much more troublesome. For example, syntax-directed editors present a model in which the user

[10] The character "*" is actually the default visual *appearance* of the flag whose internal name is `Object-Modified?` Like much of *Pan*'s user interface, both the presence and appearance of flags is configurable.

manipulates a tree; every operation carries the additional cognitive overhead of understanding the complex relationship between the tree and its two-dimensional textual presentation. This creates a serious clash between the editing model (what the user can do) and the presentation model (what the user sees).

### 1.5.1 Document presentation

One of *Pan*'s goals for document presentation (as distinct from editing, which will be described in more detail in Sections 1.7 and 1.9) is to guarantee the presence of an editing model (possibly in addition to other language-oriented editing models) that is structurally similar to the textual presentation. In other words, if a document appears as text, then it can be treated as text despite the presence of language-oriented information derived from it. Even as we explore more elaborate methods for transforming language-based structure into text (e.g. program *unparsing* (Garlan, 1985)), we would like the user to be able to think of the display as just the text it appears to be.

In *Pan I* that consistency has been a natural side effect of our first implementation strategy and our deferral of more interesting presentation services. The textual presentation in the current prototype is based on a text stream in the usual way, and the only whitespace is that which the user inserts explicitly. *Pan*'s recently added pretty-printing mode complicates this a bit by re-arranging user-supplied whitespace, but it does so in the context of a familiar and predictable service similar to indenting modes in language-based text editors like EMACS.

*Pan*'s use of fonts, point size and color does not undercut this consistency because these visual attributes are independent of the text-editing interface and are not controlled directy by the user. New text created by the user typically has default attributes (standard font, no enhancements); the system changes the attributes to display information as requested by the user.

### 1.5.2 Displaying language-based information

So complete is the reliance on the textual presentation that it might not be apparent at all when language-based information is present. Rather than hide the fact completely, *Pan*'s default configuration adds the panel flag “⅄” (shown in Figure 1.5) when structural information is being maintained.

Although a user might edit a document textually without thinking of language-based information at all, one group of *Pan*'s language-based services uses typographical *presentation enhancements* to superimpose language-based information on the text presentation. For example, *Pan* currently supports two familiar enhancements for program documents:

- Following established custom (Baecker and Marcus, 1990), *font shifts* reveal the lexical category of text: language keywords, identifiers and comments. It has been our experience that font shifts contribute significantly to program readability, but that the choice of fonts must be customized for each language.
- When *pretty-printing* is in effect, *Pan* re-arranges document whitespace after each analysis, using indentation in familiar ways to reveal syntactic nesting levels. More advanced forms of pretty-printing for program documents, including semantically-driven elision, are under development (Black, 1990).

A general form of *structural highlighting* supports other enhancements. Arbitrary groups of structural components may have their associated text rendered with one of several (generally independent) special effects: background color, stipple patterns and ink color. For example, the user might see the answer to a particular query presented as a shift to blue ink for some text, independent of yellow background shading that might be continuously highlighting components of interest for other reasons (e.g. those described in Section 1.8).

*Pan*'s use of color in document presentations is consistent with the general recommendations for program typesetting suggested by Baecker and Marcus (1990). However, we have found one of the options they describe, the use of color to reveal lexical status, to be inappropriate. We believe that documents should normally be rendered in black and white, with color reserved for various exceptional situations, under user control, as always.

### 1.5.3 Open issues

As we extend *Pan*'s presentation model to include more elaborate display mechanisms, the natural consistency between presentation and editing model enjoyed by the prototype becomes more difficult to support. For example, as blank areas in the presentation become less coupled to whitespace characters inserted explicitly by the user, it becomes less clear how to support smooth text-oriented editing in those blank areas. We expect simple heuristics to suffice for this situation.

A related but slightly more interesting problem arises when certain text has special properties, such as the *placeholders* displayed at unexpanded structural sites by syntax-directed editors. Those editors typically do not permit the user to edit text that represents placeholders. Preliminary experiments with this style of editing suggest that the restriction is not necessary in *Pan*. Textual placeholders can be added implicitly to a language description, following some simple spelling convention such as "@statement" for an unexpanded component of class "statement". The system may add these when inserting templates, but the user is also free to type them literally. A mistyped or misplaced placeholder becomes just another variance relative to the language description (see Section 1.8).

Elision is more problematic yet, since some characters (like the conventional ". . .", for example) can represent hidden information. Here it is harder to define a natural model for text-oriented editing, since the characters are artifacts of the presentation only. In this case an attempt to edit the elision symbol might trigger expansion of the elided portion, perhaps after confirmation by the user.

A related situation arises when the keywords in a language may be treated as just another aspect of language presentation (e.g. if keywords are presented visually in the user's choice of natural language; perhaps French instead of English). In this case the editing/presentation correspondence can be preserved, but the implementation becomes more complex.

Structure-oriented editors encounter many of these problems, but they typically lack *Pan*'s commitment to smooth integration with the textual presentation model. Our follow-on project is exploring these issues, both with implementation strategies and with experimentation to discover what presentation models "feel natural".

## 1.6 User models of document structure

User and editor must be able to communicate about what is being edited. When presentation and structure are similar enough, as they are for familiar text-oriented editing, simple kinds of navigation, like pointing, are obvious and effective. In contrast, language-based document structure is both less familiar to users and less evident from textual presentations. Even so, the design of many language-based editing systems presumes that user interaction can usefully be modelled as tree operations; *Pan*'s design is not based on this assumption.

Making language-based structure intelligible requires that the system present structure in terms of coherent conceptual models. The design of such models is a user-interface problem, applied here to the structure of formal languages. This section describes *Pan*'s models for the simplest kind of language-based structure: decomposition into syntactic components. Examples of more complex structures arise later, for example in the discussion of ill-formed documents in Section 1.8.

Ignoring, for the moment, problems that arise from the mixture of textual and structural models (see Sections 1.7 and 1.9), there are two important aspects of this user model: how documents decompose into components and how those components are named.

### 1.6.1 Internal representation

Most language-based editors present a document model that is based on the way they represent documents internally, typically as a tree. For example, many early editors that provided unrestricted text editing maintained a complete parse tree (Horton, 1981; Kirslis, 1986), whereas syntax-directed editors usually maintain a reduced or *abstract* syntax tree (an example of each appears in Figure 1.4). Arguments for

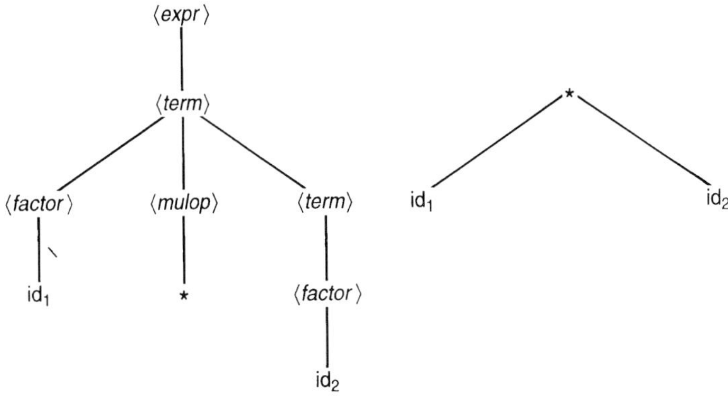


**Figure 1.4** Parse vs abstract tree representations.

the latter representation include the observation that it is a more "natural" model for user interaction, since the extra information in the parse tree is only an artifact of parsing technology. *Pan*'s internal representation is based on an abstract tree,[11] and *Pan*'s language description mechanism permits considerable flexibility in the design of this representation for each language.

Experience with language descriptions for *Pan* suggests that criticism of the parse tree representation as a user model applies to the abstract tree as well. The design of tree representations is typically driven by many issues concerning language description and analysis. Few of these issues address the need for a coherent conceptual model for documents in the language.

### 1.6.2 Names for components

Related to structural decomposition is the choice of terminology by which document components can be named. Simple structure editors (especially when designed to support multiple languages) may not name components at all, requiring the user to think in terms of representation-oriented commands (e.g. `Left`, `Right`, `In`, `Out` and `Delete-Subtree`). A single-language editor can provide more natural language-oriented commands (e.g. `Next-Function`, `Previous-Declaration` or `Delete-Statement`), but this approach doesn't generalize across languages. In some cases appropriate names overlap and depend on context; for example, a structural component internally called "variable" in a programming language representation might be conceptually both a "variable" and an "expression" in some contexts but only a "variable" in others.

### 1.6.3 A separate model for users

*Pan* decouples conceptual models of document structure from internal representations. The language description mechanism provides a loose framework in which the author of each language description is expected to design a model. This framework is based on two assumptions about how people understand syntactic structure.

The first assumption is that people generally think not in terms of

[11] Combining incremental parsing with an abstract tree representation demanded novel specification and implementation techniques (Ballance *et al.*, 1988a).

trees but in terms of familiar structural components, the ones that might be described in an informal language description: procedures, declarations, statements and the like. People understand nesting but only weakly. For example, even though natural language permits arbitrary nesting, people find sentences with even three nested levels difficult to understand.

The second assumption is that people think of structural components in the specific terms and concepts of the language being edited, "statements" for example, and not in generic structural terms like "subtree". This is an instance of the observation that users' perceptions of systems are much more sensitive to surface representations than to underlying structure.[12]

### 1.6.4 Operand classes

*Pan*'s document model is driven by a separate component of each language description. Aspects of each language that will be revealed to the user are specified in terms of *operand classes*. When used for structure, operand classes are arbitrary, possibly overlapping collections of components in the abstract representation. Each operand class has a name by which it is known to the user. Our standard Modula-2 description includes classes named "`Expression`", "`Statement`", "`Declaration`" and "`Procedure`"; some correspond to more than one kind of internal structural component, when the distinction is judged unimportant or potentially confusing to users.

That operand classes may overlap distinguishes them from operator-phyla (Kahn *et al.*, 1983); this becomes important with more complex kinds of operand classes. The operand class "`Syntax Error`" (Section 1.8) overlaps many of those classes; for example, a single internal structure might represent both a "`Statement`" and a "`Syntax Error`".

Furthermore, operand classes need not be defined for all possible internal structures. Internal structures not defined within any operand class are essentially invisible to users. This possibility, combined with *Pan*'s incremental parser, represents an alternate solution to the

[12] For example, the subjects in the evaluation of command languages for text editors failed to detect the equivalence of two editors that differed only in details of their command languages (Ledgard *et al.*, 1981).

"intermediate node" problem that vexes many syntax-directed editors (Lerner, 1992).

### 1.6.5 The operand level

Each *Pan* window has a current *operand level*, which the user selects from a menu of classes defined by the underlying language description. The operand level is a very weak input mode that modulates the operation of a few generic commands; at present these include `Next`, `Previous`, `Select`, `Extend Selection` and `Delete`. The operand level affects no other commands, and a user may choose (via menu- and key-bindings) not to use the level-sensitive versions at all. In particular, the operand level neither inhibits nor modulates text-oriented editing at any time.

When the operand level is "`Statement`", for example, the user may press the left mouse button anywhere and the "nearest" (heuristically speaking) structural component that meets the definition of operand class "`Statement`" will be selected (see Figure 1.5). In terms of implementation, the `Next` and `Previous` commands perform a tree walk, stopping only at subtrees of the appropriate operand class.

The menu of operand levels available while editing a Modula-2 document using the standard description includes "`Character`", "`Word`", "`Line`", "`Lexeme`", "`Expression`", "`Statement`", "`Declaration`", "`Procedure`", "`Syntax Error`" and "`Unsatisfied Constraint`". The first three levels correspond to textual concepts; we find that these aren't much used, since most users prefer familiar key bindings that are specific to words and lines. The last two are also predefined and are described in Section 1.8.

We have found the operand level a very useful mechanism, but we have yet to explore its full potential. We are now extending the implementation to permit more general definitions in terms of arbitrary predicates on structural components, drawing on language-oriented data in the database.

## 1.7 Inconsistency

Any situation where one kind of information is derived dynamically from another invites *inconsistency* between the two. The syntax-

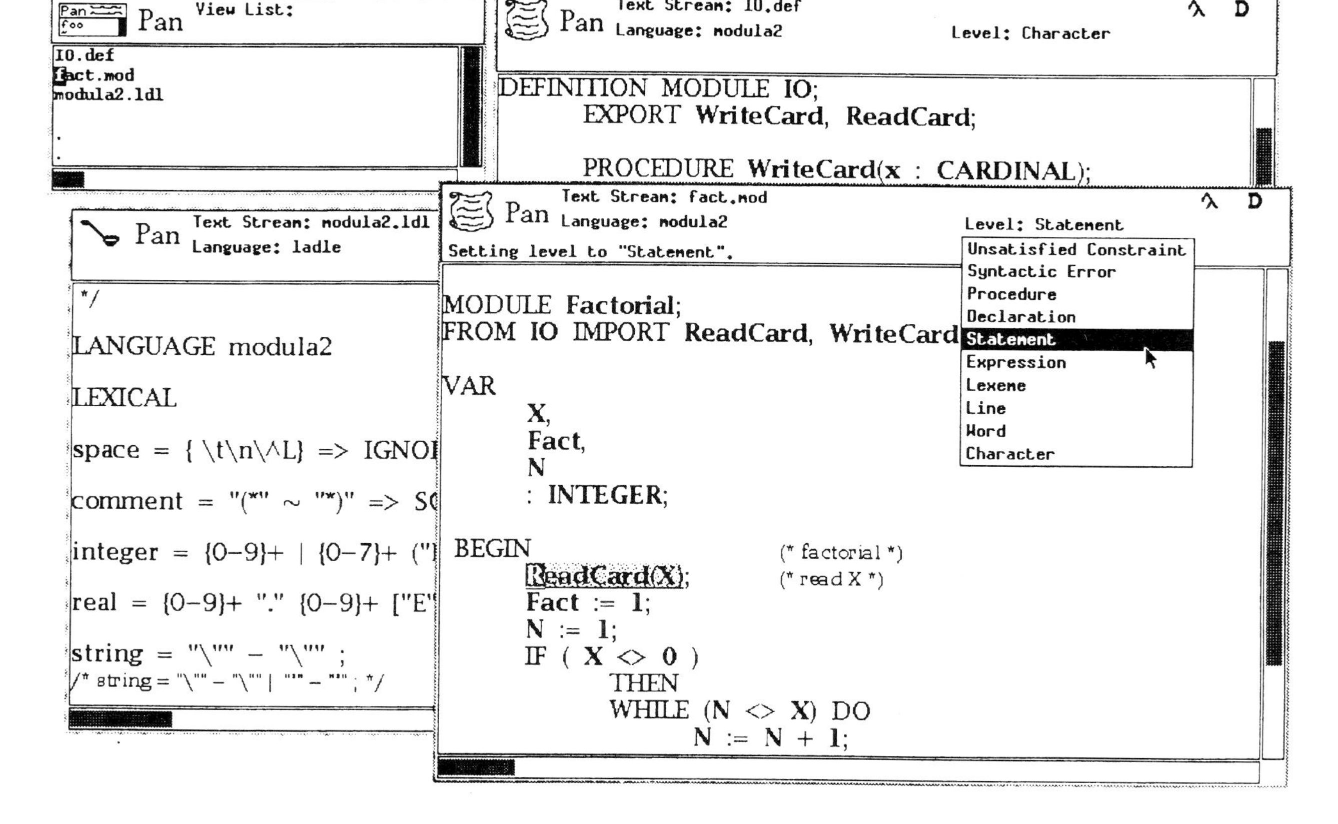


**Figure 1.5** Editing a program using *Pan I*.

recognizing approach, where language-based information may be derived from text, is no exception. During text-oriented editing, derived information maintained by the system will sometimes disagree with what the user sees. For example, font shifts revealing lexical categories become incorrect immediately after a statement has been transformed textually into a comment.

The central issue is how often and under what circumstances to attempt re-analysis, knowing that the mechanisms involved have the potential to confuse the user, to degrade performance, and to make the system's behavior unpredictable. Workable solutions demand delicate compromises involving user interface issues, analysis methods and system performance.

Inconsistency between text and derived information should not be confused with a related but different issue: the well-formedness of a document with respect to its underlying language. Many language-based editors that permit text-oriented editing are able to resolve inconsistency only for well-formed documents. A user of such a system who has edited a document textually is restrained from editing elsewhere in the document until those changes meet the system's requirements for syntactic well-formedness. *Pan* removes this restriction, using mechanisms for document analysis that *never fail*. It is possible for the text of a document in *Pan* to be well-formed with respect to its language but inconsistent with respect to the system's derived information; it is equally possible for the document to be ill-formed but consistent, in which case the derived data includes diagnoses. Section 1.8 discusses well-formedness in more detail.

### 1.7.1 Inaccurate information

Of the potential problems with inconsistency, the most serious would be to mislead the user with obsolete derived data.

*Pan*'s first defense against this prospect is the policy that language-oriented interaction may take place *only* when text and structure are consistent. This policy is implemented by *automatic re-analysis*: every command that requires derived information first ensures consistency, triggering re-analysis if needed, before proceeding. Since analysis always succeeds, this policy cannot restrict the user, although it may cause delay.

This policy is compromised when presentation enhancements (font shifts, highlighting and the like) depend on derived information but persist visually during periods of inconsistency. A conservative policy would prohibit all such enhancements in the presence of inconsistency, since any or all of them could be incorrect. In practice such a policy would lead to unpleasant visual effects after the first text-oriented change following an analysis, as the system removed all visual enhancements. More importantly, it would result in misleading feedback, since most changes ultimately have only local effects.

*Pan*'s solution rests on the assumption that the user can judge the implications of inconsistency[13] as long as it is apparent. The presence of inconsistency is revealed by the panel flag "ʎ"; this flag, otherwise displayed in boldface, appears gray during periods of inconsistency. The extent of any inconsistency is revealed partially by use of a special font for newly entered (unanalysed) text, distinct from the fonts used for lexical categories of the language. To reveal completely the extent of inconsistency would require that recently deleted text be somehow visible, but the potential benefit of that extreme approach does not seem to justify the costs in implementation and potential confusion.

### 1.7.2 Analysis strategies

Two pervasive implementation strategies help minimize analysis time. The first is complete reliance on *incremental analysis*. *Pan* only recomputes information that might change, based on a record of what text has changed since prior analysis. For many kinds of textual changes, especially simple localized changes, analysis time is roughly proportional to the extent of the changes. Unfortunately, the non-local nature of document analysis means that any change creates the potential for extensive re-analysis. We expect, and have so far confirmed, that *Pan* users will develop an intuition about the kinds of textual changes that have extensive implications and will not perceive variations in analysis time as caprice on the part of the system.

The second implementation strategy includes techniques and heuristics to retain language-based information across analyses. This strategy helps prevent information loss caused by the transient appearance of minor language violations (Ballance *et al.*, 1990).

[13] After all, the user is responsible for the changes that introduce inconsistency.

### 1.7.3 Analysis policies

Even well-designed incremental analysers can incur perceptible delays on present-day workstations, so a careful policy is important. An overly ambitious policy, attempting re-analysis after every character insertion or deletion, encounters serious problems. First, analysis at this granularity would find documents nearly always ill-formed. There is little to be gained by insisting that the system perform useful analysis on documents that are in intentionally meaningless states. Second, this drain of resources degrades performance in ways beyond the user's control. Third, over-eager update of the display is visually distracting.

Two somewhat less ambitious policies are possible. First, the system might *guess* when the document is reasonably well-formed and suitable for analysis. This policy shares the disadvantages of the more ambitious policy – the potential for unpredictable and uncontrollable behavior. Second, the system might restrict text editing to a bounded context (or *focus*) based on some internal structure, performing analysis automatically when the user attempts to leave the context in some way. This policy can create confusion about the exact nature of the context and is incompatible with *Pan*'s general insistence on unrestricted text editing with no overhead on shifts of perspective.

*Pan* uses a lazy policy, based upon the assumption that the user understands the general state of the document and can judge the trade-offs involved. Incremental analysis is only performed when requested by the user, either *implicitly* by invoking an operation that triggers automatic re-analysis (described earlier in this section) or *explicitly* by invoking the command `Analyze-Changes`. Nothing prevents a user from typing an entire document without once invoking analysis. *Pan*'s design encourages frequent analysis by making it cost-effective: both efficient and beneficial for the user.

## 1.8 Ill-formed documents

Inherent in the syntax-recognizing approach is a certainty that documents being modified are more often than not *ill-formed*: at variance with an underlying language definition. Ill-formed documents are often said to contain "errors", a pejorative term reflecting the limitations of many analysis methods. Many language-based editors that permit text-oriented editing inherit this bias. Unable to analyze ill-formed documents, these editors insist that the user correct any newly introduced "errors" before proceeding. Often justified as a service, because it limits the extent and duration of "errors", this treatment has unpleasant side-effects.

- It narrows options available to the user, who may prefer to delay trivial repairs while dealing with more important issues. An "error" may often be part of an elaborate textual transformation.
- It implies that derived information is only available and accurate when documents are well-formed, again constraining the user.
- It implies that the user has done something wrong, when in fact the system is simply unable to understand what the user is doing (Lewis and Norman, 1986).

*Pan*'s approach to the treatment of ill-formed documents pervades the system. It is an entirely normal state in *Pan* for documents to be ill-formed; every attempt is made to provide all services in the presence of any such *variances* with respect to the constraints imposed by the language description. In fact, information about variances is an important and useful kind of derived information, available whenever the user wants it.

*Pan*'s approach decouples ill-formedness from inconsistency between the textual and language-based aspects of a document, discussed in Section 1.7. An inconsistent document may or may not be well-formed (the system cannot know in this case), and a consistent document likewise may or may not be well-formed.

### 1.8.1 Variance

*Pan*'s design includes two special mechanisms for handling variances, reflecting the two layers in which the author of a *Pan* language description defines well-formedness: a context-free grammar and contextual constraints. These layers are reflected in turn by separate analysis techniques, reflecting the traditional division of analysis in language-processors into syntactic and static-semantic analysis.

*Pan* builds internal tree representations as specified by the relevant part of each language description. *Pan* implicitly extends this specification to include special document components, created automatically during incremental syntax analysis to represent instances of *syntactic variance*. Each of these special components is named after a specific kind of variance, for example "malformed statement". These components may retain well-formed subcomponents produced during prior analysis. For example, a "malformed block" might still contain well-formed statements that are accessible to the user as instances of "`Statement`". As much derived information as possible is retained along with these subcomponents for later analysis. This is an important implementation strategy for bounding the effects of minor or ephemeral syntactic variances.

*Pan*'s semantic analyzer attempts to prove that every description-defined constraint on each document component is satisfied.[14] For example, many programming languages require that all variables be declared. In a *Pan* language description, this requirement is enforced by placing an appropriate constraint on each document component where a variable can be used. If an undeclared variable appears, a *constraint variance* is added to the relevant component as one of many properties that may be recorded there.

Well-formed documents in *Pan* differ from ill-formed documents only by the absence of variances. In particular, both text- and language-based operations may proceed in the presence of variance. *Pan* offers the user three different ways to communicate *about* variances: announcement, highlighting and navigation. As with most of *Pan*'s interface, these are generally optional and under user control.

### 1.8.2 Announcement

At the conclusion of each incremental analysis, an ephemeral and unobtrusive message notes the number of variances present. Special panel flags also appear, signalling the presence of different kinds of variances. When none of these flags is present (assuming consistency), the document is well formed.

### 1.8.3 Highlighting

A user may request that the text associated with specified kinds of variances always be presented specially, in red ink for example, or with

[14] A *Pan* language description may include useful additional constraints that do not derive from the language definition, but rather from local and possibly personal conventions for the *use* of each language.

pale background shading. These forms of highlighting are designed to maintain document legibility but to give rapid feedback about the location of variances. Experienced programmers often diagnose problems at a glance, once attention is drawn to them.

### 1.8.4 Navigation

The user may request more details by selecting and moving among individual variances. After setting the operand level to one of the classes concerning variances (there are two at present: "`Syntax Error`" and "`Unsatisfied Constraint`"), the user presses the left mouse button anywhere in the display. As with other operand levels, this action sets the edit cursor at the nearest component of the appropriate class, underlines associated text and positions the text cursor at its beginning. Navigation at these operand levels has one additional side-effect; a diagnostic message associated with the particular variance appears in the panel, "malformed statement" for example.[15] Commands `Next` and `Previous` traverse instances of variances, announcing the diagnosis at each.

### 1.8.5 Classes of variance

Like most language-based systems, the distinction between syntactic analysis and contextual constraint checking is fundamental to *Pan*'s language description and analysis mechanisms. Earlier versions of *Pan* inappropriately exposed this distinction to the user in the form of the two predefined operand levels "`Syntax Error`" and "`Unsatisfied Constraint`".

A more general mechanism under development permits the author of each language description to specify one or more named operand classes for variances, each of which may contain arbitrary subsets of possible variances. For example, a simple language description might reflect the way naive users think about errors; the single operand class "`Error`" would include all possible instances of syntactic errors and unsatisfied constraints. On the other hand, experienced users seem to place variances into categories according to various criteria, for exam-

[15] A "malformed statement" is in both the "`Statement`" and "`Syntax Error`" operand classes. Recall that operand levels are defined as potentially *overlapping* classes of structural components.

ple severity, non-locality or perhaps even level of surprise; this may likewise be modelled by appropriately defined and named operand classes. This generality becomes even more important as we experiment with extra-lingual constraints, for example the imposition of locally defined, stylistic conventions.

## 1.9 Mixed-mode editing

*Pan*'s fundamental approach is to broaden rather than narrow the user's options. The user should be able to edit textually any time, any place in the document presentation; it should be equally possible to edit in terms of the language any time, any place. This section introduces *Pan*'s simple language-oriented mechanisms, with special attention to their coexistence with text-oriented editing.

### 1.9.1 Shifting perspectives

A text-oriented operation followed by a structure-oriented operation implies a shift of perspective about the document, on the part of both user and system. Humans are adept at shifting perspective, and do so frequently to suit the cognitive task of the moment. For example, studies of experienced programmers reveal that both reading and authoring activities involve a variety of fine-grained cognitive tasks, with rapid switching among them (Letovsky, 1986; Rist, 1986). *Pan* supports these activities by being ready to operate in either perspective at any time.

In contrast, many language-based editors that provide mixed text- and structure-oriented operations require user activity, both mental and physical, to shift the system's perspective. This can distract, slow down and possibly confuse the user. For example, some editors require that a structural component be specially selected for text-oriented editing; the textual content of the component then becomes the *focus* until the user wishes to resort to structure-oriented editing, at which time another explicit action may be required. Some editors provide a textual focus only in a separate window, preventing the user from editing textually in the natural visual context.

### 1.9.2 Mixing commands

Any *Pan* editing command, text- or structure-oriented, may be invoked without prerequisite. Two mechanisms make this work. The first, automatic re-analysis (Section 1.7) ensures that derived information is consistent with the text before performing any operations that require it. In many editors a similar transition can be interrrupted by analysis that encounters ill-formed text; Section 1.8 described how *Pan* avoids this problem. The second mechanism is the dual nature of *Pan*'s edit cursor.

### 1.9.3 A dual aspect cursor

*Pan*'s edit cursor has two aspects. It always has a textual location, displayed as an inverted box (Figure 1.5). It may also have a location corresponding to some structural component. In the current implementation, the cursor's structural location is revealed by turning the component's textual presentation into the current text selection. Any operation that sets the structural cursor also positions the text cursor at the first character in the structure's textual presentation. In Figure 1.5 a "`Statement`" has been selected structurally.

Editing operations that need a cursor location simply use the appropriate aspect. If the cursor has no structural aspect, then one is inferred from the text location using the same mechanism invoked when the user selects a structural component by pointing with the mouse. This design resolves the "point vs. extended cursor" problem (Teitelbaum *et al.*, 1981) by providing both behaviors simultaneously.

People seem to have little difficulty with the ambiguity, presumably for the same reasons that they can shift perspectives themselves so effectively and can manage complex, possibly ambiguous domains (deSessa, 1986a). Raskin (1989) reports that a dual-aspect text cursor, where the behavior depends on the next user action, solved a particular user-interface problem: the crucial point being that both aspects are visible and predictable.

### 1.9.4 The operand level revisited

One challenge for *Pan*'s user interface is to make language-based commands available as conveniently as the familiar text-oriented ones – via bindings to keyboard sequences, mouse buttons and menus. Other than a few generic commands (such as `Analyze-Changes`) most language-based commands operate in terms of specific languages. For example, the command `Select-Declaration` should be available when editing languages in which the concept of "declaration" has been defined; it should not be available for languages in which the concept is not defined. The command-based approach does not generalize, however, since a confusing proliferation of commands would result from *Pan*'s support for many languages.

*Pan* introduced operand levels, a level of indirection added to the basic command dispatch mechanism. Generic versions of basic commands, such as `Next`, `Select` or `Delete`,[16] use the current level to determine their actual operands. For example, when the left mouse button is bound to the generic version of `Select` and when the current operand level is "`Statement`", a left mouse button press invokes a structural command that selects the statement nearest the mouse location. The user may specify the current operand level in each window by menu selection (or key binding), where the language-specific options derive from the language description (e.g. "`Expression`", "`Statement`", "`Declaration`" and "`Procedure`").

The operand level mechanism for command dispatch is a very weak input mode. It modulates the effect of a few basic commands but has no other effects and implies no restrictions on user actions. For example, the user may edit textually, independent of the current operand level.

We sometimes find it convenient to violate the model presented above by strengthening slightly the effect of the operand level. With this adjustment, any change to the operand level causes an implicit `Select` operation at the new level. This has the visible effect of moving the edit cursor in some cases. It is not clear yet whether the convenience justifies the possible confusion.

### 1.9.5 Simple editing

The prototype implementation supports no user commands that modify internal document structure directly. A `Delete` command, invoked with a structural selection, removes the text associated with the

[16] The generic versions of these commands are called `Oplevel-Next` etc. and are bindable just like any other *Pan* command.

selected component. The internal structure corresponding to the deleted component persists until the next re-analysis, but it is invisible to the user because automatic re-analysis will delete it before any commands can use it. `Cut` places text in the clipboard, and `Paste` simply inserts text from the clipboard. If the context is appropriate, subsequent incremental analysis derives the equivalent structural information quickly.

This implementation costs a small amount of analysis time by discarding derived information when the user moves structural components. On the other hand, it guarantees the integrity and well-formedness of the document's internal representation, since the language definition is already built into *Pan*'s parser.

Complex mechanisms for direct structural editing can be a source of confusion to the user, since those editing operations may fail, something *Pan* commands never do. Worse, they may fail for the kinds of reasons we attempt to hide from the user, such as the presence of "intermediate nodes". For example, it seems reasonable to copy the list of identifiers appearing in the formal parameter list of a procedure definition and paste it into a call to that procedure. Although the two lists of identifiers might appear identically and be closely related conceptually, there may be sound implementation reasons for different internal representations in the two contexts. We prefer to avoid strategies that involve guessing the user's intent.

When a structurally inspired `Cut` and `Paste` sequence in *Pan* violates the underlying language definition, the operations succeed anyway and the problem is diagnosed by precisely the same mechanism that handles other language violations (discussed in Section 1.8).

The only cost at present of this text-based implementation is the loss of non-derivable annotations on document components during `Cut` and `Paste` sequences. We have developed, but not yet added, a strategy that avoids this information loss and provides functionality that is fully equivalent to direct structural operations. The successor to *Pan I* addresses the problem in more fundamental ways.

### 1.9.6 Other language-based operations

The ultimate advantage of language-oriented editing lies in a useful and open-ended collection of services that draw upon a rich repository of information to assist users with commonly performed tasks. One collection of these services deals with the location and diagnosis of variances (Section 1.8). This section describes other simple examples, emphasizing first that services are implemented by combining generic and language-specific components, and second that the user need not be aware of the complex internal representations needed to make them work.

One of the few forms of query supported by ordinary text editors is textual search. Searching in *Pan* can draw on *any* derived information. For example, one command locates in a document the declaration and all uses of a programming language variable, which the user identifies by pointing. The results of this and other language-based queries are made available by an interface similar to the one used for variances (described in Section 1.8). Text associated with all components of the current query result appears highlighted. The user may choose the operand level "`Query Result`" and navigate through the components. This particular command is implemented by using a query defined as part of the underlying language description. The same underlying query supports other services, for example a command that moves the cursor to the declaration of a variable pointed to by the user; this is only one example of a navigation command that follows hypertext-like links defined by the underlying language.

Like all powerful text editors, *Pan* supports textual replacement based on regular expression matching. However, one sometimes wants replacements to depend on the language structure, not on the textual structure, even when the two are similar. For example, whole-word replacement (where replacing substrings of longer words is not desired) in natural language documents is difficult to specify using patterns. One variant of *Pan*'s replacement command matches patterns only against words (lexemes) as defined by the particular language. Another variant renames variable instances in programs, drawing on information in *Pan*'s database to avoid renaming enclosed variable definitions that have the same lexical name but which are logically different variables.

## 1.10 Future directions

In addition to the general language-based services already described, we have built a number of special-purpose services that add to *Pan*'s general utility. These include a browsing interface to the file system in the form of a "directory editor", an integral help system for both naive users and developers that provides access to documentation and system source code, and a hypertext-like browser for UNIX[17] online manual pages.

We are currently working with prototypes for general language-based queries, a syntax-directed style of editing, and other new language-based services. Architectural revisions in progress include support for multiple views (or projections) per document, graphical views for tree- or graph-structured data, and the application of language-based techniques to provide visual access to internal data structures. New language description techniques are also under development.

Ongoing research projects use the leverage gained from the *Pan* system. Projects near completion include a study of program presentation that uses derived information to provide elision, alternate textual representations and mappings to hardcopy using typesetting (Black, 1990); the development of advanced document analysis techniques to specify and control user-centered program viewing; and integration with a persistent database.

An important group of issues concerns extensions to *Pan*'s language description mechanisms. First, we wish to add new descriptive layers for various purposes. For example, we intend to build browsers suitable for program call graphs and inheritance hierarchies (lattices) for languages where this is meaningful. A new description layer might define how the information is to be derived from the database and supplied to a graphical viewer. Although the present language-description facilities provide simple support for description layers, the problem of layering language descriptions requires further work.

Second, while the current implementation supports sharing among documents written in a common base language, it should be extended to deal with both documents written in multiple languages and with sharing among documents written in different languages.

Finally, we hope to move beyond the limitations in our language-description techniques that restrict *Pan* to "classical" computer languages – languages having a well-defined (and simple) syntax together with some set of well-formedness constraints. This would require adaptation of our techniques to other classes of languages: those with non-standard context-sensitive syntactic aspects, graphical languages and even natural languages.

[17] UNIX is a trademark of AT&T Bell Laboratories.

Continuing work at UC Berkeley, part of the *Ensemble* project, is generalizing *Pan*'s approach in three ways:

1. Much richer mappings among document structure, presentations and specification of appearance, building heavily on the experience gained from the $V_{OR}T_{E}X$ document system (Chen *et al.*, 1986; Chen and Harrison, 1988).
2. The extension of editing and viewing to a wide range of media – text, graphics, sound and video.
3. Integrated support for compound documents, where different languages and document types may be composed.

## Acknowledgements

*Pan* is the work of many individuals besides the authors. Christina Black, Jacob Butcher, Bruce Forstall, Mark Hastings and Darrin Lane have all made substantial contributions. The encouragement, suggestions and support of Bill Scherlis have also played a major role.

Research sponsored in part by the Defense Advanced Research Projects Agency (DoD), monitored by Space and Naval Warfare Systems Command under Contracts N00039–84–C–0089 and N00039–88–C–0292, by IBM under IBM Research Contract No. 564516, by a gift from Apple Computer, Inc. and by the State of California MICRO Fellowship Program.

## References

Apple (1984). *MacWrite Manual.* Apple Computer, Inc., Cupertino, CA.

Baecker, R. M. and Marcus, A. (1990). *Human Factors and Typography for More Readable Programs.* ACM Press, New York.

Bahlke, R. and Snelting, G. (1986). The PSG system: from formal language definitions to interactive programming environments. *ACM Transactions on Programming Languages and Systems*, **8**: 547–576.

Bahlke, R. and Snelting, G. (1992). Design and structure of a semantics-based programming environment. *International Journal of Man–Machine Studies*, **37**: 467–479.

Ballance, R. A. (1989). Syntactic and semantic checking in language-based editing systems. PhD thesis. Computer Science Division (EECS), University of California, Berkeley, CA, 94720 USA. Available as Technical Report No. UCB/CSD 89/548.

Ballance, R. A. and Van De Vanter, M. L. (1988). Pan I: an introduction for users. Technical Report No. UCB/CSD 88/410, Computer Science Division (EECS), University of California, Berkeley, CA, 94720, USA.

Ballance, R. A., Butcher, J. and Graham, S. L. (1988a). Grammatical abstraction and incremental syntax analysis in a language-based editor. In *Proceedings of the SIGPLAN '88 Conference on Programming Language Design and Implementation*, pp. 185–198. Atlanta, GA. ACM. Appeared as *SIGPLAN Notices*, **23**, July 1988.

Ballance, R. A., Graham, S. L. and Van De Vanter, M. L. (1990). The Pan language-based editing system for integrated development environments. In Taylor, R. N. (editor), *SIGSOFT '90 Proceedings of the Fourth Symposium on Software Development Environments*, pp. 77–93, Irvine, CA, USA. Appeared as *SIGSOFT Software Engineering Notes*, **15**, December 1990.

Ballance, R. A., Van De Vanter, M. L. and Graham, S. L. (1988b). The architecture of Pan I. Technical Report No. UCB/CSD 88/409, Computer Science Division (EECS), University of California, Berkeley, CA, 94720, USA.

Black, C. L. (1990). PPP: the Pan Program Presenter. Master's thesis, Computer Science Division (EECS), University of California, Berkeley, CA, 94720, USA. Available as Technical Report No. UCB/CSD 90/589.

Borras, P., Clément, D., Despeyroux, T., Incerpi, J., Kahn, G., Lang, B. and Pascual, V. (1988). CENTAUR: the system. In Henderson, P. (editor). *ACM SIGSOFT '88: Third Symposium on Software Development Environments*, pp. 14–24, Boston, MA, USA. Appeared as *SIGSOFT Software Engineering Notes*, **13**(5): November, 1988.

Budinsky, F. J., Holt, R. C. and Zaky, S. G. (1985). SRE – a syntax-recognizing editor. *Software – Practice & Experience*, **15**(5): 489–497.

Butcher, J. (1989). Ladle. Master's thesis. Computer Science Division (EECS), University of California, Berkeley, CA 94720, USA. Available as Technical Report No. UCB/CSD 89/519.

Chandhok, R., Miller, P., Pane, J. and Meter, G. (1990). Structure editing: evolution towards appropriate use. *CHI'90 Workshop on Structure Editors*. Presentation.

Chen, P. and Harrison, M. A. (1988). Multiple representation document development. *IEEE Computer*, **21**: 15–31.

Chen, P., Coker, J., Harrison, M. A., McCarrell, J. W. and Proctor, S. (1986). The $\mathrm{V_{OR}T_{E}X}$ document preparation environment. In *Proceedings of the 2nd European Conference on $T_EX$ for Scientific Documentation, Lecture Notes in Computer Science*, **236**: 24–32, Springer Verlag, Strasbourg.

Constable, R. L., Allen, S. F., Bromley, H. M., Cleaveland, W. R., Cremer, J. F., Harper, R. W., Howe, D. J., Knoblock, T. B., Mendler, N. P., Panangaden, P., Sasaki, J. T. and Smith, S. F. (1986). *Implementing Mathematics with the Nuprl Proof Development System.* Prentice-Hall, Englewood Cliffs, NJ.

DiSessa, A. A. (1986a). Models of computation. In Norman, D. A. and Draper, S. W. (editors) *User Centered System Design: New Perspectives on Human–Computer Interaction*, pp. 201–218. Lawrence Erlbaum Associates, Hillsdale, NJ.

DiSessa, A. A. (1986b). Notes on the future of programming: breaking the utility barrier. In Norman, D. A. and Draper, S. W. (editors), *User Centered System Design: New Perspectives on Human–Computer Interaction*, pp. 125–152. Lawrence Erlbaum Associates, Hillsdale, NJ.

Donzeau-Gouge, V., Huet, G., Kahn, G. and Lang, B. (1984a). Programming environments based on structured editors: the MENTOR experience. In Barstow, D. R., Shrobe, H. E. and Sandewall, E. (editors). *Interactive Programming Environments*, Chapter 7, pp. 128–140. McGraw-Hill, New York.

Donzeau-Gouge, V., Kahn, G., Lang, B. and Mélèse, B. (1984b). Document structure and modularity in Mentor. In Henderson, P. (editor), *Proceedings of the ACM SIGSOFT/SIGPLAN Software Engineering Symposium on Practical Software Development Environments*, pp. 141–148. Pittsburgh, PA, USA. Appeared as *SIGSOFT Software Engineering Notes*, **9** (3): May 1984.

Douglas, S. A. and Moran, T. P. (1983). Learning text editor semantics by analogy. In Janda, A. (editor), *Proceedings of the CHI'83 Conference Human Factors in Computing Systems*, pp. 207–211. Boston, MA, USA. ACM, New York.

Fraser, C. W. and Lopez, A. A. (1981). Editing data structures. *ACM Transactions on Programming Languages and Systems*, **3**: 115–125.

Gansner, E. R., Horgan, J. R., Moore, D. J., Surko, P. T., Swartout, D. E. and Reppy, J. H. (1983). SYNED – a language-based editor for an interactive programming environment. *IEEE Spring Compcon '83*, pp. 406–410.

Garlan, D. (1985). Flexible unparsing in a structure editing environment. Technical Report CMU-CS–85–129, Computer Science Department, Carnegie-Mellon University, Pittsburgh, PA, USA.

Goldberg, A. (1987). Programmer as reader. *IEEE Software*, **4**(5): 62–70.

Halme, H. and Heinänen, J. (1988). GNU EMACS as a dynamically extensible programming environment. *Software – Practice and Experience*, **18**: 999–1009.

Hansen, W. J. (1971). Creation of hierarchic text with a computer display. PhD thesis, Stanford University, Stanford, CA, USA.

Hansen, W. J. (1984). User engineering principles for interactive systems. In Barstow, D. R., Shrobe, H. E. and Sandwell, E. (editors), *Interactive*

*Programming Environments*. Chapter 11, pp. 217–231. McGraw-Hill, New York.

Hilfinger, P.N. and Colella, P. (1989). FIDIL: a language for scientific programming. In Grossman, R. (editor), *Symbolic Computation: Applications to Scientific Computing*, pp. 97–138. Society for Industrial and Applied Mathematics, Philadelphia, PA.

Holt, R. W., Boehm-Davis, D. A. and Schultz, A. C. (1987). Mental representations of programs for student and professional programmers. In Olson, G. M., Sheppard, S. and Soloway, E. (editors), *Empirical Studies of Programmers: Second Workshop*, pp. 33–46. Ablex Publishing, Norwood, NJ.

Horton, M. R. (1981). Design of a multi-language editor with static error detection capabilities. PhD thesis, Computer Science Division (EECS), University of California, Berkeley, CA 94720, USA.

Kahn, G., Lang, B., Mélèse, B. and Morcos, E. (1983). Metal: a formalism to specify formalisms. *Science of Computer Programming*, **3**: 151–188.

Kirslis, P. A. C. (1986). The SAGA editor: a language-oriented editor based on an incremental LR(1) parser. PhD thesis, University of Illinois at Urbana-Champaign, Urbana, IL, USA.

Lampson, B. W. (1978). *Bravo Users Manual.* Xerox Corp., Palo Alto, CA.

Lang, B. (1986). On the usefulness of syntax-directed editors. In Conradi, R., Didriksen, T. M. and Wanvik, D. (editors), *Advanced Programming Environments*, Lecture Notes in Computer Science, **244**: 47–51. Springer-Verlag, Heidelberg, New York.

Ledgard, H., Singer, A. and Whiteside, J. (1981). *Directions in Human Factors for Interactive Systems*. Springer-Verlag, Berlin, Heidelberg, New York.

Lerner, B. S. (1992). Automated customization of structure editors. *International Journal of Man–Machine Studies*, **37**: 529–563.

Letovsky, S. (1986). Cognitive processes in program comprehension. In Soloway, E. and Iyengar, S. (editors), *Empirical Studies of Programmers*, pp. 58–79. Ablex Publishing, Norwood, NJ.

Letovsky, S. and Soloway, E. (1986). Delocalized plans and program comprehension. *IEEE Software*, **3**: 41–49.

Lewis, C. and Norman, D. A. (1986). Designing for error. In Norman, D. A. and Draper, S. W. (editors), *User Centered System Design: New Perspectives on Human–Computer Interaction*, pp. 411–432. Lawrence Erlbaum Associates, Hillsdale, NJ.

Mack, R. L., Lewis, C. H. and Caroll, J. M. (1983). Learning to use word processors: problems and prospects. *ACM Transactions on Office Information Systems*, **1**(3): 254–271.

Medina-Mora, R. and Feiler, P. H. (1981). An incremental programming environment. *IEEE Transactions on Software Engineering*. **SE-7**(5): 472–481.

Meyrowitz, N. and Van Dam, A. (1982). Interactive editing systems: parts I and II. *ACM Computing Surveys*, **14**: 321–416.

Minör, S. (1990). On structure-oriented editing. PhD thesis, Department of Computer Science, Lund University, Sweden. LUTEDX/(TECS-1002)/1–198/(1990).

Neal, L. R. (1987). Cognition-sensitive design and user modelling for syntax-directed editors. In *Proceedings of the CHI'87 Conference on Human Factors in Computing Systems and Graphics Interface*, pp. 99–102, Toronto, Canada. ACM, New York.

Notkin, D. (1984). Interactive structure-oriented computing. PhD thesis, Carnegie Mellon University, Pittsburgh, PA 15213, USA.

Notkin, D. (1985). The GANDALF project. *Journal of Systems and Software*, **5**(2): 91–105.

Oman, P. and Cook, C. R. (1990). Typographic style is more than cosmetic. *Communications of the ACM*, **33**: 506–520.

Raskin, J. (1989). Systemic implications of leap and an improved two-part cursor: a case study. In *Proceedings of the CHI'89 Conference Human Factors in Computing Systems*, pp. 167–170, Austin, TX, USA. ACM, New York.

Reps, T. and Teitelbaum, T. (1984). The synthesizer generator. In Henderson, P. (editor), *Proceedings of the ACM SIGSOFT/SIGPLAN Software Engineering Symposium on Practical Software Development Environments*, pp. 42–48, Pittsburgh, PA, USA. Appeared as *SIGSOFT Software Engineering Notes*, **9**(3): May 1984.

Rich, C. and Waters, R. C. (1988). The programmer's apprentice: a research overview. *IEEE Computer*, **21**: 10–25.

Rist, R. S. (1986). Plans in programming: definition, demonstration and development. In Soloway, E. and Iyengar, S. (editors), *Empirical Studies of Programmers*, pp. 28–45. Ablex Publishing, Norwood, NJ.

Rose, C. (1985). *Inside Macintosh*. Addison-Wesley, Reading, MA.

Soloway, E. and Ehrlich, K. (1984). Empirical studies of programming knowledge. *IEEE Transactions on Software Engineering*, **SE-10**(5): 595–609.

Stallman, R. M. (1981). EMACS: the extensible, customizable, self-documenting display editor. In *Proceedings of the ACM SIGPLAN SIGOA Symposium on Text Manipulation*, pp. 147 156, ACM, Portland, OR. Appeared as *SIGPLAN Notices*, **16**(6): June 1981.

Strömfors, O. (1986). Editing large programs using a structure-oriented text editor. In Conradi, R., Didriksen, T. M. and Wanvik, D. (editors), *Advanced Programming Environments, Lecture Notes in Computer Science*, **244**, pp. 39–46. Springer-Verlag, Berlin, Heidelberg, New York.

Teitelbaum, T. and Reps, T. (1981). The Cornell Program Synthesizer: a syntax-directed programming environment. *Communications of the ACM*, **24**(9): 563–573.

Teitelbaum, T., Reps, T. and Horwitz, S. (1981). The why and wherefore of the Cornell Program Synthesizer. In *Proceedings of the ACM SIGPLAN SIGOA Symposium on Text Manipulation*, pp. 8–16, ACM, Portland, OR. Appeared as *SIGPLAN Notices*, **16**(6), June 1981.

Teitelman, W. (1985). A tour through Cedar. *IEEE Transactions on Software Engineering*, **SE-11**: 285–302.

Van De Vanter, M. L. (1989). Error management and debugging in Pan I. Technical Report No. UCB/CSD 89/554, Computer Science Division (EECS), University of California, Berkeley, CA 94720, USA.

Waters, R. C. (1982). Program editors should not abandon text oriented commands. *SIGPLAN Notices*, **17**(7): 39–46.

Welsh, J. and Toleman, M. (1992). Conceptual issues in language based editor design. *International Journal of Man–Machines Studies,* **37**:419–430.

Winograd, T. (1979). Beyond programming languages. *Communications of the ACM*, **22**(7): 391–401.

Wood, S. R. (1981). Z – the 95% program editor. In *Proceedings of the ACM SIGPLAN SIGOA Symposium on Text Manipulation*, pp. 1–7, ACM, Portland, OR. Appeared as *SIGPLAN Notices*, **16**: June 1981.

# CHAPTER 2

# Using Mjølner Orm as a structure-based meta environment

*Sten Minör and Boris Magnusson*

## Abstract

Mjølner Orm is a structure-based environment supporting incremental software and language development. The environment functionality is specified by grammars developed to support object-oriented languages. These grammars are interpreted by the environment allowing both the syntax and semantics to be changed on the fly. This has been shown useful for language customization, design, and maintenance. This chapter gives an overview of the environment and focuses on Orm used as a meta environment.

## 2.1 Introduction

The problems with developing software systems are widely acknowledged. After several decades of software development experience there is still no satisfactory solution to the "software crisis". There are no "simple" solutions or shortcuts to these problems and it has to be accepted that development of large software systems is a complex and time-consuming task. We do, however, believe that providing appropriate support for software engineers will make the software development process far more efficient than it is today.

The overall goal in the Mjølner project is to provide support for object-oriented software development in general. The project started in 1986 as a joint Nordic project involving universities and industry in Denmark, Finland, Norway, and Sweden. The activities within the project have dealt with three different aspects of software development:

Structure-based Editors and Environments
ISBN 0-12-681890-8

- *Methods* for object-oriented software development.
- Object-oriented *languages*.
- *Environments* for object-oriented development.

The method activity has worked with support for specification. It has resulted in the design of an object-oriented extension to SDL, called OSDL (Møller-Pedersen *et al.*, 1987). OSDL has after smaller modifications been accepted by CCITT as the new standard, SDL 92.

The language aspect has been covered in the development of the language Beta (Kristensen *et al.*, 1987). Beta, a simple but powerful object-oriented language, is a successor of Simula (SIS, 1987), generalizing its constructs for abstraction and inheritance.

The environment activity includes the development of the Mjølner Orm environment and the Mjølner Beta environment. The Beta environment (Knudsen *et al.*, 1989) has been developed along with the work with the Beta language at Aarhus University in Denmark. The focus of this chapter is the Orm environment, developed at Lund University in Sweden. Orm is a generic structure-oriented environment supporting object-oriented languages.

The work with the Orm environment has covered a number of topics within the area of software development environment research, e.g. structure-oriented architectures, structure editing, incremental compilation techniques, configuration control, user interfaces, grammar formalisms, and grammar interpretation. In Section 2.2, we will give an overview of the environment. After an introduction, the system is presented from the user's point of view in Section 2.2.1. Section 2.2.2 shows the user's view of the environment used for program development. The functionality of the environment and the architecture is presented in Section 2.2.3. Orm as a meta environment is discussed in Section 2.3. Section 2.3.1 shows the user's view of grammars and grammar editing. Section 2.3.2 describes the grammar interpretive technique which is the basis for the meta capabilities. The practical consequences of the grammar interpretive technique are discussed in Section 2.3.3. In Section 2.4 the implementation status and experience from using Orm in practice is presented, followed by some topics for future research in Section 2.5 and a summary in Section 2.6.

Mjølner (pronounced something like "Miolner") is a name from the ancient Nordic mythology. It is the name of the hammer of the god Thor. Mjølner is his most powerful tool. It always hits its target, and afterwards it returns to Thor's hand. When Thor uses his hammer, it is noticed by humans as thunder and lightning. Orm is the name of the main character in a famous Swedish saga of the viking age. He was a great admirer of Thor, even after he eventually, against his will, was christened.

## 2.2 The Orm environment

The common interest in the Mjølner project has been to strive to support industrial large scale software development and object-oriented programming. The participants in the project have a long tradition of using object-oriented languages, even before the term was invented and became popular. In the Scandinavian tradition, block-structured, strongly typed object-oriented languages play an essential role (Madsen *et al.*, 1990). Supporting early error detection, such as static typing, is important when developing high-quality software. Turnaround time in systems for such, usually compiled, languages has, however, often been a major problem. This is in contrast to interactive environments based on interpretation which often show an attractive, at least short term, decrease in development time. They do, however, not support early error checking. In the Orm system we have set out to provide both: supporting strongly typed languages with an incremental compiling environment.

Another problem in traditional, so called "tool based", environments is the proliferation of tools, all with different user interfaces. In the Orm environment we have designed a direct manipulation graphical user interface (Hutchins *et al.*, 1986) used consistently in program manipulation, observation of program execution, and for the definition of language grammars. In the Orm environment the functional components are integrated into one single tool. The components use a shared program representation, which is built on program structures rather than source text. This makes it possible to add functionality to the system, sharing the structured view of programs. Extensibility is essential since we have certainly not yet developed all the functionality we would like to see in a full blown industrial strength programming environment. The structural view is also crucial for the development of fine-grained incremental compilation techniques and more suitable as a basis for language-specific support such as structured browsing, navigation tools, as well as syntax-based and semantics-based editing than a textual representation.

The Orm environment is generic and can be specialized to support different languages. The technique used for specialization is interpretation of grammars defining the actual programming languages. The grammars have been developed with Orm itself which shows to us the value of interactive development in language implementation as well. A modification of one of the grammars takes effect immediately. The use of grammar interpretation also makes the system flexible and easy to customize by adding or removing language constructs, supporting special syntax, or even using Orm as a language laboratory.

Several structure-oriented environments have addressed similar problems before. Mentor (Donzeau-Gouge *et al.*, 1980) is an early example. Other structure-oriented environments incude The Synthesizer Generator (Reps and Teitelbaum, 1984), Gandalf (Notkin, 1985), PSG (Bahlke and Snelting, 1992), DOSE (Feiler *et al.*, 1986), Muir (Winograd, 1987), Centaur (Borras *et al.*, 1988), and *Pan* (Chapter 1 this volume).

The major contributions of the Orm environment are:

- Consistent use of a hierarchical object-oriented browser to represent different tree-structures such as: nested program constructs, runtime data structures, file system directories, and grammar aspects.
- Direct manipulation of objects in an integrated environment rather than tool interaction.
- Incremental compilation.
- Graphical object-oriented execution observation.
- Interactive development of language grammars with immediate feedback.
- Use of grammar interpretation for providing language dependent support.
- Structuring of grammars after different aspects: abstract and concrete syntax as well as static and dynamic semantics.

### 2.2.1 A structured view of the system

In order to give an overview of the system, we start with the user's perspective. The Orm user interface is based on two major principles: (1) direct manipulation and (2) visualization of context by means of a hierarchical window system. These principles are applied consistently throughout the system to present the objects of major interest in the system: programs, program executions, and grammars.

Objects of major interest, e.g. programs and their internal parts like procedures and classes, are represented as windows on the screen. The user can *identify* a window with the object it represents. This identification is a basic prerequisite for direct manipulation. Each window has its own set of operations which are relevant to the object, and operations such as moving, cutting, and pasting can be performed directly.

In a direct manipulation system the physical object structure is of great importance. An object always resides in a particular location, which is usually significant for its meaning. For instance, the meaning of a procedure (method) "draw" depends on which class it resides in. For some kinds of objects, the location is more of organizational than of semantic importance such as the organization of files into different folders. This physical structure is visualized directly in the Orm browser by means of a generalized window system. In traditional window systems the windows reside on the screen, overlap and can be iconized to save screen space. In the Orm window system this principle is generalized to allow not only the screen to contain windows, but *any* window can contain a local set of overlapping iconizable windows. This gives a hierarchical organization of windows which is employed to match the physical object structure. For instance, a class containing two procedures "draw" and "scale" is visualized by a class window which contains two procedure windows. Moving the "draw" window to another class window is an editing operation on the program. The context of a particular object is always clear due to the hierarchical organization of windows.

In the snapshot from an Orm session shown in Figure 2.1, the outermost window represents a UNIX directory ("ormdemo") containing a grammar object ("demo.gram"), a program object ("thing.mjol") and the I/O window for an executing program ("Program Execution Window"). The program window contains the source code with procedures nested inside classes. Different shades of grey in the window backgrounds enhance the hierarchical window structure.

The browser provides a basic navigation aid in terms of the physical object structure: an object can be located by opening its surrounding windows. The status of inner windows (e.g. icon/window state, placement within the outer window, and size) is preserved as a window is iconized or moved. When iconizing the program window, all its parts (main block, local classes and procedures) will disappear from view,

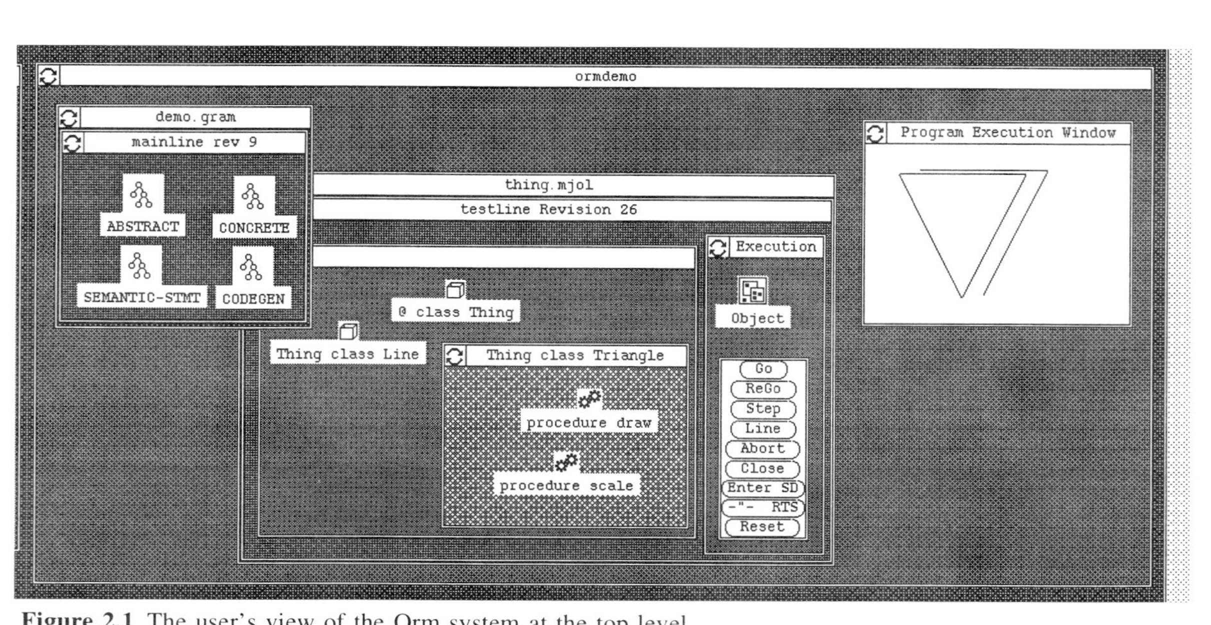


**Figure 2.1** The user's view of the Orm system at the top level.

but reappear in exactly the same state when the program window is opened again. Such information is preserved also between sessions.

In addition to the structural navigation supported by the browser, there are many important relations along which the user may want to navigate. In particular, there are many such relations in programs, e.g. super/subclass relations between classes and declare/use relations between identifiers in a program. Such relations are handled by means of *hyperlinking*. When the user clicks on a link button, the target window is made visible (opening outer windows if needed) and visual feedback is given on the target window by flashing its contents. Thus, the object structure can be seen as basically hierarchical, but with possibilities to follow links to bring another part of the hierarchical structure into view quickly.

The direct manipulation interface of Orm is similar in flavor to the Macintosh finder. It differs from the Macintosh finder by the introduction of a hierarchical windows feature which always shows an object (window) in its context. We consider this a particularly valuable property in programming environments for object-oriented languages where the context has strong semantic importance. Earlier programming environments utilizing graphics and windows usually follow the tool-based approach where the windows represent tools rather than objects, e.g. the browser and inspector tools in the Smalltalk environment (Goldberg, 1984).

### 2.2.2 A structured view of a program

The main objects of interest to a programmer are programs and program executions. A program is internally composed of blocks, e.g. classes and procedures. A program execution is an incarnation of a program and is composed of block instances, e.g. class instances (objects) and procedure activations. Documentation is viewed as part of the program rather than as a separate document.

Figure 2.2 shows an example of editing and executing a program. The source program contains classes modelling graphical "things" and the execution window contains instances of these classes. The program execution is halfway through drawing some of the "things" and the output drawing appears in the I/O window.

The goal has been to allow the user to focus on the program and execution objects, rather than tools. Thus, compilation activities, such as semantic checking, code generation, linking, and loading, are all

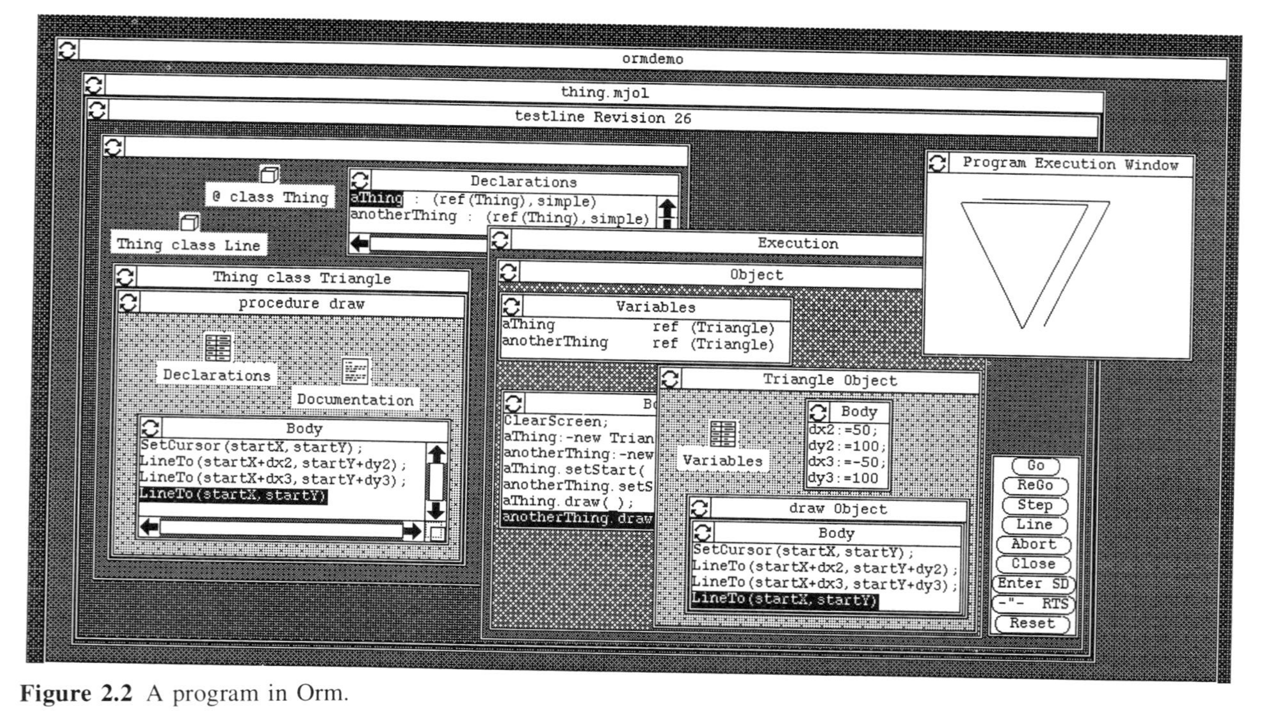


**Figure 2.2** A program in Orm.

done automatically as needed by the system without need for initiation by the user.

*Source program*

Program blocks, represented as windows and icons, can freely be laid out in two dimensions. Each block contains its "locals" (i.e. parameters and variable declarations) and its "body" (statements). These are presented as a textual unparsing of syntax trees and are edited in a structure-oriented way. The usual view of a program as a linear text has thus been abandoned.

Static semantic checking is performed incrementally as the user edits the program and continuous monitoring of static semantic errors is provided. Errors are marked unobtrusively and can be corrected at any time. An explanation of the error can be obtained via a menu command. Both erroneous and incomplete programs can be executed, giving the effect of a breakpoint at the point of the error.

*Program execution*

In the execution window, objects and procedure activations are shown nested according to the block structure. The present state shows an activation of the draw procedure of a Triangle object. The draw activation belongs to the Triangle object and is therefore located within the object. Inside the draw activation, its locals and body windows are shown. These windows are essentially instances of their templates in the program source, but the locals window contains the values of the variables rather than the types as is the case in the source. The code window shows the current (re-)activation point of the draw activation. For reference (pointer) variables, the value field contains the actual qualification of the referenced object, and it works like a link button which links to the referenced object. When the user clicks on the variable "anotherThing" for instance, the "Triangle" object's window flashes. For simple variables, such as integers and booleans, the value is shown directly in the local variables window.

The link facility for reference variables provides a method of navigating within object structures. In our example, the variable, "aThing" references a Triangle object other than the one currently shown. This object will appear when the user clicks on the variable. Another way to bring up windows is via the call stack (not shown in the example). The user thus has full control over what objects are shown in the execution window. He can iconize and also remove windows as he wishes. The reference variables act as starting points for exploration by the user.

### 2.2.3 Functionality in Orm

Orm is a structure-based environment, representing programs as well as grammars as attributed abstract syntax trees. The representation is shared among a number of tightly coupled components. The components together form the functionality of the environment. The components are integrated in two ways. First, they are communicating using a scheme for notifications, which allow incremental updates of changes to the shared representation to be distributed to other components. The notifications minimize a component's knowledge of other components in the environment, resulting in an open-ended environment where components can be added and removed without changes to the existing components (Minör, 1990). Second, the components share the same user interface, appearing as one single, seamless tool to the user.

Figure 2.3 shows a simplified view of the components in the environment. Both grammars and programs are represented as abstract syntax trees. The abstract syntax trees are divided into two levels. The higher level represents *blocks*; a block is typically an abstraction in a programming language, e.g. a class, a method or a procedure, or an

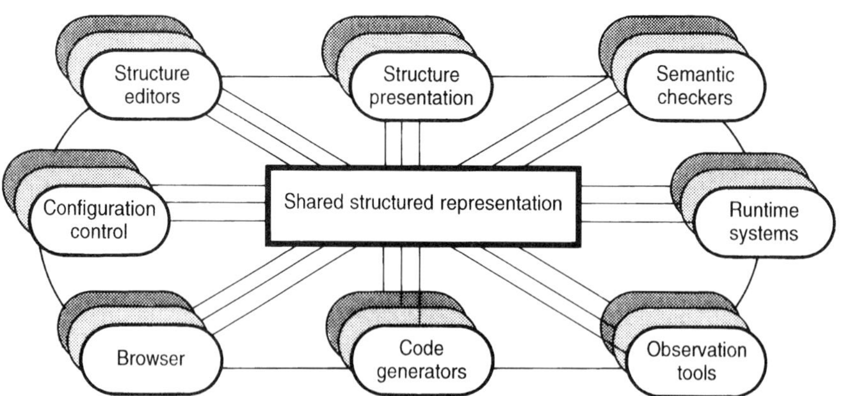


**Figure 2.3** The components of the Orm environment.

aspect of a grammar. The lower level represents the contents of a block, e.g. a class body, the contents of a method or procedure, or the contents of a grammar aspect.

The division of the abstract syntax tree has been used throughout the system. The block level is the basic means for navigation in the system using the hierarchical browser. It is also used by the incremental compiler for maintaining symbol tables and defines the granularity of the code generation. A program is organized as blocks also when stored in persistent storage in a program database. Individual components of a block, such as the symbol table, or the abstract syntax tree for the locals, can be retrieved from the database in one single operation. This grain of storage fits the access patterns of the user and of the compilation components. The database (Gustavsson, 1990) is tree structured and allows all kinds of data to be stored, e.g. syntax trees, symbol tables, window status, documentation, and generated code. Finally, the block level has been the foundation for the architecture of the environment. Each block has bound to it a cluster of communicating components operating on its content, e.g. a structure editor, a presentation component, a static semantic checker, and a code generator. The clusters of components bound to different blocks are executing in parallel, allowing multiple simultaneous activities in the different blocks, such as simultaneous editing of different classes and procedures.

The remainder of this section will give a short overview of the research areas involved in the Orm environment.

### *Structure editing*

Since Orm is a structure-oriented environment, both programs and grammars are edited by means of structure editing. There are two levels of structure editing corresponding to the division of the representation into the block level and the block-contents level.

At the block level, a structure is viewed as a hierarchy of windows in the browser (Hedin and Magnusson, 1988). The browser is essentially a structure editor with a semi-graphical presentation. The browser is used for editing the structure in accordance with a block grammar. The block grammar is defining the top level of a language's syntax. A program, for instance, consists of a number of classes and procedures and one body. A class, in turn, consists of instance variables, a class body and a number of procedures. The browser allows the user to navigate in a program in terms of abstractions of the target language. The hierarchy of the abstractions are made explicit by the window hierarchy. In contrast to browsers in most other environments, e.g. the Smalltalk browser which is activity-oriented, the Orm browser is object-oriented, allowing an arbitrary number of blocks to be open and edited simultaneously.

Within a block, a nested structure editor is used for editing. The structure editor uses textual presentation of the block's contents. The contents are edited in accordance with the syntax of the language which is expressed in language grammars. The main contributions of the Orm structure editor, compared to other structure editors, is the consistent representation of grammars and programs as abstract syntax trees and the grammar-interpreting technique utilized for defining language dependence. This is described in more detail in (Minör, 1990) and we will expand on it somewhat in Section 2.3. The editor is basically a "pure" structure editor using the "synthesizer approach", although we currently provide also conventional parsing of expressions as an alternative. However, we work on finding alternative ways to improve the interaction in pure structure editors and avoid parsing of text altogether (Minör, 1992).

The editor supports *context-sensitive editing* (Hedin, 1991). Based on the semantic information available in the semantic attributes in the abstract syntax tree, e.g. symbol tables, scope rules, and types, the editor presents variables, classes, and procedures visible in the current scope in a menu. Typed references in statically typed object-oriented languages make it possible to determine the legal set of messages to an object. The messages are made available in submenus and will, if selected, be inserted in the program with the correct number of parameters. This facility has proved to be very practical and is appreciated by the users. Particularly when using class libraries or tool-boxes, context-sensitive editing relieves the user from remembering the class names, their methods, and the protocols for calling them.

*Incremental compilation*

Orm is based on incremental compilation, for two reasons. One reason is efficiency. Since the techniques used in Orm are intended to be useful in "real" applications, the efficiency of the executing program is important. Thus compilation to native code is required. The turn-around time from editing to execution must not be too long in an interactive system. Thus fine-grained incremental techniques are used. The second reason for incremental compilation is early error detection. If the user makes an error, such as creating a type system inconsistency, the environment should report this error as soon as possible, preferably immediately after the editing operation causing the error. The user can then either correct the error or leave it as it is for later correction. The incremental compilation is divided into two components: the incremental static semantic checker and the code generator.

A semantic checker incrementally maintains syntax-tree attributes and symbol tables after each modification performed by an editor component. Incremental attribute evaluation techniques are used which consider attribute dependencies within a local abstract syntax tree. Non-local dependencies, e.g. dependencies between the syntax tree attributes and the symbol table, and between different symbol tables in the block tree, are handled by general scope mechanisms which are interfaced to the attribute evaluation mechanism. This has been found a good solution, as opposed to the pure attribute grammar approach which does not work well for incremental scope handling. We have found other solutions to these problems (e.g. Hoover and Teitelbaum, 1986) to be insufficient for the combined inheritance and nesting scopes found in object-oriented languages. The semantic checker interprets evaluation plans which are generated from a semantic grammar. More details on the semantic checking in Orm and the attribute grammar formalism are given in Hedin (1992).

Code generation is done incrementally with the granularity of a block, i.e. a class, procedure or equivalent. Each block results in a template (a structural description of the block) and a code object (a sequence of instructions). Since there is no need for feedback to the user, it can be run as a low-priority background process, keeping the executable code consistent with the source representation. The medium-grained incremental code generation lies in between the traditional separate compilation of modules and the fine-grained incremental compilation technique presented in Fritzson (1984). The code generation is divided into a front-end translating syntax trees annotated with semantic attributes to an intermediate code and a back-end translating intermediate code to native code. The front-end interprets attribute evaluation plans generated from the code generation grammar. The back-end is created using a traditional generative approach. Hence, it is possible to modify the target language interactively whereas generating code for a new target machine requires re-generation of the back-end.

*Runtime environment*

The runtime environment is not based on formal descriptions. It is hand-crafted using more traditional techniques. However, it has been our purpose to make the design general enough for runtime representation of programs in a class of block-structured object-oriented languages, such as Simula (SIS 87), Beta (Kristensen *et al.*, 1987), Eiffel (Meyer, 1987), and C++ (Stroustrup, 1986).

The runtime representation of a program is a *template tree* where each template corresponds to a block in the source representation. The dynamic state of the running program is represented as a collection of block instances which are all heap allocated and subject to garbage collection. Block calls (e.g. procedure calls, and class "new"s) are performed via indirect pointers in the template tree. The code for a block call is thus independent of the position of the called block. The flexibility of the template tree runtime representation is of great importance in the incremental setting. Individual templates and code objects can be incrementally loaded without affecting existing code in the system (Magnusson, 1983). This is utilized to support incremental execution, i.e. allowing continued execution after changes to the program (Hedin and Magnusson, 1986).

The executing program runs as a UNIX process separate from the development system, enabling program development to take place in a host-target setting. This is an essential requirement for real-time systems which usually run on dedicated machines.

Runtime observation (i.e. viewing the objects and procedure activations as described in Section 2.2.2) and incremental loading is done by a small kernel embedded in the runtime system. The bulk of the code implementing this functionality is part of the development system.

Since Orm is planned to handle programs with strict real-time requirements, the garbage collector has to work incrementally and guarantee free memory to be supplied at a required rate. Such demands are difficult to meet at a low cost. At the expense of a somewhat longer guaranteed response time, the garbage collection efficiency can, however, be improved considerably by using a generation-based algorithm (Bengtsson, 1990).

*Configuration control*

The configuration management component of Orm basically has two purposes. It supports the user in controlling the evolution of (large) software systems, and it defines and implements a data model used for persistent storage of programs and grammar structures.

The purpose of the data model is to store the block tree and the contents of the blocks, i.e. syntax trees and block attributes, on persistent storage. The block attributes may contain information such as symbol tables and generated intermediate code created by various components in the environment. Attributes of the abstract syntax tree nodes may refer to the block attributes, and the block attributes may refer to other blocks in the block tree. The data model has to deal with a graph structure, even though the basic structure is tree-like. The physical storage of the structure is implemented as sequences of bytes stored in Unix files. The representation on persistent storage is a byte representation of the program *structure* which has little in common with traditional textual representation of programs. Furthermore, the configuration control component handles loading of program and grammar structures from persistent storage. In this way the space requirements for the abstract syntax tree representation are reduced since only the parts of the representation actually used are loaded.

The evolution control supports the user in keeping track of *versions* and *alternatives* of program and grammar structures. An alternative of a program is understood as a series of versions that represent one evolution line among other alternatives. The configuration manager supports the user in creating different alternative lines of evolution and their respective versions. It also supports several different binding rules between program modules and between programs and grammars. It guarantees mutual exclusion, i.e. modifications are not done by two users simultaneously in an uncontrolled manner. It also offers the ability to merge two alternative evolution lines into one. Different specializations of such objects make it possible to maintain program structures and to support organization of software objects by membership and by composition in a uniform way. This is described in detail in Gustavsson (1990).

## 2.3 Orm as a meta environment

Several components of the Orm environment are generic. Their language dependence is defined in grammars which are interpreted by the components. Both grammars and programs are represented and stored as abstract syntax trees in Orm, enabling the use of the environment for language development as well as software development. When working on language grammars, the Orm components are interpreting meta grammars. In this section we explain how to use Orm as a meta environment. More details are found in Minör (1990).

### 2.3.1 A structured view of grammars

In Section 2.2, the editing and execution of a program object within Orm was shown from the end user's perspective. Here, we will show the grammars from the language implementor's perspective.

Grammars are manipulated by "opening" a grammar connected to a program. Editing a grammar includes development of the syntax as well as the static and dynamic semantics of the language. Like a program, a grammar in Orm has an internal structure. It consists of four *aspects* of a language, each specified in a separate formalism. The division of the aspects into different grammars using different formalisms is a consequence of the goal to make Orm open-ended. This includes the possibility to add new language aspects if necessary and to make configurations of Orm where some aspects are not present. The architecture allows experiments with different formalisms for a certain language aspect without affecting the other aspects.

- The *abstract grammar* defines the abstract syntax.
- The *concrete grammar* defines the concrete syntax.
- The *semantic grammar* defines the static semantics.
- The *code-generation grammar* defines the translation to intermediate code.

Figure 2.4 shows editing of the grammar "demo.gram". The four aspects constituting the internal structure of the grammar have all been opened and may thus be manipulated by means of structure-oriented editing. The fact that the grammar is interpreted means that the various grammars may be modified incrementally and that each

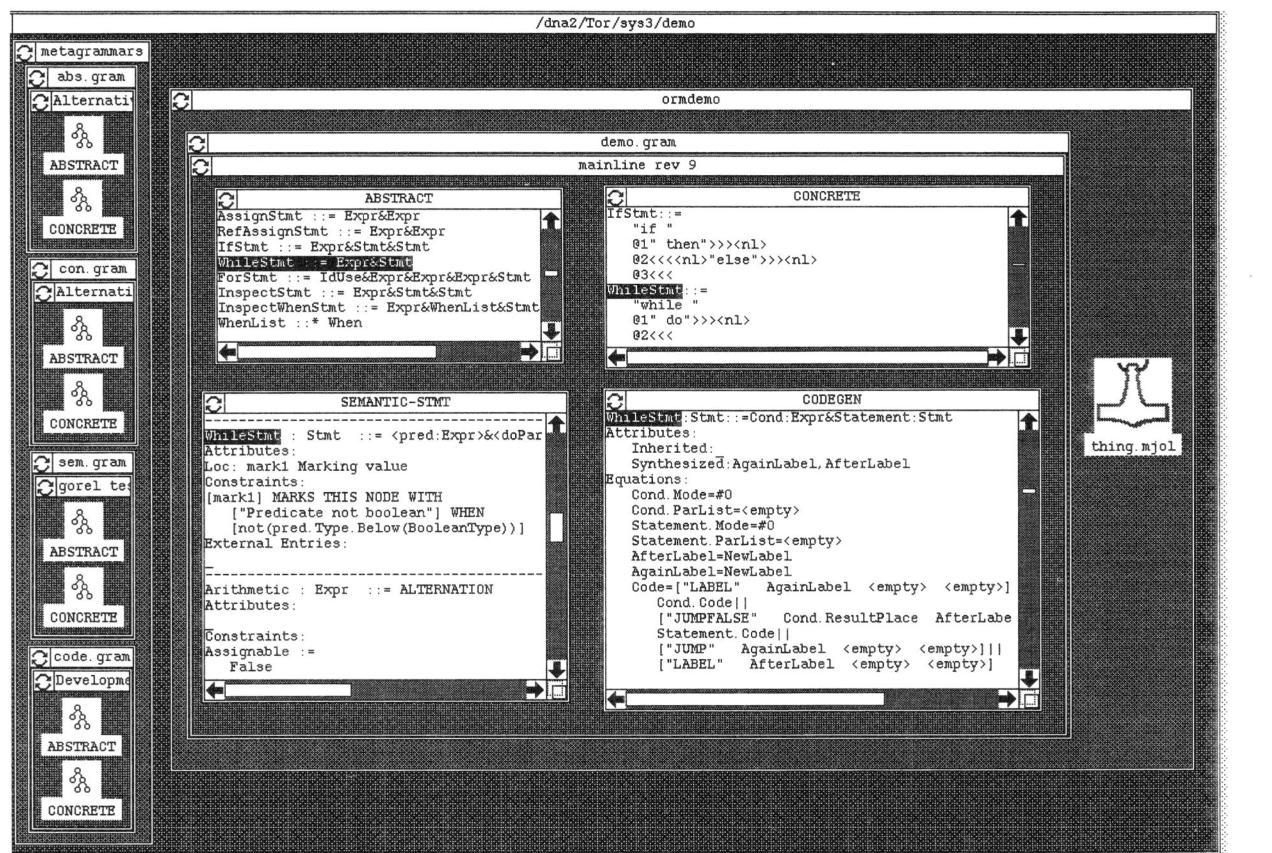


**Figure 2.4** Editing a grammar.

change comes into effect instantly. For instance, adding a new production in the respective grammar automatically means that the program may be manipulated and executed according to the new version of the language, i.e. without leaving the environment or re-generating parts of the environment.

Grammars and programs are edited with the same structure editor. The editor interprets meta grammars when editing these language grammars, e.g. the meta grammar "abs.gram" in Figure 2.4 describes different aspects of abstract grammars. Similarly, the other language grammars are described by a corresponding meta grammar. The use of the same editor and browser for both programs and grammars gives a consistent user interface throughout the environment.

### 2.3.2 Grammar interpretation

Earlier structure-oriented environments, such as the Cornell Synthesizer (Teitelbaum and Reps, 1981), were "hand-crafted" for one specific language. They were followed by systems allowing structure-oriented editors or environments to be automatically generated from a language description. ALOE (Medina-Mora, 1982) and the Synthesizer Generator (Reps and Teitelbaum, 1984) are two examples of such systems. In environments generated from language grammars the implementor alone is able to modify the grammars. Grammar modifications in generative systems are typically done in a "batch"-oriented manner with a considerable turnaround time from the grammar modification until a new generated environment is obtained.

Orm differs from most other structure-oriented environments by its grammar interpretive approach. Orm *is* a generic environment that is parameterized with language specific information by interpreting language grammars dynamically. A consequence of the approach is that the environment need not be re-generated in order to handle various languages or to adapt to a modification in the language specification. Interactive modifications of grammars are supported explicitly in the Orm environment. When a grammar has been modified the changes come into effect in the program, depending on the grammar instantly. Thus Orm supports interactive modifications and construction of both programs and *languages* within the environment. The ability to develop or modify grammars can either be utilized by a language implementor or by the end user.

Figure 2.5 illustrates the differences between a hand-crafted envir-

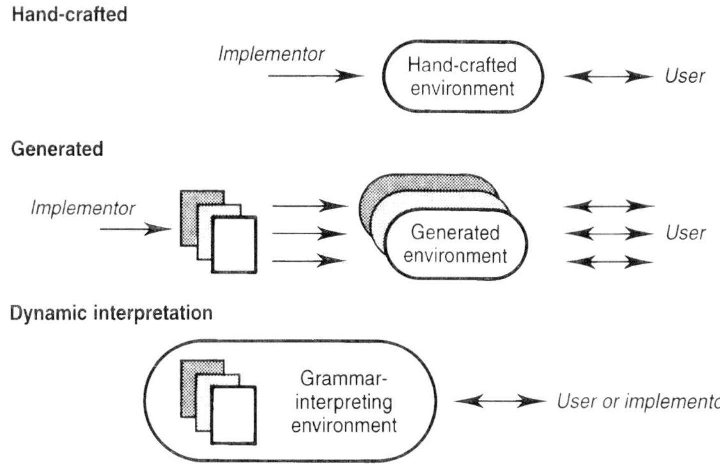


**Figure 2.5** Different kinds of environments.

onment tailored for one language, environments generated from language grammars, and a generic grammar-interpreting environment.

Other environments based on interpretation of language descriptions are DOSE, developed at Carnegie-Mellon University within the Gandalf project, and Muir, developed at Stanford University. DOSE (Feiler *et al.*, 1986), an acronym for Display-Oriented multi-purpose Structure Editor system, is a system mainly intended for language design. DOSE interprets grammars for syntax definitions and uses an action routine for the semantics. The environment is divided into the implementor's environment which is generated from language grammars and the user's environment interpreting the language description.

Muir (Winograd, 1987) is a structure-oriented language development environment. The environment interprets syntax descriptions both at the target language level and at the meta level. The environment has functionality for transformation between abstract syntax trees derived from different versions of a grammar (Nørmark, 1987).

Orm, to our knowledge, is the only system with syntax, static semantics and code generation based on grammar interpretation. Programs and grammars are consistently represented as abstract syntax trees, and the same generic components are used, including the meta level.

*Grammar interpretation in Orm*

As mentioned in the previous section, a language is described by four grammar aspects. Each aspect is interpreted by different components in the environment. Figure 2.6 shows the components and the grammar aspects interpreted. In the block grammar it is defined how the grammar is divided into higher level blocks, as discussed in Section 2.2.3. Below, we will sometimes call the abstract aspect of a language grammar the "abstract grammar" for short, and likewise the concrete aspect of a grammar the "concrete grammar", etc.

- The *block grammar* describes the structure of the higher levels of the abstract grammar, such as classes, procedures, etc.
- The *abstract grammar* defines the structure of the language. It is specified in a structured variant of extended BNF with all concrete syntax elements removed.

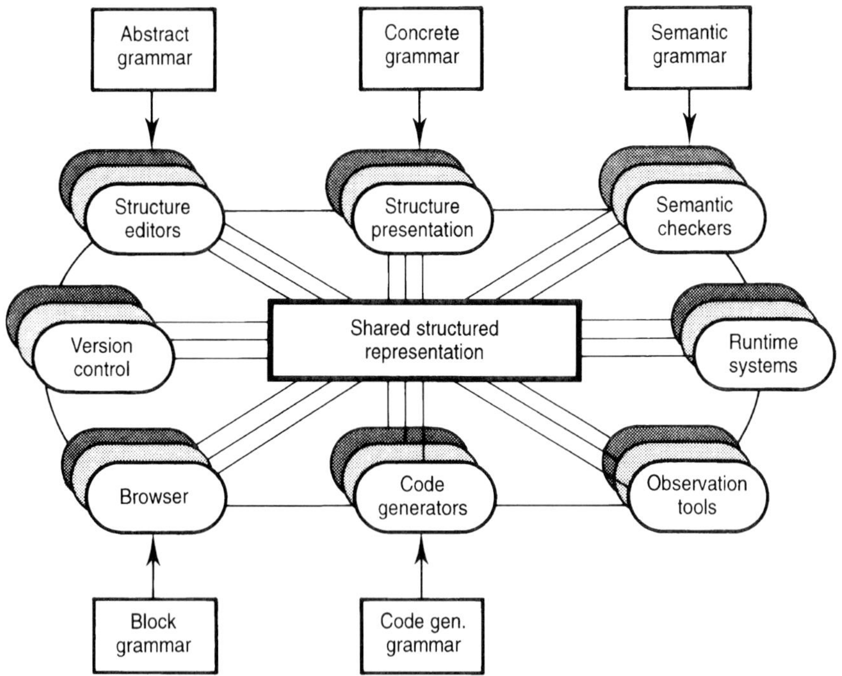


**Figure 2.6** Grammar aspects interpreted by Orm components.

- The *concrete grammar* defines the textual presentation of the language constructs, as defined in the abstract grammar. This includes the specification of the "syntactic sugar" and presentation layout for each language construct.
- The *semantic grammar* defines the static semantics, such as scope-rules, identifier bindings, and type checking. It is expressed in an object-oriented variant of attribute grammars, called door attribute grammars (Hedin, 1992), allowing attributes and equations to be inherited between productions. This allows the behavior to be specified at the appropriate level of generalization. Furthermore, they allow side-effects in a controlled manner which facilitate efficient incremental attribute evaluation (Hedin, 1988).
- The *code-generation grammar* translates the language constructs into an intermediate language expressing the dynamic semantics of the language. In this case an object-oriented variant of attribute grammars is used as well.

The browser interprets block grammars. It traverses the block grammar representation in order to check the editing operations at the block level. The block grammar is also the basis for constructing the language menu used for editing in the browser.

The abstract grammar is interpreted by the structure editor. The operations in the editor are: *expansion*, i.e. replacement of a meta node by a single new node; *pasting*, i.e. the replacement of a meta node by existing subtrees; *cutting*, i.e. the removal of subtrees; and syntactical *transformation*. In all these operations the abstract syntax tree representing the abstract grammar is traversed in order to check the consistency of the operation with the grammar. The menus used for structure editing are initially derived from the abstract grammar and the concrete grammar.

Interpretation of the concrete grammar is done by consulting the concrete grammar representation whenever the target structure has been modified by the editor. This is done by a presentation component, which is responsible for keeping the screen representation updated incrementally.

The semantic grammar is used by the incremental semantic checker components. A semantic checker maintains the attributes in the target tree incrementally after each modification according to the attribute grammar. Since attribute grammars are dependent on plans for attribute evaluation, the semantic grammar is preprocessed, resulting in evaluation sequences. The evaluation sequences are interpreted during attribute evaluation.

Finally, the code generator components generate code for a block at a time. The translation to intermediate code is based on interpretation of code generation grammars, which are variants of attribute grammars.

*Editing grammar structures*

Since all grammars, or the aspects of a grammar, are represented in the same way as programs, they may be edited by the same structure editor. When editing a grammar, the editor is interpreting a grammar description of a grammar, i.e. a meta grammar. The meta grammars are also represented as abstract syntax trees and may be edited by the structure editor as well. Figure 2.7 shows the target program, the language grammars (or grammar aspects) and the meta grammars. It also shows the structure editors used for editing the language grammars. Notice that the structure editors in Figure 2.7 consist of both an editor component interpreting abstract grammars and a presentation component interpreting concrete grammars.

In Figure 2.7, the target program structure is shown to the right. The target program is edited and executed by means of the Orm components, which are interpreting the language grammars, represented by arrows in the figure. Simultaneously as the language grammars are interpreted they may be edited by structure editors having the language grammar as target structure while interpreting meta grammars. For instance, the abstract grammar may be edited by a structure editor interpreting an abstract grammar describing the structure of abstract grammars and a concrete grammar describing the presentation of abstract grammars. In the same way any of the language grammars (or the grammar aspects) may be edited in accordance with the respective abstract and concrete meta grammars.

The figure is not symmetric, since the language grammars only are manipulated by the editor and presentation component according to the abstract and concrete grammars. In future versions of the environment we would like to include all grammar aspects also in the meta grammars resulting in five meta grammar aspects for each language grammar aspect. The block grammar could be used for browsing among productions. The semantic grammar could be used for checking static semantics in grammars, for instance if an attribute in an attribute

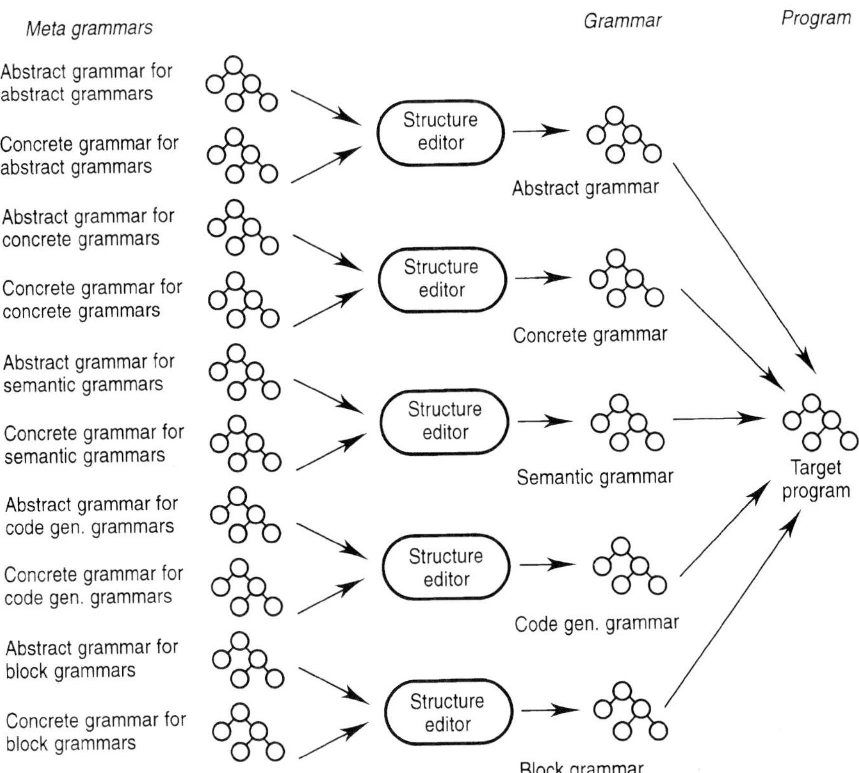


**Figure 2.7** Meta grammars, grammars and target program.

grammar is declared in that production. The code generation grammar might be used for preprocessing in the attribute grammars generating plans for attribute evaluation.

### *The meta grammars*

The meta grammars are defined and presented in terms of themselves. Figure 2.8 shows the relationships between the meta grammars, i.e. how meta grammars define and present each other.

Since Orm can be used for editing any hierarchical structure defined by grammars, it is possible to edit any of the meta grammars. However, it is only sensible to edit some of them. The concrete grammars for any

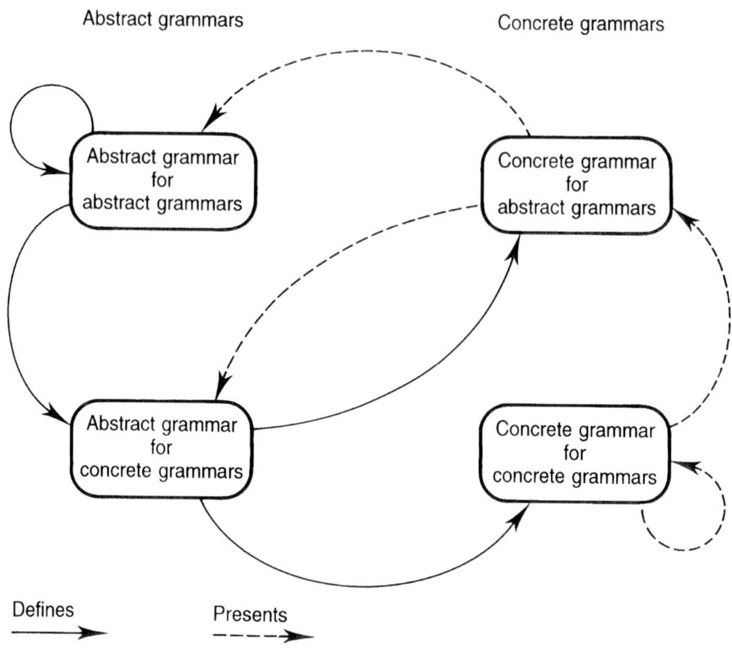


**Figure 2.8** The relationship between abstract and concrete meta grammars

of the language aspects may be modified freely, which make it possible to modify the presentation of grammars and meta grammars. The abstract grammars may, however, not be modified at all. This is due to the fact that the structure of the grammars is defining the language understood by the Orm components. They must not be modified without requiring corresponding modifications of the Orm program components.

### 2.3.3 Using Orm for language design, maintenance, and customization

The fact that all the aspects of the language grammars are interpreted results in a high degree of flexibility for the user to modify both the structure and the surface syntax of a language as well as the static and dynamic semantics. In environments generated from language grammars, only the implementor, i.e. the person generating an instance of the environment from a grammar, has this ability. The end user of the generated environment has no ability to influence it.

*Customizing surface syntax*

By changing the concrete grammars of a programming language, the surface syntax may be changed completely. The potential of this ability should, however, not be overestimated. The concrete grammars in Orm are rather primitive, but even with more elaborate presentation schemes the possibilities are limited. A programming language is to a large extent dependent on the abstract structure of the language. Changing the concrete grammars can, however, be useful. Users often have different opinions about how a program should be formatted in order to gain in readability and aesthetic qualities. It is partly a matter of taste, but a familiar layout certainly makes a program easier to read and comprehend. This has led companies to develop standards for program layout in order to support different programmers to read each other's programs. Through the use of a system where the user may specify his or her own personal layout, such discussions become irrelevant. The layout is not a property of the individual program. Various presentations may easily be created without affecting the abstract representation of the program. The program layout is thus a matter of personalization. The personalization is defined interactively by the user himself, only limited by the expressive power of the concrete grammars. If the user wishes, the surface syntax may be totally redefined.

*Command language interface*

In Minör (1987), the use of the Orm structure editor as a command language interface is discussed. For command interaction it is reasonable to provide different surface syntax of the command language for different user categories. Beginners may prefer a more verbose dialog with explanations, whereas expert users prefer a more compact surface syntax. The different presentations of the command structures may be set up in Orm in several alternative concrete grammars for different levels of expertise. The language used in the presentation (Swedish or English for instance) or translation between languages used in the command interaction can also be done in a straightforward and well defined manner. Furthermore, badly designed command languages with incompehensible commands may be improved, at least at the surface level. The ability to personalize the command language of an interface interactively is, of course, desirable.

*Text processing*

Modifications of concrete grammars may be of use in text processing as well. The text contents, paragraphs, chapters, and different styles are represented in the abstract syntax tree according to the abstract grammar. The concrete grammar specifies the actual layout chosen. As an example consider reference lists. A reference typically is a structure with author(s), title, magazine or book or proceedings, volume, number, month, and year. The typography used in different reference lists varies although the contents are identical. Converting from one format to another may be a tedious task. An abstract representation of a document together with a presentation that is modifiable by the user interactively would be desirable. However, the example requires a concrete grammar more powerful than the one provided in Orm at present, with support for different fonts, italic, underline, etc.

*Language design and maintenance*

Language design is an area where the ability to modify the syntax and semantics interactively is interesting. Designing a programming language is a long term process where several design decisions have to be considered carefully. However, language design is not restricted to the design of traditional programming languages. In many cases application-oriented languages, or *little languages* (Bentley, 1986), are designed as an interface to an application, e.g. command languages, database query languages, specification languages, configuration languages, and grammar formalisms. The ability to create and modify grammars interactively supports rapid prototyping of languages. A modification of the grammar will directly come into effect and the result becomes visible for evaluation instantly. The importance of *iterative design* in general has been emphasized, e.g. in Brooks (1987). The interpretation of grammars supports iterative design of both the syntax and semantics of languages. The language designer is allowed to experiment with the language and to evaluate it repeatedly while it evolves.

A related area is software environment maintenance. Since most of the language specific information is expressed in grammars in Orm, a large deal of the language-related maintenance in traditional environments becomes grammar maintenance in Orm. Like programs, grammars are not correct from the beginning. In Orm grammar errors may be corrected and the result may be evaluated without the need of a time-consuming regeneration of the environment. Grammars may be "debugged" interactively. The environment also makes it easy to adapt a grammar to a new version of a language or to a specific dialect.

A general problem when having a language description and an instantiation of it, e.g. a program, is the inconsistency of the program and the new description after modifying the description. The problem applies to software environments in general, but it is particularly evident in grammar-based environments. The grammar interpretive technique itself is not causing the inconsistencies. Systems generating environments from grammars are subjected to the same problem. However, the ability to modify grammars interactively encourages frequent modifications and hence potentially more inconsistencies. Inconsistencies resulting from changes of concrete, semantic and code-generation grammars are easy to deal with since the screen representation, the semantic attributes, and the code can easily be regenerated even though it may take some time. The abstract grammar, however, defining the structure of the representation is a harder problem.

A rather simple approach to deal with inconsistencies is used in DOSE (Feiler *et al.*, 1986). Target structures inconsistent with the grammar are simply discarded. The approach is, of course, not satisfactory, even though it actually results in consistent grammars and programs. More elaborate approaches based on semi-automatic transformations have been suggested in Nørmark (1987) and Garlan *et al.* (1987). None of the approaches allows target structures to be transformed automatically after complex grammar modifications. Orm follows a different strategy. Inconsistent programs have to be transformed "manually" after any modification of the abstract grammar causing inconsistencies. However, not all modifications result in inconsistencies. An abstract grammar can be extended, for instance, without causing inconsistencies.

## 2.4 Orm in practical use

The development of Orm has served several purposes. One was to actually implement and try out the techniques mentioned above.

Another was to establish a platform to serve as a basis for the integration of more advanced features needed for large-scale programming and covering a broad range of the software life cycle. An alternative would have been to stick to the traditional textual representation. This is currently the dominating trend for commercial systems which are often built on existing tools as a way to quickly get useful results. We are, however, convinced that in the long run, incremental techniques based on a structured representation have a much higher potential for achieving powerful program development systems.

Orm is implemented in Simula and runs on Sun SPARCStations. As a programming environment it is in the test state and has been used for implementation of small programs (up to 1000 lines or so). As a language laboratory, the system has proven highly attractive although it still has some "rough edges". The largest grammar implemented is the one for Simula, consisting of about 100 productions in each grammar aspect. Printed out, the grammar is about 4000 lines of sparse text. In handling full programming languages there are currently shortcomings, mainly due to some semantic functions not being general enough. For this reason it has not yet been possible to implement, for instance, a full Beta system.

The response times of the environment as a whole are typically fractions of a second. When executed on a modern workstation, most operations such as edit operations, semantic checking, and changes to the language grammars, are fast enough for interactive use. The response times have not degraded noticeably when programs and grammars grow to the size we have used, as described above. As a matter of fact, the underlying window system at present is one of the most time-consuming components. We are thus confident that the approaches and techniques used scale up to support development of large programs in industrial applications.

The grammar editing facilities have been valuable within the Mjølner project. They have been used for developing the language grammars for Simula and partly for Beta. All the grammar aspects for these languages have been developed and "debugged" interactively.

The grammar formalisms for the block, abstract, concrete, semantic, and code generation grammars have also been developed using Orm. This has been done by developing the meta grammars, i.e. the abstract and concrete grammars, for each grammar. The first versions of the meta grammars for abstract and concrete grammars were, however, initially entered using other tools in order to bootstrap the grammars.

The meta grammars for the block, semantic and code generation grammars have been developed using Orm exclusively. The grammars have evolved successively and modifications have been frequent. Orm has shown to be a practicable tool for experimenting with the formalisms.

Outside the project, Orm has been used as a meta environment in teaching the compiler construction course at our department for several years with good experience. The students have been able to experiment with grammars and languages, and it has been possible to evaluate the results of grammar modifications immediately. Several companies have used Orm in the same way for prototyping language syntax.

Orm also has been used for "little" languages. At a company, grammars for the Unix *make* language were developed. Make is known to have a cryptic syntax which is hard to learn and use. The support offered by the structure editor has proved to be useful. An alternative more verbose and readable concrete syntax was defined. The atlernative concrete syntax was shown to users while the original concrete syntax was used when the makefile "program" was sent to make for execution.

Orm has been used by Swedish telecom companies for language prototyping in several projects. The ability to obtain a structure editor in short time and to evaluate and refine the language constructs has been reported to be very useful. It was also used to demonstrate ideas of a prototype language to a larger group of users. The users were given Orm with the prototype grammar and were allowed to play around with the language in order to evaluate it and give feedback. Based on the feedback, the language design was refined. The structure editing was considered to be very useful when exposing users to a new language design, since the users did not have to be experts in the syntax. The structure editor support allowed them to construct, e.g. programs or database schemes for evaluation without prior training.

Structure editing in general has proved useful when using Orm as a meta environment. Since the grammar formalisms used are not well-known, which is often the case with programming languages, the suport of the structure editor has been appreciated. Even among the people who have designed the notations themselves, the support was needed. One reason for this is that the users are "occasional users", i.e. they do not use the notations daily, and after a month or two they needed support when editing a grammar. Another reason is that the notations sometimes have changed from day to day, particularly in the initial phases of the project. The structure editors have helped in presenting the up-to-date version of the grammar notation.

Orm has been used in practice as a programming environment in teaching the first programming course at the authors' department. A group of 15 students used the Orm environment when the rest of the students used a traditional compiler/debugger. On the course, object-oriented programming was taught using Simula. In general, the experiences using Orm were good. There was no examination at the end of the course, making it hard to evaluate if the student using Orm was better or worse than the other students. A fairly comprehensive questionnaire was filled in by the students using Orm, enabling a rough evaluation to be made. The questionnaire was handed out after the students had used the traditional Simula system on the next course, enabling them to compare Orm and a conventional environment. Most students had additional experience from using other systems from outside the university.

Most complaints about Orm concerned execution speed. On the course, the system had to be used on machines too small and too slow to give Orm a fair chance as an interactive system. Despite the bad response times, the students were surprisingly positive: 80% of the students considered Orm to be better or much better than a traditional environment; all students thought the system was fairly easy to learn; over 80% felt that they had control over what happened in the environment. The integrated incremental nature of the environment was considered to be of great value by all students. Almost all students had "fun" using the system.

Concerning structure editing, no student was unhappy with constructing programs by means of menus. When asked whether they preferred using menus or editing program fragments as text, 20% answered that they used both to equal degree and 80% preferred using the menus. All students considered the context-sensitive editing technique to be very useful and 93% of the students preferred using this mode to entering identifiers as text. Furthermore, the hierarchical browser was received positively by all students. On the question whether they had preferred to use a conventional text editor instead of the structure editor, 93% answered "no" or "absolutely not" (7% answered "maybe"). On the questons about other parts of the environment, the students were almost exclusively positive.

Finally, a surprising observation about programming style was made during the programming of assignments. Several students using Orm placed the declarations of variables at the wrong block level. Usually the variables were declared globally instead of locally. Apparently, the Orm browser makes the (geometrical) distance between blocks smaller: a global declaration is made as easy as a local one. The browser did not, as it was intended to, convey the hierarchical scopes of the language to the extent that such mistakes were avoided. None of the students using the conventional system made that kind of error.

Besides the use of Orm reported above, the system has been distributed to other universities and research labs for evaluation and experimenting. At the time of writing, the system has been distributed to some 30 sites. We hope to get feedback on the system and gain experience to improve it.

## 2.5 Future research

Orm has been developed to a fairly complete and working prototype. However, it is still far from being a system that can be used in "real" applications, e.g. in industrial settings. But the system may be used as a platform for further research concerning both the language and software development support.

Concerning language development, we would like to extend the functionality and the meta grammars to give more support for browsing in grammars and check semantic consistency, as described in Section 2.3.3. We also would like to give more support for language "debugging", e.g. the possibility to follow the attribute evaluation in the semantic and code generation grammars. Better support for transformations of abstract syntax trees between different versions of the grammar is also needed.

Support for "little languages" is an interesting area to explore further. The integration of the Orm meta environment to different environments using the little language as interface has to be further developed. Another area in this context is grammar modularization. When developing different little languages, as well as different programming languages, a general problem is that parts of different languages are identical or similar, e.g. the syntax and semantics for arithmetic and boolean expressions. The ability to share grammars or grammar parts between different languages would be a solution to this problem. In the same manner as code is shared and reused by using modules and separate compilation in software environments, grammars could be reused. In a system supporting *grammar reuse*, it would be possible to define "production libraries" from which the syntax and semantics of different language constructs could be imported and further specialized.

Concerning software development, semantic support is an area needing further investigation. The context-sensitive editing in Orm has shown to be very useful. The idea to use the semantic information stored in the abstract syntax trees for supporting the user to read, understand, and browse in a software system is a more powerful way of working than that available in conventional text-based systems. By improving this kind of facility, the power for giving valuable language-based support in structure-based environments could be demonstrated.

Programming-in-the-many, or cooperative software development, is an interesting and demanding area for future research. Software development is frequently teamwork, but commonly used systems are weak in supporting groups working together. Orm is at present a single-user environment, even though its revision control system together with its module handling gives some support for developing software and grammars in a project. We believe the environment is a suitable platform for developing an environment supporting multiple cooperating users. This belief is based on the following properties of the environment:

- The structure-oriented architecture.
- The integrated configuration control system.
- The graphical browser.

First, the structure-oriented architecture facilitates full control in the system. All information about a program is collected in one representation, which facilitates sharing of the information among multiple users. If some part of a program is changed by one user, it is possible for any tool to keep track of its consequences and notify other users if desired. Second, the integrated configuration control system keeps track of versions and revisions. It also handles alternative development lines of a program, which is crucial when developing software in a project. Finally, the graphical browser is used for navigation in the program by means of a hierarchy of windows reflecting the program structure. Since the presentation is structured, we believe it will facilitate visualization of several users working on one software component.

## 2.6 Summary

We have presented Orm, a structure-oriented environment for program and language development. A significant characteristic of Orm is its consistent user interface based on direct manipulation. Direct manipulation relieves the user of cognitive burdens and has proven highly successful within the office automation area. We think it is extremely useful also for simplifying the complex activity of software development.

To make mundane activities such as compiling, linking, and loading fast, automatic, and invisible to the user, Orm employs incremental techniques which are triggered by need. Similar ideas have also been pursued in other environments such as Gandalf and the Cornell Synthesizer, but have been developed further in Orm. Orm also differs from these environments in focusing on object-oriented languages which require more complex compilation technology, and by providing more powerful support during execution.

As a language development environment, Orm offers similar advantages as for programming: grammars are developed by means of direct manipulation and new or changed language constructs can be immediately tested on a target program. This applies not only to developing programming languages, but to developing any formal language, including the meta formalisms themselves. Development of formalisms will certainly be an increasing area of importance within software development, in particular for the design and development of small special purpose languages.

For many important issues, like programming-in-the-many and programming-in-the-large, we have only scratched the surface. However, we believe it much more beneficial to tackle these problems within a highly interactive and structure-based environment like Orm, than within traditional text-based environments.

## Acknowledgements

The Orm environment has been developed within the Department of Computer Science at Lund University. Several people have been involved in the work described in this chapter. Mats Bengtsson has been working with the run-time environment and garbage collection;

Lars-Ove Dahlin has worked with incremental semantic checking together with Görel Hedin who also designed the attribute grammar formalism; Göran Fries has been working with the back-end code generation generator; Anders Gustavsson has designed the configuration control system; Dan Oscarsson has supplied the hierarchical window system; Magnus Taube made the run-time observation tool; Sten Minör has been involved in the structure editor, the grammar interpretation, and the front-end code generator; and finally, Boris Magnusson has been project leader of the Orm subproject within the Mjølner project.

The work reported in this chapter has been carried out as part of the Nordic Mjølner project. The project is supported by The Nordic Fund for Industrial Development and The Swedish National Fund for Technical Development (STU–NUTEK).

## References

Bahlke, R. and Snelting, G. (1992). Design and structure of a semantic-based programming environment. *International Journal of Man–Machine Studies*, **37**(4): 467–480.

Bengtsson, M. (1990). Real-time compacting garbage collection algorithms. Licentiate thesis, Department of Compuer Science, Lund University.

Bentley, J. (1986) Little languages. *Communications of the ACM*, **29**(8): 711–721.

Borras, P., Clément, D., Despeyroux, T., Incerpi, J., Kahn, G., Lang, B. and Pascual, V. (1988) CENTAUR: the system. In *Proceedings of ACM SIGSOFT'88: Third Symposium on Software Development Environments*, November.

Brooks, F. P. Jr. (1987) No silver bullet – essence and accidents of software engineering. *IEEE Computer*, April: 10–19.

Donzeau-Gouge, V., Huet, G., Kahn, G. and Lang, B. (1980). Programming environments based on structure editors: the MENTOR experience. INRIA Research Report, No. 26. Also in Barstow *et al.* (editors), *Interactive Programming Environments*, McGraw-Hill, 1984.

Feiler, P. H., Jalili, F. and Schlichter, J. H. (1986) An Interactive Prototyping Environment for Language Design. In *Proceedings of the Nineteenth Annual Hawaii International Conference on System Sciences.*

Fritzson, P. (1984). Towards a distributed programming environment based on incremental compilation, Ph.D. thesis, Dissertation no. 109, Department of Computer and Information Science, Linköping University.

Garlan, D., Krueger, C. W. and Staudt, B. J. (1987). A structural approach to the maintenance of structure-oriented environments. In *Proceedings of the ACM SIGSOFT/SIGPLAN Software Engineering Symposium on Practical Software Development Environments, SIGPLAN Notices*, **22**(1): 160–170.

Goldberg, A. (1984). *Smalltalk-80: The Interactive Programming Environment*, Addison-Wesley Publishing Company.

Gustavsson, A. (1990). Software configuration management in an integrated environment, Thesis LU-CS-TR:90–52, Department of Computer Science, Lund University.

Hedin, G. (1988). Incremental attribute evaluation with side-effects. In *Proceedings of the Workshop on Compiler Compiler and High Speed Compilation*, Berlin GDR.

Hedin, G. (1991). Context-sensitive editing in Orm, Department of Computer Science, Lund University.

Hedin, G. (1992). Incremental semantic analysis, Ph.D. thesis, Department of Computer Science, Lund University.

Hedin, G., and Magusson, B. (1986). Incremental execution in a programming environment based on compilation, In *Proceedings of the 19th Hawaii International Conference on System Sciences.*

Hedin, G. and Magnusson, B. (1988). The Mjølner environment: direct interaction with abstractions. In *Proceedings of ECOOP'88 European Conference on Object-Oriented Programming*, Lecture Notes in Computer Science, Vol. 322, Springer-Verlag.

Hoover, R. and Teitelbaum, T. (1986). Efficient incremental evaluation of aggregate values in attribute grammars. In *Proceedings of the ACM SIGPLAN Symposium on Compiler Construction.*

Hutchins, E. L., Hollan, J. D. and Norman, D. A. (1986). Direct manipulation interfaces. In Norman, D. A. and Draper, S. W. (editors), *User Centered System Design*, pp. 87–124, Lawrence Erlbaum Associates.

Knudsen, J. L., Madsen, O. L., Nørgaard, C., Petersen, L. B. and Sandvad, E. (1989). *An Overview of the Mjølner BETA System*, Mjølner Informatics ApS, Aarhus.

Kristensen, B. B., Madsen, O. L., Møller-Pedersen, B. and Nygaard, K. (1987). The BETA programming language. In Shriver, B. and Wegner, P. (editors), *Research Directions in Object-Oriented Programming*, The MIT Press, Cambridge MA.

Madsen, O. L., Magnusson, B. and Møller-Pedersen, B. (1990). Strong typing of object-oriented languages revisited. In *Proceedings of OOPSLA '90 Conference on Object-Oriented Systems, Languages, and Applications.*

Magnusson, B. (1983). Code-objects: a support for incremental compilation, Department of Computer Science, Lund University.

Medina-Mora, R. (1982). Syntax-directed editing: towards integrated programming environments, Ph.D. thesis, Department of Computer Science, Carnegie-Mellon University.

Meyer, B. (1987). *Object-Oriented Software Construction*, Prentice Hall.

Minör, S. (1987). Structured command interaction based on a grammar interpreting synthesizer. In *Proceedings of the Second IFIP Conference on Human–Computer Interaction*, North-Holland.

Minör, S. (1990) On structure-oriented editing, Ph.D. thesis, Department of Computer Science, Lund University.

Minör, S. (1992). Interacting with structure-oriented editors, *International Journal of Man–Machine Studies*, **37**(4): 399–418.

Møller-Pedersen, B., Belsnes, D. and Dahle, H. P. (1987). Rationale and tutorial on OSDL: an object-oriented extension of SDL, *Computer Networks*, **13**(2): 97–117.

Notkin, D. (1985). The GANDALF Project, *The Journal of Systems and Software*, **5**(2): 91–105.

Nørmark, K. (1987). Transformations and abstract presentations in a language development environment, Ph.D. thesis, Computer Science Department, Aarhus University.

Reps, T. and Teitelbaum, T. (1984). The Synthesizer Generator. In *Proceedings of the ACM SIGSOFT/SIGPLAN Software Engineering Symposium on Practical Software Environments.*

SIS (1987) *Programming Languages – SIMULA*, Swedish Standard, SIS, Stockholm.

Stroustrup, B. (1986). *The C++ Programming Language*, Addison-Wesley.

Teitelbaum, T. and Reps, T. (1981). The Cornell Program Synthesizer: a syntax-directed programming environment, *Communications of the ACM*, **24**(9): 563–573.

Winograd, T. A. (1987). Muir: a tool for language design, Report No. STAN-CS-87–1159, Department of Computer Science, Stanford University.

# CHAPTER 3

# An OCCAM programming environment

*M. Filali, Y. Jamoussi, A. Knani and J. C. Maurize*

## Abstract

We present a programming environment for the concurrent language OCCAM. Its main feature is program transformations. The basic transformations for finite programs have been extended to a class of infinite programs: programs designed within the client–server model. Distribution transformations are considered, and advanced features of distributed programming are discussed.

## 3.1 Introduction

The OCCAM language has been defined formally as a process algebra (Hoare, 1985); calculus rules and equivalence relations between terms of this algebra have been established. In order to let a programmer manipulate his OCCAM programs with these well defined concepts (algebra terms, term transformations), we have built an OCCAM programming environment.

The main feature of this programming environment is a syntax directed editor which provides program transformations, where the transformation consists of replacing a term by an *equivalent* term; consequently, although the program is modified textually, its semantics remain the same. In our programming environment, we have implemented some of the basic transformations for finite programs (Roscoe and Hoare, 1988) and other transformations which concern infinite programs: programs designed within the client–server model, for which we propose distribution transformations. A distribution transformation consists of replacing a *server* process with a composition of processes; the motivation of such a transformation is to adapt a given program to some underlying distributed architecture.

Structure-based Editors and Environments
ISBN 0–12–681890–8

In the remainder of this chapter, we successively present the tool with which we have built our programming environment and the OCCAM language. Then, we consider the basic transformations of that language and the new transformations considered: *distribution transformations*. In the last section, we discuss some extensions: debugging features and *superposition* that should be easily considered in the framework of our programming environment.

## 3.2 A syntax editor generator: the Cornell synthesizer generator

In this section, we will give a brief overview of the Cornell synthesizer generator (henceforth called CSG), we have used for building our environment; for a detailed description and examples the reader is referred to Reps and Teitelbaum (1989).

The CSG is a tool for specifying and manipulating an abstract syntax. The programming language of the CSG is SSL; mainly, this language allows the description and the manipulation of abstract syntax trees. Another important feature of the CSG is the definition of inherited and/or synthesized attributes (Deransart *et al.*, 1988). For instance, these attributes can be used for semantic checks.

In this section, we will try to present the CSG as a tool for the definition of an algebra and the manipulation of its terms. Other descriptions and uses of the CSG may be found, for instance, in Van Eijk (1990) where the CSG has been used for the development of LOTOS tools, or in Vogt *et al.* (1990) where the goal research issue was the development of program transformation systems.

### 3.2.1 Definition of an algebra

An abstract syntax can be considered as an algebra; each language construction being the application of an operator to algebra terms. In SSL, a *phylum* defines such a set of terms. There exist some predefined phyla like strings (*STR*), integers (*INT*), etc.

We can build new phyla, for instance, in a recursive way, let the phylum *process*[1] defined as follows:

[1] The word "process" employed here is different from the "process" of operating systems; here, a process may be composed of one or more operating system processes executing on one or more processors.

```
process  :  Skip()
         |  Stop()
         |  Par(process process)
         |  Seq(process process)
         ;
```

The phylum *process* is then defined from syntax constructs or operators:

- of arity zero: *Skip*, *Stop*;
- of arity two: *Par*, *Seq*;

and recursively the phylum *process*.

Then the set of terms described are the following:

- The phylum *process* is the set:
  *process* = {*Skip*(), *Stop*(), *Par*(*Skip*(),*Skip*()), *Par*(*Skip*(), *Stop*()), . . .
- The syntax construct *Skip* defines the one element set {*Skip*()}.
- The syntax construct *Par* defines the following process subset:
  {*Par*(*Skip*(),*Skip*()),*Par*(*Skip*(),*Stop*()), . . .
  . . .

### 3.2.2 Manipulation of terms

In SSL, each term is *typed* by a phylum; basic operators to manipulate phyla are:

- term construction;
- typed term assignment;
- definition of term sets by a pattern matching mechanism;
- conditional expressions;
- functions.

Since our aim is not to give a detailed presentation of the SSL language, we will instead present the main features of that language by means of examples.

*Examples*

A conditional expression is an expression whose value depends on the pattern of its argument. The following example illustrates such a feature; the value of the conditional expression built is the process resulting from its reduction; that reduction is done according to its pattern:

```
with    / *argument* / (p)(
        / *pattern*/         Par(Skip(),p0)        :p0,
                             Par(Stop(),Stop())    :Stop(),
                             Seq(p1,Skip())        :p1,
                             Seq(Skip(),p2)        :p2,
                             default               :p
                             )
```

SSL functions are like usual functions with parameters and a result which belongs to a phylum. The body of an SSL function consists of an expression, whose value will be the result of the function. For instance, the following function called *reduction* returns the process obtained by recursive reduction of the process parameter:

```
process   reduction   (process p){
          with        (p)(
                      Par(Skip(),p0)        :reduction(p0),
                      Par(Stop(),Stop())    :Stop(),
                      Seq(p1,Skip())        :reduction(p1),
                      Seq(Skip(),p2)        :reduction(p2),
                      default               :p
                      )
          }
```

### 3.2.3 Phylum editing and transformation

In order to interact with terms, the CSG defines:

- how to input a term;
- how to visualize a term by means of unparsing rules;
- how to transform a term by means of transformation commands.

*Term input*

In the CSG, a term can be constructed by input from the environment; a term can be constructed from a stream of characters through the UNIX compilation tools LEX (Lesk, 1975) and YACC (Johnson, 1978).

*Unparsing rules*

An unparsing schema can be considered as a visualization function defined on a term. In SSL, such functions have their own syntax which is well suited to special editing aspects (tabulations, fonts, . . . ). A noteworthy feature of these functions is the *view* concept through which a given term can be edited in several ways (for instance, a statement can have an extended and resumed unparsing schema).

*Term transformation*

Moreover, the user can transform a term interactively. These transformations substitute a term by another one of the same phylum.

For the phylum *process* defined in Section 3.2.1, we can define the following editing transformations:

```
transform  process         on
           "SKIP"          < process >   :Skip(),
           "STOP"          < process >   :Stop(),
           " < SEQ > "     < process >   :Seq([process],[process]),
           " < PAR > "     < process >   :Par([process],[process]);
```

Figure 3.1 illustrates the construction of terms associated with the phylum *process* through editing transformations.

We note that these transformations could be used for other transformations than syntactic transformations. For instance, we could define a semantic transformation by the replacement of a term by an equivalent term according to some equivalence relation. If we consider the preceding process example, we can specify the following semantic transformations:

```
transform  process                 on
           "SKIP||P = P"           Par(Skip(),p0)         :p0,
           "STOP||STOP = STOP"     Par(Stop(),Stop())     :Stop(),
```

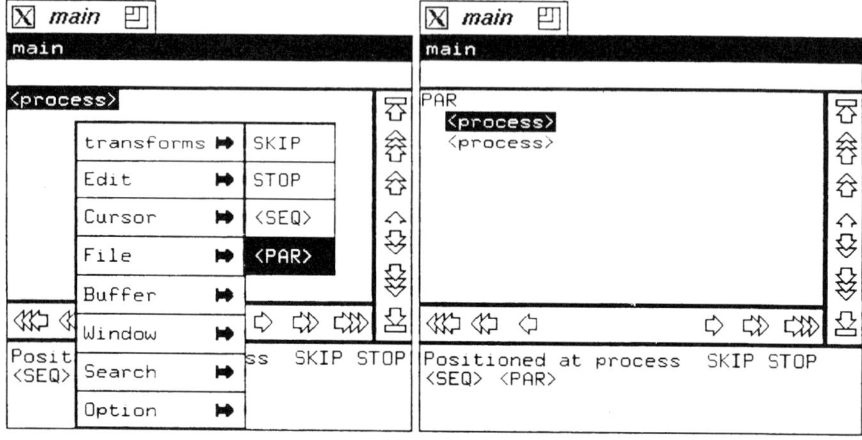


**Figure 3.1** Syntactic transformations of the term *process*.

| | | |
|---|---|---|
| *"P;SKIP = P"* | *Seq(p1,Skip())* | *:p1,* |
| *"SKIP;P = P"* | *Seq(Skip(),p2)* | *:p2;* |

# 3.3 The OCCAM language

## 3.3.1 Informal description

In the OCCAM language, an application is structured as a network of communicating processes. In this section, we give a brief overview of the language. OCCAM processes are either primitive or constructed. The primitive processes are as follows:

- **SKIP** is the process that does nothing but terminates.
- **STOP** is the process that does nothing but does not terminate (such a process is generally used when an error is encountered).
- The assignment $v := e$.
- The input process $c?v$ where $c$ is a channel and $v$ is a variable, consists of reading from channel $c$ into variable $v$.
- The output process $c!e$, where $c$ is a channel and $e$ is an expression, consists of writing to the channel $c$ the value of $e$.

In the following, we present the original constructors **PAR**, **ALT** only, the other ones, **SEQ**, **IF WHILE**, being well known.

- The **PAR** constructor defines a process whose execution is the concurrent execution of a list of processes.
- The **ALT** constructor defines a process whose execution is a non-deterministic choice (the choice can depend on the variables of the process but also on its environment, i.e. the state of its channels) between a list of processes.

In OCCAM, the communication on channels follows the rendezvous protocol: the first process sending (respectively receiving) a message on a channel has to wait for the second process receiving (respectively sending) a message on that channel; then both processes will execute concurrently.

Finally, we mention that OCCAM has facilities to define typed data and iterated constructs.

*Examples*

We illustrate some original aspects of the language by means of two well known examples:

- a merge process (Figure 3.2),

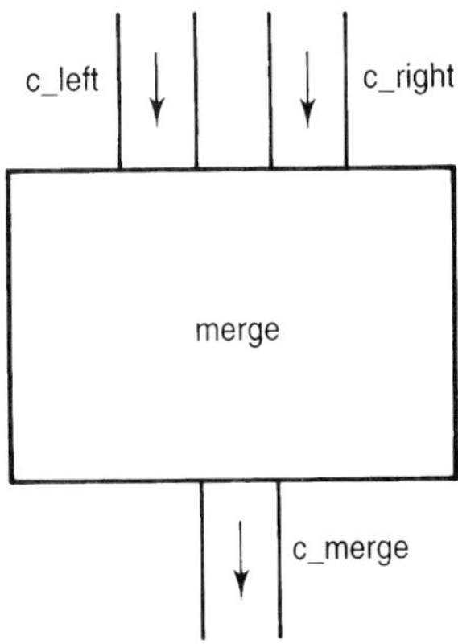


**Figure 3.2** A merge process.

```
CHAN OF INT c_left, c_right, c_merge:
INT m:
WHILE TRUE
  ALT -- non deterministic choice
    c_left ? m
      c_merge ! m
    c_right ? m
      c_merge ! m
```

- a concurrent buffer (Figure 3.3),

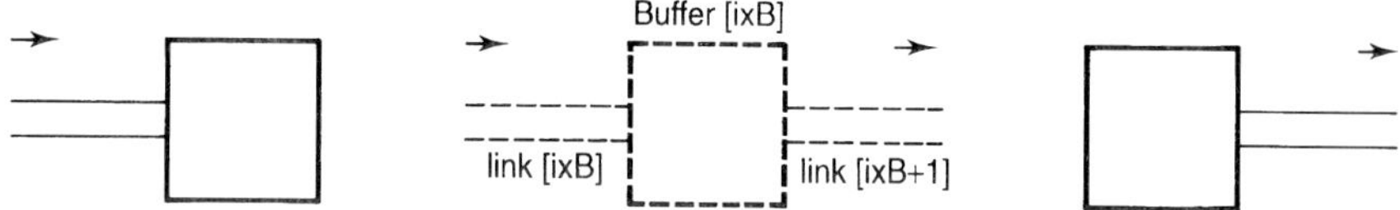


**Figure 3.3** A concurrent buffer process.

```
VAL INT length IS 10:
[length+1] CHAN OF INT link:
PAR ixB = 0 FOR length
  INT data:
  WHILE TRUE
    SEQ
      link[ ixB ] ? data
      link[ ixB + 1 ] ! data
```

### 3.3.2 Syntax description

In this section, we present the OCCAM language algebraically (Hoare, 1985): we present the set of terms to which belongs an OCCAM program. This presentation may seem unusual, but it results from the fact that we do not want to write OCCAM programs but we want to *transform* them. Finally, we do not consider the language aspects concerning time and physical distribution over processors.

An OCCAM program is a term of the process algebra. That algebra is described by the following SSL specification (the phyla declaration `decl`, expression `expr` and element `elem` are supposed known;

moreover the notation `List(x)` denotes a phylum `list_x` which consists of a list whose elements are of type `x`):

```
List(process) /* list_process */
process:  NullProcess()
        | Bloc(decl process)
        | Skip()
        | Stop()
        | Inp(elem list_inp)
        | Out(elem list_out)
        | Assignment(list_elem list_expr)
        | Seq(list_process)
        | SeqRepl(repl process)
        | If(list_if)
        | IfRepl(repl if)
        | Loop(expr process)
        | Par(list_process)
        | ParRepl(repl process)
        | Alt(list_alt)
        | AltRepl(repl alt)
        | Instance(name list_expr)
        | Comment(comment process)
        ;

List(inp) /* list_inp */
inp: TheInIt(elem);

List(out) /* list_out */
out: TheOutIt(expr);

repl: Therepl(name expr expr);

List(if) /* list_if */
if: NullIf()
   | GuardedIf(expr process)
   |IfIf(list_if)
   | IfIfRepl(repl if)
   ;

List(alt) /* list_alt    */
```

```
alt: NullAlt()
   | AltAlt(list-alt)
   | AltAltRepl(repl alt)
   | GuardInp(elem list_inp process)
   | GuardBoolInp(expr   elem list_inp process)
   | GuardOut(elem list_out process)
   | GuardBoolOut(expr   elem list_out process)
   | GuardBoolskip(expr   process)
   ;
```

We note that the main differences from a sequential language are:

- the phyla which express concurrency constructed by means of the operator *Par* and its iterated version *ParRepl*;
- communication terms constructed by means of the operators *Inp* and *Out*;
- the phyla which express non-determinism, constructed by means of the operators *Alt* and *AltRepl*.

Note that, in order to consider the whole basic transformations, we have extended the basic OCCAM with the operators *GuardOut* and *GuardBoolOut* which represent respectively output commands and guarded output commands.

As an illustration, of the preceding abstract syntax, we show the first levels of the structure of the following concrete OCCAM program.

— — *Bloc*

*decl* { **CHAN OF** *INT CNS, TNS, Rep* :
*INT coffee, tea, CreditCard* :

*process* {
— — *Loop*
**WHILE** *TRUE*
*process* {
— — *Alt*
**ALT**
*list_alt* {
— — *GuardBoolInp*
*(coffee > 0)& CNS ? CreditCard*
⋮

## 3.4 Basic transformations of the OCCAM language

A basic transformation (Roscoe and Hoare, 1988) transforms a process into another *semantically* equivalent process. These semantic transformations can be used, for instance, for (Burns, 1988)

- program optimization;
- program mapping.

### 3.4.1 The basic transformations

The basic transformations are associated with the basic semantic properties of the language. Some of these properties are:

- associativity of the constructions *If*, *Seq*, *Alt*, *Par*;
- symmetry of *Par*, *Alt*;
- synchronous communication properties.

*Example*

Figure 3.4 illustrates the transformations of a guarded command:

**ALT**($b$ & $g$ $P$, $G$) = **IF**($b$ **ALT**($g$ $P$, $G$), $\neg b$ **ALT**($G$))

where

- $b$ is (coffee > 0)
- $g$ is CNS ? CreditCard
- $P$ is

```
SEQ
  coffee := (coffee − 1)
  Rep ! (CreditCard − 1)
```

- $G$ is

```
(tea > 0) & TNS ? CreditCard
  SEQ
    tea := (tea − 1)
    Rep ! (CreditCard − 1)
```

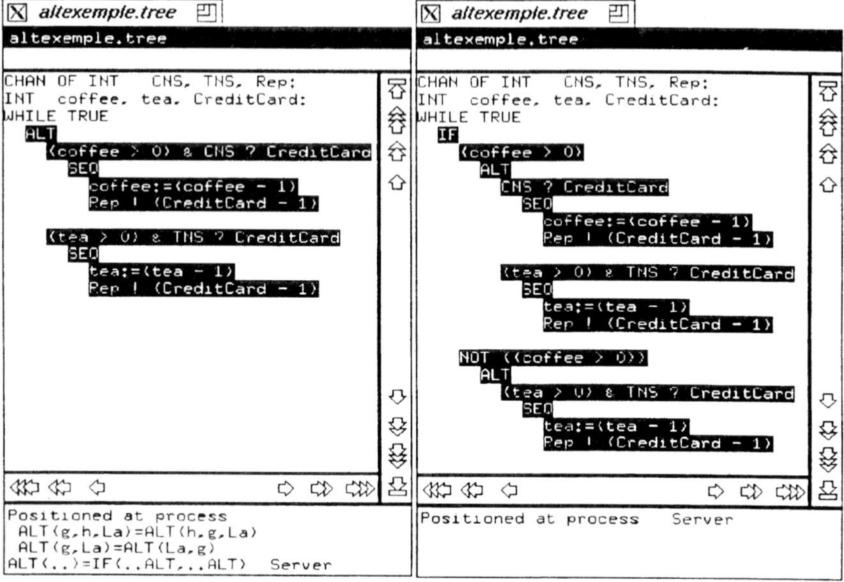


**Figure 3.4** Guarded command: before and after transformation.

## 3.4.2 The transformation kernel

The transformation kernel we have implemented is built on top of a syntax directed editor for OCCAM; it contains most of the basic transformations. As a transformation example, we give the SSL text of some of the transformations related to the *Par* construction.

The following two transformations express the *Par* commutativity:

**transform** process[2]

**on** "P||Q=Q||P"
    Par(Theprocesss(p,Theprocesss(q,listR))) : Par(q::p::listR),
**on** "P||(Ps)=(Ps)||P"
    Par(Theprocesss(p,listQ)) : Par(listQ@Theprocesss (p,Nullprocesss));

[2] In SSL, @ is the infix symbol for concatenating two lists, :: is the infix symbol for inserting an element at the beginning of a list.

In a single transformation, the user can only interchange the first two processes or rotate the process list by position; actually, these transformations are sufficient to commute any two elements of a process list. Then, these two transformations express the following property:

$$\forall\ \pi\ \textit{permutation of}\ \{1, \ldots, n\} PAR^{n}_{i=1}\ p_i = PAR^{n}_{i=1}\ p_{\pi(i)}$$

In summary, the basic transformations can change a process into another one without modifying its semantics. Elaborated semantic transformations can be constructed by composing these basic transformations. In the next section, we present another kind of transformation: transformations for particular process patterns. The basic transformations could be used to shape a process into a suitable pattern for such transformations.

## 3.5 Client–server model transformations

The transformations presented in the preceding section concerned the language constructs. In that section, we considered transformations of program patterns obtained by applying some design methodology. The design methodology considered here is the client–server one.

### 3.5.1 The client–server model

The client–server model illustrated by Figure 3.5 is a design model well suited for distributed applications; for instance, it is the underlying model of the SUNOS distributed operating system. In this design

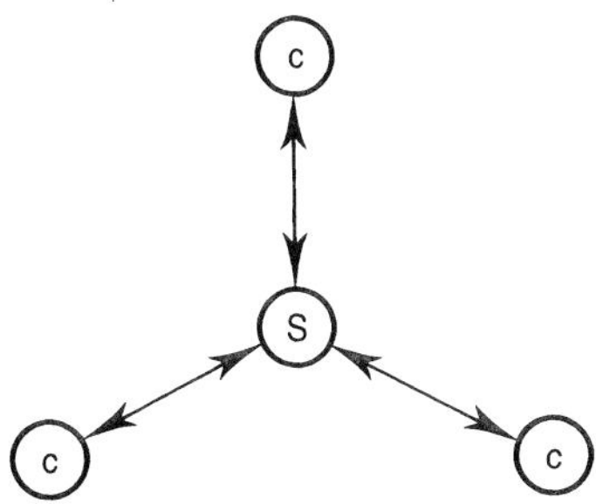


**Figure 3.5** Client–server model.

model, each function of the system is encapsulated within a service; this service is implemented by a server, local or remote users (i.e. on the same node or not) transmit their request transparently to the server; the basic execution mechanism for that request is the remote procedure call (Coulouris and Dollimore, 1988).

This model has the following properties:

- simple concept;
- easy implementation;
- distribution transparency.

However, we think that such a model is not well suited for execution on highly parallel machines:

- inside a server, there is no concurrency;
- being on one node, the server can become a system bottleneck.

In the following sections, we present transformations for programs designed within the client–server model; the aim of these transformations is to easily adapt such programs for their execution on highly parallel machines.

In order to illustrate such transformations, we will consider the well known example of the drinks machine.

### 3.5.2 Server primitive form

In this section, we characterize the server processes (henceforth called servers) that we will consider. Such servers are described as follows.

- A server is characterized by a set of typed requests and a static set of users. For each user–request pair, a request channel and a response channel are assigned.
- The behavior of a server consists of an infinite loop, of which each iteration consists of receiving a request on a request channel, processing the request and transmitting the results on the corresponding response channel.
- A server has private memory. Moreover, servers do not communicate with each other; server processings are atomic.

The concrete OCCAM form of a server is then:

```
<service static variables>
WHILE TRUE
  <service parameters>
  ALT ix = 0 FOR n — — n: number of clients
    ALT
      :
      request.k[ix] ? <parameter> — — request for service k
        SEQ
          <Processing.k>
          response.k[ix] ! <results> — — end of service k
      :
```

Figure 3.6 represents the OCCAM program for the drinks machine.

### 3.5.3 Server transformation principles

The key idea of a server transformation is to split the original server of the design model into a set of local servers, the role of each local server

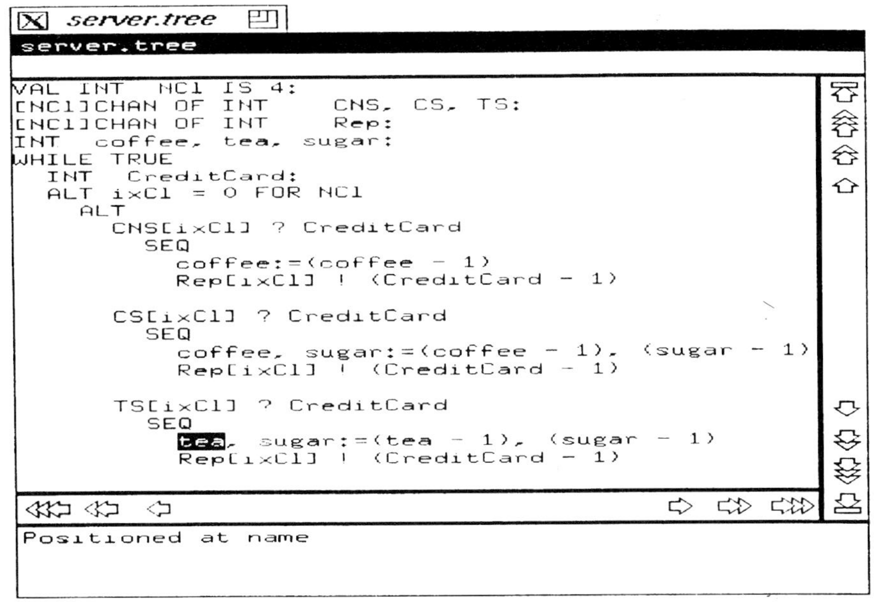


**Figure 3.6** OCCAM drinks machine program.

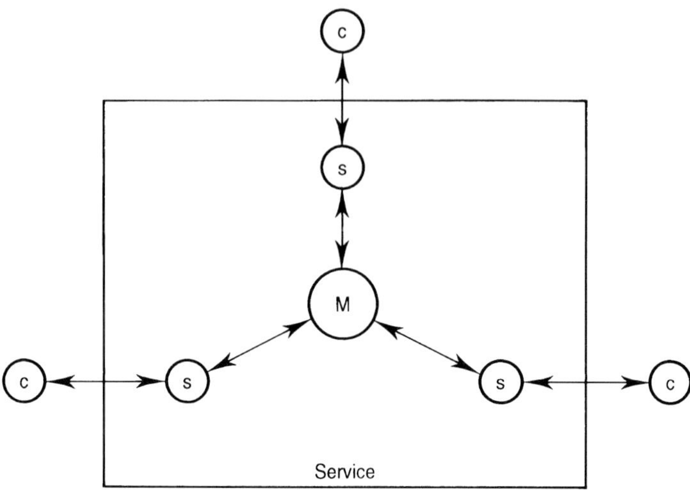


**Figure 3.7** Distributed client–server model.

being to handle the requests of a subset of clients (in our study, we have considered a simple case: each local server handles the requests of a unique client). In order to serve their assigned clients, the local servers will share the memory of the initial server; such a memory will be implemented by a memory process.

Formally, the server transformation, illustrated by Figure 3.7, is the following:

$$t(S) = (\|_{i=1}^{n} s_i) \| M$$

where $S$ is the server process, $M$ is the memory process and $s_i$ is a *local* process which handles request of client $i$.

Distribution transformations will consist in proposing a process composition equivalent to $S$. Then, the process of this composition could be placed on different physical nodes.

### 3.5.4 Server term

In this section, we formalize a server by assigning to it an operator of the process algebra. A server term will be characterized by:

- its global variables, first *list_decl*;
- its request parameters, second *list_decl*;
- its fixed set of clients, replicator *repl*;
- the processing of each request, *list_alt*.

Then the new operator introduced for the phylum *process is:*

*Server(list_decl list_decl repl list_alt*)

## 3.6 Process–server transformation

In this section, we explain how a given process term of the process algebra is recognized as a server. A process term is a server if it has the following pattern:

```
p=
    Bloc(d1, }
    :         } global variables
    Bloc(dn, }

Loop(Btrue,                              }
                                         }
  Bloc(d1, }                             }
  :         } service variables          } service infinite
  Bloc(dn, }                             } loop
                                         }
  AltRepl(rep,                           }

    AltAlt(ListAlt)
```

Then this process term could be transformed into the following term:

Server(d1:: . . . ::dn::Nulldecls, v1:: . . . ::vn::Nulldecls, rep, ListAlt)

In order to proceed to that transformation, we have defined the following SSL functions on a process *p*:

- *ModelServer* to test if a process has a server form or not;
- *DeclGlob* to calculate the list of global variables of *p*;

- *DeclParam* to calculate the list of service parameters;
- *Replicator* to calculate the AltRepl replicator of *p*;
- *ListAlt* to the list of alternatives of *p*.

Finally, the process–server transformation is expressed by the following SSL text:

```
transform process on "server"
    p : ModelServer(p)? /* transformation test*/
        Server(DeclGlob(p),DeclParam(p),Replicator(p),ListAlt(p))
        p
```

## 3.7 Some types of server distribution

In this section, we present some types of server distribution. The aim of these distributions is to provide an equivalent server to the initially centralized server and to reuse existing algorithms well suited to given underlying topologies. More precisely, the reused algorithms are distributed mutual exclusion and distributed access to a set of distributed resources (Raynal, 1985).

### 3.7.1 Distribution with a global memory

In this kind of distribution, memory is globally accessed. Such a distribution could be used if the underlying network is a ring network; then the global memory will be exchanged between the local servers instantiated on each node of the ring.

For processsing each request, all global variables are first taken; these variables are released at the end of the processing. In fact, we have a circulating memory among the different local servers. The memory access protocol is the following:

- request the current copy of the global variables;
- wait until the variables arrive;
- process the request with the obtained copy;
- release the new variables copy.

In order to implement such a protocol, we use for each local server the following set of channels:

- MemGlob.Req variables access request channel;
- MemGlob.Acq variables transfer channel;
- MemGlob.Rel variables release channel.

The behavior of a local server is described as follows:

```
MemGlob.Req[ix] ! any —— variables request access
MemGlob.Acq[ix] ? <memory.variables>
<Processing.j>
MemGlob.Rel[ix] ! <memory.variables> —— release new memory
copy
```

The centralized server is then transformed into the following process:

```
PAR
  MemGlob(< protocol channels >)
  PAR ix = 0 FOR n —— local servers
    WHILE TRUE
      <service parameters>
      < decl. of mem. vars >
      ALT
        :
        request.k[ix] ? <parameters> —— request for service k
        SEQ
          MemGlob.Req[ix] ! any —— mem. access request
          MemGlob.Acq[ix] ? <mem vars>
          <processing.k> —— processing of request k
          MemGlob.Rel[ix] ! <mem vars> —— end of mem. access
          response.k[ix] ! <results> —— end of service
        :
```

The process *MemGlob* ensures that the concurrent accesses made by the local servers are equivalent to accesses made by a centralized server. Basically, *MemGlob* is a mutual exclusion process so several mutual exclusion alorithms could be reused here. Figure 3.8 illustrates such a distribution.

```
server.tree
server.tree

PAR
  MemGlob(MemGlob.Req, MemGlob.Acq, MemGlob.Rel)
  PAR ixCl = 0 FOR NCl
    WHILE TRUE
      INT  CreditCard:
      INT  coffee, tea, sugar:
      ALT
        CNS[ixCl] ? CreditCard
          SEQ
            MemGlob.Req[ixCl] ! any
            MemGlob.Acq[ixCl] ? coffee; tea; sugar
            coffee:=(coffee - 1)
            MemGlob.Rel[ixCl] ! coffee; tea; sugar
            Rep[ixCl] ! (CreditCard - 1)

        CS[ixCl] ? CreditCard
          SEQ
            MemGlob.Req[ixCl] ! any
            MemGlob.Acq[ixCl] ? coffee; tea; sugar
            coffee, sugar:=(coffee - 1), (sugar - 1)
            MemGlob.Rel[ixCl] ! coffee; tea; sugar
            Rep[ixCl] ! (CreditCard - 1)

        TS[ixCl] ? CreditCard
          SEQ
            MemGlob.Req[ixCl] ! any
            MemGlob.Acq[ixCl] ? coffee; tea; sugar
            tea, sugar:=(tea - 1), (sugar - 1)
]           MemGlob.Rel[ixCl] ! coffee; tea; sugar
            Rep[ixCl] ! (CreditCard - 1)

Positioned at process  < BLOC >  <SEQ>    <SEQ FOR> <PAR>    <PAR FOR>
<IF>      <IF FOR>  <WHILE>   <ALT>     <ALT FOR> <SKIP>    <STOP>    <--Comm>
<INPUT>   <OUTPUT>  <ASSIGN>  <INSTANCE>  Server
```

**Figure 3.8** Global memory drinks machine.

### 3.7.2 Distribution with a distributed split memory

In this kind of distribution, memory is split into pieces. Such a distribution could be used if some requests need disjoint sets of pieces and to be served concurrently.

Now, for each request, a local server can ask only for the pieces it needs. As in the preceding case, a *MemSplit* process ensures that the concurrent accesses made by the local servers are equivalent to accesses made by a centralized server. The basic problem underlying this distribution is the generalized dining philosophers' problem: each local server is a philosopher and the pieces he needs are the forks; so we can reuse, for instance, the implementation given in Chandy and Misra (1988). Figure 3.9 illustrates such a distribution.

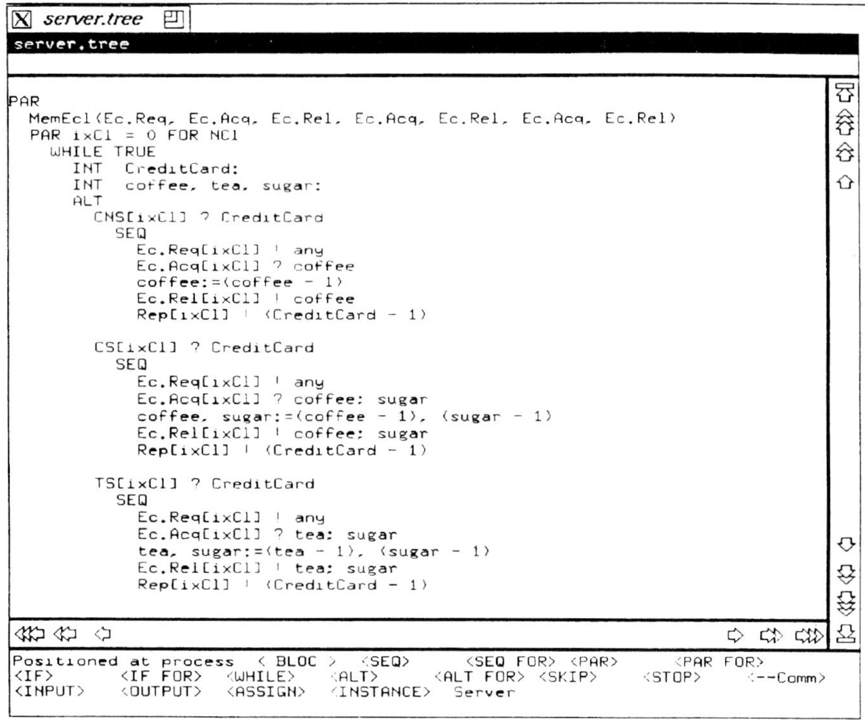


**Figure 3.9** Split memory drinks machine.

## 3.8 Extensions

In this section, we discuss some aspects that should be easily considered in the framework of the environment kernel for OCCAM programming. However, it should be noted that these aspects are just discussed and have not been implemented.

First, we discuss debugging issues. Then we consider an elaborated programming concept: namely *superposition*; we present how it could be understood in a message passing model and implemented within our programming environment for OCCAM.

### 3.8.1 Debugging issues

A basic aspect of debugging is tracing; tracing is generally considered at two levels:

- data tracing;
- control tracing.

In the following sections, we show how basic tracing can be implemented by a transformation of the abstract syntax tree; in Section 3.8.2, we will present a high level mechanism that could be used to implement elaborated traces.

*Data tracing*

Within our environment, data tracing is easily implemented by a transformation of the abstract syntax tree; suppose we want to trace all the changes made to a variable $x$, then we can transform all the nodes where $x$ is modified, e.g. assignments to $x$, reception in $x$, . . . into new nodes containing a tracing instruction. In the case of assignments, the shape of these transformations will be the following:

```
x :=e →     SEQ
                x := e
                chan_x ! x
```

where chan_$x$ is a channel of the type of $x$ and is always served by a tracing process for instance a visualization process.

*Control tracing*

We can consider control tracing in the same way. As we have considered tracing the changes made to a varible, we can consider tracing the beginning and the ending of a process $p$; let Ending($p$) and Starting($p$) be two processes for the respective traces, then we can consider the following transformations:

```
  p    →     SEQ
                 Starting (p)
                 p
```

and

$p \quad \rightarrow \quad$ **SEQ**
 $p$
 Ending($p$)

More generally, we can envelope the process $p$ by the following transformation:

$p \quad \rightarrow \quad$ **SEQ**
 Starting($p$)
 $p$
 Ending($p$)

### 3.8.2 Superposition

In the field of distributed programming, an application is often structured as a set of layers, each layer relying on the services provided by the lower layers. For instance, the OSI structure obeys such a programming discipline. *Superposition* (Bougé and Francez, 1987; Chandy and Misra, 1988) is a programming concept intended to establish such a discipline; in their formalism, called UNITY, Chandy and Misra (1988) consider superposition as a program transformation and describe it by the following two rules:

1. Augmentation rule: a statement $u$ in the underlying program may be transformed into a statement $u \| s$ where $s$ does not assign to the underlying variables.
2. Restriction rule: a statement $s$ may be added to the underlying program provided that $s$ does not assign to the underlying variables.

In our programming environment, such rules could be easily implemented by establishing restrictions on the modifications of a superposed syntax tree.

However, in the case of OCCAM programming, the preceding rules call for an appropriate interpretation:

- Rule 2 can be directly adapted with the following rules:
  (a) message sending and receiving cannot be done on the underlying channels;
  (b) we canot put a received message in an underlying variable.

- Rule 1 can also be directly adapted; however, we can give it another interpretation in the case of OCCAM programming. Let us consider more closely the transformation

$$u\_v := u\_e \quad \rightarrow \quad u\_v, s\_v := u\_e, s\_v^{3}$$

Considering the assignment transformation of OCCAM:

```
v := e ↔     PAR
               c ! e
               c ? v
```

then, rule 1 can be interpreted as the following OCCAM transformation:

```
PAR                    PAR
  c ! u_e      →         c ! u_e; s_e⁴
  c ? u_v                c ? u_v;s_v
```

Consequently, such a transformation tells us that we can:

- transform the protocol of a channel by extending it;
- modify the emission and reception statements on this channel provided that the received variable *s_v* is not an underlying variable. A similar mechanism has been studied in (Lokpo and Padiou, 1992), however it was not considered as a special case of superposition and proposed as a transformation.

### *Discussion*

We note that in Section 3.8.1 on debugging, we presented only basic mechanisms; however, elaborated debugging aspects such as the detection of stable properties, e.g. deadlock, termination, or snapshot recording could be programmed by means of the superposition mechanism.

[3] The statement *u* has been rewritten as *u_v:=u_e* and *s* as *s_v:=s_e* and *u* || *s* as *u_v,s_v:=u_e,s_v*.

[4] The declaration of *c* should be transformed to satisfy the new use of *c*.

## 3.9 Conclusion

Our programming environment is built on top of a syntax-directed editor. However, its original feature is to propose semantic program transformations. More precisely, we have been concerned with distribution transformations for programs designed within the client–server model. Finally, we have seen that some debugging features and advanced aspects of distributed programming could be added to our environment.

Concerning the OCCAM language, it has usually been considered as a primitive programming language; so, in most projects, it has been replaced by more conventional languages; e.g. PASCAL or C, with concurrency constructs, have been used. On that point, we think that OCCAM requires a new programming style: programming by derivation where the program and its proof are derived together (Chandy and Misra, 1988). In that framework, program transformations (Gries and Volpano, 1990) are an important aspect to be supported by a programming environment. We should note that this programming style is different from modular programming where what is reused is considered as a black box, whereas with transformations, what is reused remains visible but semantically protected.

## References

Bougé L. and Francez, N. (1987). A compositional approach to superimposition. Technical Report 87–13, LIENS.

Burns, A. (1988) *Programming in Occam 2.* Addison-Wesley.

Chandy, K. M. and Misra, J. (1988). *Parallel Program Design, a Foundation.* Addison-Wesley.

Coulouris, F. and Dollimore, J. (1988). *Distributed Systems Concepts and Design.* Addison Wesley.

Deransart, P., Jourdan, M. and Lorho, B. (1988). *Attribute Grammars: Definitions and Bibliography.* Springer-Verlag.

Gries, D. and Volpano, D. (1990). The transform – a new language construct. *Structured Programming*, **11**(1): 1–10.

Hoare, C. A. R. (1985). *Communicating Sequential Processes.* Prentice Hall.

Johnson, S. C. (1978). Yacc: Yet another compiler-compiler. Technical report, Bell Laboratories, Murray Hill.

Lesk, M. E. (1975). Lex – a lexical analyzer generator. Technical report, Bell Laboratories, Murray Hill.

Lokpo, I. and Padiou, G. (1992). Reusability in the occam language. *Structured Programming*, **13**(2): 65–74.

Raynal, M. (1985). *Algorithmes Distribués et Protocoles*. EY-ROLLES.

Reps, T. W. and Teitelbaum, T. (1989). *The Synthesizer Generator A System for Constructing Language-based Editors*. Springer-Verlag.

Roscoe, A. W. and Hoare, C. A. R. (1988). The laws of occam programming. *Theoretical Computer Science*, **60**(2): 110–229.

Van Eijk, P. (1990). Attribute grammar applications in prototyping LOTOS tools. In Deransart, P. and Jourdan, M. (editors), *Attribute Grammars and Their Applications*, pp. 91–100. Springer-Verlag.

Vogt, H., Berg, A. V. D. and Freije, A. (1990). Rapid development of a program transformation system with attribute grammars and dynamic transformations. In Deransart, P. and Jourdan, M. (editors), *Attribute Grammars and Their Applications*, pp. 101–115. Springer-Verlag.

# Part II Evolution of structure–based environments



The chapters in Part I all dealt with structure editors which offer a range of capabilities to support the programming process. In some cases, user testing was done or the design and implementation was based on a description of the intended user population. User testing and evaluation can play an important role for the feedback it provides. More importantly, evaluation can lead to system enhancements in response to a clear notion of the problems that occur and the benefits that result during use of a system.

While this is true for systems in general, it is of special importance with structure-based environments for two reasons. Firstly, the field has suffered from a perception that tools meant to aid in the programming tasks were so restrictive that their benefits could rarely be realized. Secondly, the programming task is very complex for the range of users from novices to highly experienced, and tools offering support need to provide the forms of support required without a great cost to the user. This applies to the support for syntax semantics, user interface customization, configuration management, reuse, etc.

The common theme of the three chapters in Part II is that all evaluate the usefulness of structure-based environments. However, different techniques are used and the studies have different foci. In the first two chapters, extensive user studies are reported upon and enhancements are made to the systems reflecting the findings; the third uses an analytical approach.

Chapter 4, by Lerner, examines the interfaces of environments generated by the Gandalf program environment generator. Studies showed problems that occurred during system use which could be removed by automated customization. Mechanisms were added to Gandalf to allow this feature in generated editors; one such environment, for ARL, a description language for structure editor semantic operations, was studied to determine the effectiveness of the concept.

Chapter 5, by Neal, describes a system which incorporates example programs which are used for education and program construction. The original concept of example-based programming came from studies of structure editor use. The system met its expectations; additional benefits became clear. For example, the system showed greatest usefulness with experienced programmers who were using new languages. Such benefits were shown through evaluation of the system with users having a range of programming experience as well as differing in other respects.

Chapter 6, by Robillard *et al.*, examines the question of the usefulness of structure editors. This question can be examined using numerous techniques; the authors chose, rather than to study multiple projects or a variety of users, to use software metrics applied to one project in order to minimize the number of influential factors. Software metrics were used to measure structure-based program editing vs. complexity of program text.

# CHAPTER 4

# Automated customization of structure editors

*Barbara Staudt Lerner*

## Abstract

A common method of developing structure editors is by generating them using an environment generation tool, such as Gandalf or the Synthesizer Generator. One weakness of this approach is that the user interfaces to the generated structure editors tend to be difficult to customize to individual languages and users. Customization is typically left to the user with macros and mode settings. An alternative described in this chapter is automated customization. To support automated customization, the system must determine what customizations to perform, when to perform them, and how to evaluate them without user intervention. This chapter reports on mechanisms added to the Gandalf environment generation system to support automated customization, as well as results of experimentation with these mechanisms.

The major results of experimentation are the following. Automated customization resulted in a 7% decrease in the number of commands required to complete a task, and up to 25% reduction in the number of errors encountered. In addition the evaluation mechanism performed well, correctly evaluating 95% of the automated actions.

## 4.1 Motivation

Many program generation systems produce programs that use structure editors as their user interface. Use of a program generator benefits users not only because of the speed with which programs can be developed, but also because each program generated by the same system will have the same "look and feel".

Unfortunately, such generated structure editors are not without

Structure-based Editors and Environments
ISBN 0–12–681890–8

problems. While program generators typically provide facilities that make it easy for a programmer to describe the behavioral characteristics of the program, they rarely allow the programmer to modify the user interface itself. Instead the structure editor is derived directly from a grammatical description of the program's structures. To change the user interface, the programmer must therefore change the grammar. Since the grammar is typically also used as the internal representation of the structures maintained by the underlying program, there is an inherent tension between creating a grammar that is easy for the programmer to manage, and one that is easy for users to interact with.

Since the programmer cannot easily modify the user interface, the user must perform the user interface tailoring explicitly using mode settings and macros. This forces the user to understand what customization he wants and how to express them. In contrast, this chapter describes research in completely automated customization. The system learns what customizations to apply, applies them and evaluates their usefulness without direct user intervention.

Customizations are learned by observing patterns in the user's behavior and automating those patterns. Automated evaluation is supported with a mechanism called success/failure criteria. When a customization is automatically performed, success/failure criteria monitor subsequent user actions to determine if the automated customization was acceptable to the user, or if the user undid the effects of the customization. The results of this evaluation are used as feedback to the customization process. The ability to evaluate customizations automatically is critical to the entire process, since it allows the system to learn which customizations are inappropriate for a particular user so that they can be disabled. Without such evaluation, it would be infeasible to automate customizations without first requesting confirmation from the user.

The mechanisms developed in this research allow the user interface designer to write heuristics that provide automatic customization of user interfaces. These mechanisms support the development of heuristics as part of the general-purpose utilities that are re-used by all generated editors, as well as specialized heuristics appropriate only for a single editor. While the concepts presented in this paper are relevant for a variety of applications, the specific mechanisms have been designed for use in the generation of Gandalf programming environments (Notkin, 1985).

Section 4.2 provides an overview of the problem and solution.

Section 4.3 briefly describes the mechanisms developed to support automatic customization. Section 4.4 describes a prototype system in which automated customization was implemented. Section 4.5 discusses the experimental results. Section 4.6 suggests some areas of future research. Section 4.7 compares this work to other work in intelligent user interfaces. Finally, Section 4.8 provides some concluding remarks.

## 4.2 Overview

An intelligent user interface should treat users as individuals, and be responsive to the needs and desires of the individual. Knowledge of the user is therefore of paramount importance for an intelligent user interface. User knowledge comes in several different flavors, such as the user's understanding of the application domain, the user's goals, the user's physical abilities, the user's habits and preferences etc. User modeling research has concentrated on the first three of these types of user knowledge. The resulting user models have been most useful in help and advice-giving systems (such as Broverman *et al.*, 1986; Kass, 1987; Wilensky *et al.*, 1988). However, knowledge of the user's habits and preferences has been largely overlooked. This research explores how knowledge of a user's habits can be used in automatic customization.

### 4.2.1 Intended uses

Three uses for automated customizations are investigated in this research: automation of routine tasks, tolerant command interpretation and active help. In this section, these terms are defined, and examples of particular customizations are given.

*Automation of routine tasks*

Routine tasks can be automated by a heuristic that identifies patterns that are consistently used in a given context by a particular user. When the user encounters that context, the heuristic can automate the pattern. For instance, suppose whenever a programmer uses a case statement, he includes a default case. A heuristic could recognize this pattern, and automatically construct the default clause whenever the user creates a case statement. By providing such automation, the heuristic has simplified the user's job. The editor understands that the user has a specialized way of writing case statements, and the heuristic can essentially customize his case statements to always include the default case.

Of course, the user could write a macro to automate the construction, thereby achieving the same effect. But this is a fundamentally different approach, since it places the burden of automation upon the user's shoulders, requiring that the user:

- recognizes that a pattern exists;
- determines how to automate the pattern;
- remembers (how) to invoke the macro.

In contrast, if the heuristic provides the automation, the user needs to do none of the above steps. Simply issuing the standard command to construct a case statement has the desired effect.

*Tolerant command interpretation*

In addition to automating routine tasks, an intelligent user interface should also be tolerant of the idiosyncrasies of individual users. Rather than provide a strict set of commands with rigid preconditions, which all users must adhere to, the user interface should understand input in a variety of forms.

For instance, a heuristic could dynamically acquire bindings between commands and semantic operations by observing the interactions between the user and the system. In this model, the system must still define the initial bindings between commands and semantics, but the bindings should become customized to the individual user automatically. If the user issues some syntax with a known semantics, those semantic operations should be performed. However, if there is no binding available, or some precondition of the bound operation does not hold, the heuristic should attempt to infer what the user intended. To do this, the heuristic must apply the knowledge it has along various dimensions, such as common sense, the application domain, the context in which the user is operating, and also knowledge about the user.

Once again, knowledge of a user's habits is useful. For instance, if the user inputs an unbound sequence of control characters to an editor, it is unlikely that some other types of knowledge can determine what actions to take. However, if the user used the same control sequence previously, a heuristic can remember which command the user executed after the unrecognized command, and postulate the binding between the control characters and the command.

Once the heuristic has determined how to interpret the user's input using any of its knowledge sources, it should cache information about the decision. In this way it is building a model of the user's preferred interpretations. If the user repeatedly issues the same input, and the heuristic's interpretation is repeatedly evaluated positively, then the heuristic can gain enough confidence in these customizations so that it is no longer necessary to perform the time-consuming reasoning required to explain the input using the other knowledge sources.

Without tolerant command interpretation, unknown user input results in an error message. Tolerant command interpretation therefore relieves the user from the burden of:

- understanding an error message;
- determining the accepted way of accomplishing the desired action;
- incorporating the accepted method into his model of how to use the system.

### *Active help*

Finally, an intelligent user interface should recognize when the user is confused, and actively offer advice in an attempt to resolve the confusion. This advice may be in the form of a help message, examples, suggestions of what other users typically do in the same situation etc. One way that confusion is manifested is with a sequence of commands that involve performing some action and immediately undoing it. The help messages can be specific to the types of objects the user is manipulating, the types of actions the user is undertaking, and the past experiences of the user.

For example, a new user of a structure editor may be unfamiliar with the commands available to construct structures. This is particularly likely to be true if he is also new to the application domain the structure editor is for. In such situations it is common for users to try various commands in an experimental fashion in order to find the one they desire. Such experimentation is evident when the user applies one of these commands, deletes the resulting structure, applies a different command and deletes that structure. An active help heuristic could recognize the user's inability to make progress and provide detailed help describing the various commands used to construct structures.

Active help is superior to the more normal passive help, because it relieves the user of the following burdens:

- remembering how to invoke the help system;
- actually invoking the help system;
- navigating through the help system to find the appropriate information.

### 4.2.2 User patterns

These styles of automatic customization depend upon three preconditions. First, the user must have habits that are exhibited as patterns in his behaviour. Second, the system must be able to recognize those patterns. Third, the system must know when automatic application of the pattern is appropriate. Patterns have been observed in regular users of a text editor (Card *et al.*, 1983), a natural language interface to a database (Lehman and Carbonell, 1989) and a structure editor (Lerner, 1989). Given that patterns exist, this chapter concentrates on the last two issues, pattern identification and pattern automation.

Automated customization is facilitated if the following two hypotheses hold:

- *Repetitive hypothesis.* Actions that the user has taken repetitively in the past are likely to be taken in the future.
- *Recency hypothesis.* Actions that the user has taken recently in the past are likely to be taken again in the near future.

The first hypothesis simply states that if a pattern has been recognized in the past, it probably will be useful in the future. Therefore, if the system has recognized a pattern, it is worthwhile to expend some effort in determining when the pattern is applicable, so that when those conditions are seen in the future, the pattern can be automatically applied.

The second hypothesis states that users often operate in modes where similar actions are applied repetitively over a relatively short period of time. If the user operates in this manner, then the system can be more ambitious about automating activity if it has observed the user perform those actions in the recent past.

### 4.2.3 Complete automation

One of the primary goals of this research is to develop interfaces that could be customized automatically, that is, without requiring any user intervention. In particular, the system should decide what actions to take, take the actions and evaluate the usefulness of those actions without user intervention. To achieve complete automation, several measurements are associated with each customization: a confidence value, benefit–risk analysis, and a record of past performance.

- *Confidence values.* In general, an individual heuristic is capable of recognizing more than one pattern in a user's interactions. In order for customization to take place, a heuristic must identify a pattern as being a likely candidate for automation. If the user has applied a pattern frequently, then by the repetitiveness hypothesis, the heuristic can be fairly confident that the pattern will be useful in the future. Similarly, if the user used a pattern recently, then by the recency hypothesis, the heuristic can expect the pattern to be useful in the near future. The more frequently and more recently the user has exhibited a pattern, the more confident the heuristic can be that automation of that pattern would be acceptable to the user.
- *Benefit–risk.* A benefit–risk analysis is performed on potential customizations to compare the benefit of applying a correct customization to the cost of a mistake. Each operation that can be automated has a method of computing both its potential benefit and potential risk. For example, creating a complex structure has a high potential benefit, and low potential risk, while the reverse is true for deletion of the same structure. The benefit–risk value of a particular customization is determined by summing the benefits and risks of each primitive operation that occurs in the customization. An appropriate confidence threshold is set depending upon the results of the benefit–risk analysis. The riskier a customization is, the more confidence we require before automation may proceed.
- *Past performance.* Once a customization has been applied, the heuristic should evaluate whether or not the customization was acceptable to the user by monitoring the user's activity subsequent to the customization. If the user performs actions that take advan-

tage of the customization, then the user has confirmed the customization. Conversely, if the user takes actions that undo the work of the customization, the customization is inappropriate. This performance evaluation provides feedback to the heuristic. If a customization is consistently inappropriate, the heuristic should prevent that customization from being re-applied.

Automation of routine tasks, tolerant command interpretation, and active help can all be supported in a totally automated fashion by relying upon the mechanisms presented in this chapter. Each type of customization has three essential questions to address, as shown in Table 4.1. First, we need to identify what customization to perform. Second, we need to identify when the customization should be applied. Third, we need to determine how to evaluate whether the customization was actually beneficial to the user. The *what* question is answered by observing the user's actions, and looking for patterns in the user's behavior. The *when* question is answered by finding some preconditions to the pattern that accurately predict that the pattern is desired. The *OK* question is answered by identifying actions the user could perform that would indicate that the user had benefited from or been hindered by the customization.

For example, when automating a routine task, a heuristic determines what to do by identifying routine tasks in the user's behavior. When the user initiates the task, the heuristic can apply a customization that will complete it. If the user uses the work of the customization, the customization is considered successful. If the user undoes the work of the customization, the customization is considered to have failed.

In the tolerant command interpretation case, a customization is performed when the user issues a command that cannot be executed

**Table 4.1** System-level issues for automating customization

| *Type of customization* | *What?* | *When?* | *OK?* |
|---|---|---|---|
| Automating routine tasks | Task completion | Task initiated | Results used |
| Command interpretation | Alternative interpretation | Failed command | Results used |
| Active help | Contextual help | Confusion observed | User progresses |

using the underlying system's interpretation of the command. The customization will attempt to use an alternative interpretation of the command. Again, the customization is evaluated by determining if the user uses or undoes the customization.

Finally, active help is given when a heuristic sees the user is failing to make progress. The heuristic applies a customization that provides help related to the types of actions the user is attempting, and the objects the user is manipulating. The help is evaluated by observing if the user is able to make progress after receiving the help.

All three types of customization rely upon analysing the history of interactions between the user and the system. These interactions are used to determine when to provide customizations and whether customizations were successful or not. In addition the history may also be used to determine what customizations to apply. The ability to analyse the history of user interactions is essential to attain complete automation. If history was not analysed, the user would need to define the customizations manually or rely on some pre-defined customizations. Furthermore, without the ability to reason about history, it would not be possible for the system to understand when to apply the customizations, thereby requiring manual invocation of the customizations. If manual invocation is used, then the customization process becomes even simpler, because the evaluation phase is no longer needed. Instead, customizations could be assumed successful since they were explicitly requested by the user.

### 4.2.4 Example session

This section presents a few commands from an example session with Lantern, the system that was developed during this research. (Lantern is presented in more detail in Section 4.4.) The example here is intended to provide the reader with a more concrete understanding of what Lantern can automate. This example will be referred to throughout the chapter as more details of the mechanisms underlying these customizations are presented, so that the reader can understand how such automation occurs.

Lantern is an environment for programming in ARL, a special-purpose programming language designed for tree manipulation. The user interface to Lantern is a structure editor. Most commands in a structure editor are structural commands rather than the textual commands found in text editors. The editing cursor selects an entire structure, such as an if-statement, rather than a single character. If the delete command is applied when the cursor is selecting an if-statement, the entire if-statement is deleted. Similarly, programs are constructed by issuing structural commands. For instance, the *if* command in ARL builds the structure for an if-statement, and presents this structure to the user with the following template:

```
if $boolean_expr then
   $statement
$else_part
end if
```

In the scenario presented in this section, we show the operation of two heuristics and an example of the customizations they have learned for particular users. In the following, the user is editing a procedure, with the cursor currently on the structure identified as $decl:

```
procedure Example
  begin
    $decl
    $statement
  end;
end Example;
```

*$decl* is a placeholder for further constructions. With the cursor at *$decl*, the user issues the *int* command. The user is attempting to declare a variable whose type is *integer*. However, the *int* command is not legal at the *$decl* placeholder. In the standard ARL environment, the error "int unknown command" would be reported to the user. In Lantern, a tolerant command interpretation heuristic intercepts the error message. It determines that the user's command would be legal if an intermediate structure representing the declaration was first created. The heuristic creates the necessary intermediate structure and re-applies the user's command. The user gets the results he expected, without even being aware that a heuristic was executed:

```
procedure Example
  begin
    integer $variable_def;
    $statement
  end;
end Example;
```

Now, the user completes the declaration of the variable. By doing this, he is using the results of the heuristic customization. This results in a positive evaluation of the heuristic, increasing the likelihood that it will be repeated in the future.

After a few more commands, the user constructs an if-statement. In the standard ARL environment, the result of constructing an if-statement results in the template first shown. With this template, a user can create an else-statement by issuing the *else* command at the $*else_part* placeholder. Alternatively, if the user does not want an else-statement, he can issue the *empty* command at the $*else_part* placeholder. In the standard ARL environment, when the user constructs an if-statement, the else part is always a placeholder. The user must specify whether to construct an else-statement or omit it. In Lantern, there are three possible outcomes for the else construct, depending on the user's history of interactions with the system.

- If the user has used an else-statement in the vast majority of his recent if-statements, a heuristic in Lantern that automates routine tasks will automatically construct the else-statement when the user constructs the if-statement:

```
procedure Example
  begin
    pascal cursor ident;
    if $boolean_expr then
      $statement
    else
      $statement
  end;
end Example;
```

- If the user has not used an else-statement in the vast majority of his recent if-statements, a heuristic in Lantern that automated routine tasks will automatically omit the else-statement:

```
procedure Example
  begin
    pascal cursor ident;
    if [$boolean_expr] then
      $statement
  end;
end Example;
```

- If neither of the above conditions hold, that is the user has developed no pattern with regard to use of else-statements, the heuristic will do nothing, and the standard template shown previously will be presented to the user.

Suppose the else-statement is automatically constructed for the user, and in this case it is not what the user wanted. The user will then move the cursor to the else statement, and delete it. The deletion is undoing the work of the heuristic, and will result in a negative evaluation of the customization. This reduces the confidence the heuristic has in that particular customization. If the confidence falls below a threshold, the heuristic will not automate the customization again until the user manually constructs the else-statement sufficiently frequently to raise the confidence value over the threshold again.

## 4.3 Mechanisms to support automated customization

Creating user interfaces that can be automatically customized to meet the needs of individual users presents a number of problems. How are customizations defined? What types of customization are supported? How can they be allowed to vary from user to user? How can we decide when a customization is appropriate? How can we tell if the customizations are actually beneficial once they have been applied? How can customizations change as the user evolves? This section answers these questions by describing the high-level design of mechanisms that provide the desired functionality.

### 4.3.1 Heuristics

A computer user typically performs a variety of tasks. The tasks range from jobs that require human creativity to purely mechanical chores.

Consider the jobs involved in writing a program. Deciding on the proper algorithms and data structures to use requires creativity. Writing the program to a file and compiling it are examples of purely mechanical chores. Mechanical chores are often automated, while almost everything else requires explicit user actions.

Between these two extremes of creativity and mechanization, we see situations where users develop patterns of behavior. This occurs when a user can choose from a set of actions, but in practice performs the same action almost every time. For instance, using a text editor, different programmers are very likely to develop their own patterns with respect to program indentation. While it may be less obvious that similar patterns would emerge when using a structure editor, experimentation (described in Section 4.5) has shown that patterns do develop. For example, different programmers tend to build different types of structures, they navigate through the structures differently etc.

Until now, the term "heuristic" has been used in a fairly informal sense. A more precise description of the role of heuristics in this research follows. *Heuristics* are components that customize the user interface by reasoning about the user's preference and habits, the context in which he is working, and the application domain with which he is working. While the user interface designer must write the heuristics, both the development of specific customizations and their use are automated. Ideally, the user should perform his normal tasks oblivious to the presence of heuristics. Over time the heuristics should learn the user's patterns transparently by analysing his actions. They should also learn what situations the user applies each pattern in. Then when that situation is encountered, the heuristic can automatically apply a customization that performs the pattern without explicit user invocation. These customizations might provide macro-like sequences of actions, perform transformations on the objects the user is manipulating, interpret commands that are not strictly legal etc.

Typically, a heuristic is designed so that it can operate in a wide variety of contexts. The heuristics illustrated in Section 4.2.4 are more general purpose than they may seem from the simple example. For example, the heuristic that intercepted the "unknown command" error message can work in any context, not just during declarations. However, it learns different customizations for the different contexts. The example showed that it learned to build an intermediate declaration structure. Similarly, the heuristic that automatically constructed the else-statement in Section 4.2.4 operates in a wide variety of contexts. The example showed how it might operate in the context of an if-statement.

Heuristics are not an integral part of the user interface, but rather are separate components that are added to a more traditional user interface. The traditional user interface defines a complete and precise form of user interaction. Heuristics enhance the user interface by adding intelligent behavior. Each heuristic encapsulates a piece of intelligence, leaving the underlying user interface unchanged. As will be shown in Section 4.3.5, this modularity simplifies the process of adapting to individual users.

### 4.3.2 Loggers

One can imagine an intelligent user interface that gets its intelligence solely from reasoning about the application domain and the current system state, but does not apply any knowledge specific to the individual user. For instance, such an interface could be used to interpret user input that deviated somewhat from the expected form, by reasoning about how the input could be mapped into a sequence of interpretable commands. While providing intelligence to the user interface, such components would stop short of actually customizing the interface for individual users, since they would treat all users identically. This is the approach taken by Teitelman (1972) in DWIM.

To truly customize a user interface, we need knowledge of the user. Such knowledge is gained through *loggers*, which collect data about the user by analysing the user's interactions with the system. Loggers receive as input the events that occur in the underlying system. These events are structures that provide a very detailed description of the system activity. Both the number and detail of events generated by the system is enormous, and greatly exceeds what any individual heuristic requires. Loggers create abstract events from these data based upon the needs of the heuristics and present these abstract events to heuristics to reduce the amount of analysis required by each heuristic.

The types of abstraction that a logger can perform are the following. Using *selection*, a logger can collect only those events that satisfy some desired property. Using *projection*, a logger can save only certain attributes of an event. *Sequencing* allows a logger to maintain a sequence of primitive events as a single abstract event. *Calculation* allows a logger to create new attributes whose values are based upon one or more attributes of one or more events. Projection is therefore a trivial form of calculation. *Counting* indicates the number of times that the same abstract event is encountered in the user's history. *Time-stamping* indicates the last time that an abstract event was encountered in the user's history.

### 4.3.3 Conflict resolution

Heuristics observe user behavior to identify useful customizations. When the user encounters a situation in which the heuristic believes application of a customization is appropriate, the heuristics will make a proposal to an arbiter that customization be applied. A *proposal* describes the customization the heuristic wishes to take. Since heuristics generally operate independently of one another, it is possible that multiple heuristics will make proposals in response to a particular user action. In some situations the proposals will be independent, affecting different parts of the environment. In such situations, all the heuristics' proposals can be applied. In other situations, the heuristics may propose conflicting customizations whereby application of one customization would make it impossible to apply one of the other proposed customizations. In this type of situation, the arbiter must decide which of the conflicting customizations should be applied.

To customize interfaces to individual users, the ideal rule to use for conflict resolution is to choose the customization that matches what the user would like to have done. A simple and accurate solution, then, would be to ask the user which customization should be applied. However, this approach violates the goal of providing customization without user intervention. Instead we want the arbiter to perform conflict resolution without involving the user.

The solution requires each heuristic to indicate how much confidence it has in each customization it proposes. The conflict resolution algorithm can then sort the conflicting proposals by these confidence values. Only in those rare cases where there are multiple proposals with nearly equivalent values is the user asked to choose from among them (or choose none at all). Heuristics determine the confidence value for a proposal by analysing how accurately the proposal reflects the user's patterns. If heuristics are reasonably accurate in this user modeling, it is unlikely that multiple heuristics will have strong confidence in conflicting customizations. As a result the user will rarely be asked for confirmation.

### 4.3.4 Performance evaluation

Using heuristics to interpret user input is an inherently risky activity. Anytime a heuristic makes a guess about the user's intentions, there is a possibility that the guess will be incorrect. To prevent incorrect guesses from being made frequently, heuristics evaluate the performance of the customizations they perform and use these evaluations when computing confidence values. Customizations that receive positive evaluations receive increased confidence, while negative evaluations result in lowered confidence. The result is that heuristics can initially make proposals when they have some minimal confidence in those proposals. Positive evaluations result in raised confidence, encouraging the heuristics to repeat the customization. Negative evaluations lower confidence, thereby requiring more supporting evidence to be acquired before the same proposal can be used again.

Two mechanisms are provided for this performance evaluation: success/failure criteria and timeouts.

*Success/failure criteria*

Instead of requiring users to indicate their approval or disapproval of a customization explicitly via confirmation, *success/failure criteria* allow the user's approval or disapproval to be determined implicitly. In particular, success criteria identify user actions that take advantage of work performed by a customization, thereby indicating that the user found the customization useful. Failure criteria watch for user actions that effectively undo the customization, and thus signal disapproval with the customization. For instance, the construction of the else-statement performed in the example in Section 4.2.4 is considered a success if the user constructs a statement for the body of the else-statement. It is considered a failure if the user deletes the else-statement. Similarly, if a customization causes a new window to appear on the user's screen, failure is recognized if the user removes the new window. Success occurs if the user uses the new window.

*Timeouts*

Sometimes neither success nor failure criteria will be met for a long time. If after several days, the user takes advantage of the customization, causing the success criteria to be evaluated, it is reasonable to believe that the user did approve of the customization. However, if failure criteria are evaluated after several days, it is not clear that this really represents a failure of the customization. It might simply mean that the user changed his mind about some aspect of his work, and as a result is undoing the customization's work as well.

In other situations there may be no user action that would identify a customization as successful. For instance, if a customization performs spelling correction to some text the user entered, failure can be recognized by the user editing the word. However, there is no action the user could take that would indicate the correction was appropriate. Instead it is the lack of an undo action that implies success.

*Timeouts* are designed to handle the above two situations. A *context switch timeout* occurs when the user's focus of attention changes. An *activity switch timeout* occurs when the types of actions the user takes changes. A *command threshold timeout* simply counts the number of commands since the customization is applied. Finally, *system exit* can be treated as a timeout. Each heuristic decides which timeout(s) to use. When its timeout is reached, the corresponding customization is considered successful, and its success/failure criteria are disabled.

### *Developing success/failure criteria*

The ability to evaluate customizations automatically is one of the major contributions of this research. In order to make it easy for a user interface designer to provide accurate and efficient criteria, the system provides primitive success/failure criteria for each primitive action that a customization can take. A customization's success/failure criteria are compositions of the system-defined criteria for the primitive actions taken by the customization. For instance, if one of the primitive actions is *create-structure*, the corresponding failure criterion is *delete-structure* of the newly created structure, while success would be recorded if the user performed a construction within the new structure, or the structure remained unchanged until a timeout occurred. A customization that uses *create-structure* as one of its actions can then use *create-structure's* success/failure criteria as one component of the customization's success/failure criteria.

Since customizations ultimately consist of a sequence of primitive actions, a customization's success/failure criteria can be automatically constructed by combining the success/failure criteria of each primitive action. The combination technique used in the prototype (described in

Section 4.4) is the *first result rule*: a customization is considered successful (failed) if the first primitive action to be evaluated is a success (failure). For example, a customization that creates a complete subtree (one with no placeholders) in a structure editor really performs a sequence of *create-structure* operations. It would fail if any of the structures were deleted. It would only succeed if the customization received a timeout. There are no explicit success criteria in this case, since the customization created a complete structure, thereby eliminating the possibility that the user could perform a construction within the new structure.

The first result rule is extremely easy to use. It is also very efficient, because as soon as one primitive success/failure criterion is evaluated, the result of the entire customization evaluation is known. Therefore a very small amount of computation is actually required to evaluate a customization. The problem is that they might not be accurate in all situations. If a customization performs a single primitive action, then the first result rule is satisfactory. However, if a customization performs several primitive actions, determining the customization's success by the result of a single primitive might not be very accurate.

The fundamental problem with this approach is that it treats each primitive action as equally important, when in fact some actions may be much more central to the customization's performance and should therefore be weighted more highly. To solve this problem, we turn to a hybrid approach in which the system provides the primitive criteria, and the user interface designer determines how to combine them. In the hybrid approach the user interface designer identifies which of the customization's actions are *significant* with respect to the evaluation of the customization. Only the criteria for significant actions are used in the construction of the customization's criteria. The hybrid approach results in success/failure criteria that are nearly as easy to use as the automatically generated criteria. They may be more efficient than the automatically generated criteria, since it might not be necessary to monitor each primitive action taken. They should be more accurate, since the user interface designer can decide how to evaluate each automated action individually. Finally, since they are not hard-wired into the system, it is easy for the user interface designer to experiment with them and tune them to individual customizations.

### 4.3.5 Garbage collection

Since heuristics are designed to identify individual user patterns, they may act differently for different users, or even for the same user at different times. A side effect of this customization process is that heuristics that are useful for one user might not be useful for another, and customizations that were applicable at one time might not be applicable at another time. Such ineffective heuristics and customizations are referred to as *garbage*.

A heuristic may be ineffective for one of two reasons. First, it is possible that an individual user simply does not exhibit patterns that a heuristic can use effectively. Second, a user interface designer can write general purpose heuristics that can be used in a wide variety of contexts. This form of code reuse is very powerful, but it can result in heuristics being applied in situations that they are not really designed for. For instance, a spelling checker can be written that is independent of the application domain, but if it is applied to the bibliography section of a journal article, it will perform poorly due to the fact that authors' names will not be in its dictionary.

A heuristic can recognize that it is ineffective in one of two ways. Either it is unable to discern patterns in the user's behavior and thus cannot provide any useful customizations, or the customizations that it is proposing are frequently evaluated as failures. In the former case we may want to disable the heuristic completely so that it does not waste computing resources and the user's time. In the case of frequent failures, we may simply want to mark the customizations that have been counterproductive so that they are not applied in the future, thereby preventing the heuristic from repeating its mistakes, while still allowing it to develop new customizations.

Garbage may also result from changes in the habits of users. A heuristic might learn certain customizations that are beneficial to a user for some time. But as the user continues to use the system, he may develop new patterns and discard old ones. This can occur because he has become more familiar with the system, his tasks have changed, or the system has been enhanced. In any event, if heuristics are to model the user's preferences accurately, the knowledge that is encoded about the user must be updated to reflect these changing habits. As a result, customizations that no longer benefit the user should be discarded.

## 4.4 Lantern

Lantern is the prototype implementation that was developed for experimental evaluation of this research. Rather than construct a toy system with a heuristic-enhanced user interface, user interface heuristics were added to an existing system. There are two parts to the implementation. First, some underlying mechanisms were added to the Gandalf system, a programming environment generator. Second, several heuristics were added to an existing Gandalf programming environment to create an environment with an automatically-customized user interface. Lantern, like all Gandalf environments, is a programming environment with a structure editor interface.

The user interface of Gandalf environments provides an interesting base to add heuristics to. This is true for several reasons.

- *Automating routine tasks.* Users tend to develop patterns of behaviour to accomplish routine tasks. The presence of patterns in user behavior allows us to judge whether heuristics can identify and apply such patterns without requiring explicit user invocation.
- *Tolerant command interpretation.* A typical Gandalf environment supports several hundred commands. This naturally leads to a complex and potentially confusing user interface. This complexity helps us to evaluate whether heuristics are capable of understanding a user's intent when a command's preconditions are unmet, interpreting these commands correctly instead of reporting errors to the user.
- *Active help.* The structure editor supplied by Gandalf is able to provide minimal help through continuously displayed menus. However, if the command names are not familiar to the user, this type of help is not sufficient. Instead a novice user often resorts to trial-and-error style editing in order to achieve his goals. Using Gandalf, we can evaluate whether active help heuristics can provide better assistance to the user than the current passive help system.

Gandalf also provides an architecture that can be practically extended to support heuristic mechanisms. To generate a programming environment, an environment implementor[1] must supply a description of the syntax and semantics of the language for the desired environment.

[1] Throughout the remainder of this chapter the term *implementor* is used to refer to the person who develops a particular Gandalf environment, including the syntactic,

These descriptions are compiled and linked with the Gandalf kernel to produce an environment. The separation of environment development into the kernel and environment-specific specifications provides an appropriate foundation for the separation between heuristic mechanisms and heuristics needed in Lantern. The modified kernel used in Lantern also defines environment-independent heuristics, which can be used in all generated environments, while the user interface designer provides environment-specific heuristics.

The ARL environment was selected as the environment for experimentation for a number of reasons. First, in order to allow experimentation with both novice and expert users, it was necessary to choose an existing environment so that there would be experts. Of the Gandalf environments in use at Carnegie Mellon University, ARL is by far the most commonly used.

In addition to the fact that both novices and experts were readily available for the ARL environment, the language itself presented an interesting choice for experimentation. ARL is a rich programming language, but designed for a specific application domain. Many of the statements and expressions in ARL have no equivalents in other programming languages. Therefore, ARL novices can only directly use their expertise from other languages for the subset of constructs that it shares with more standard imperative programming languages.

When deciding upon the kinds of heuristics to implement in Lantern, two goals were considered. First, we wanted to test the efficacy of heuristics, loggers, and success/failure criteria in a variety of situations. Second, we wanted to provide customizations that would be most beneficial to users.

The scripts of three users of the standard ARL environment, ranging from novice to expert, were collected over the course of approximately two months. These scripts were studied to determine what customizations would be most beneficial for users. Heuristics that could learn these customizations were then designed and added to Lantern. Five heuristics were implemented, covering all three styles of customization: automation of routine tasks, tolerant command interpretation, and active help. Four of the five implemented heuristics rely heavily upon knowledge collected by observing user interactions. In particular, they are not ARL-specific, but could be reused in any Gandalf environment.

semantic and heuristic portions. It does not refer to the person who develops the Gandalf tools that enable an environment to be generated.

Their customizations are derived solely from gathering knowledge of the user. This section describes one heuristic of each customization style that was implemented in Lantern.

### 4.4.1 Automatic construction

One of the most frequent user actions in a Gandalf environment is the construction of new program fragments. Constructions are implemented by replacing a placeholder in the user's tree with a node whose structure is defined by a structure editor command. For example, the command "if" in ARL replaces a "statement" placeholder with the template for an *if-statement*

```
if $boolean_expr then
  $statement
$else_part
end if
```

The *automatic construction* (AC) heuristic was developed to automate construction whenever possible. It collects data concerning the constructions the user performs at each placeholder of each template. If the user exhibits a pattern where the same constructive command[2] is usually applied at a particular placeholder, this heuristic will automatically apply the favored constructive command at the corresponding placeholder in the future. This heuristic is coded with no knowledge of a specific environment. Nevertheless, since it collects different data for each constructive command in each environment, it behaves in an environment-dependent, context-dependent and user-dependent fashion.

In the example from Section 4.2.4, the AC heuristic learned to construct an else-statement automatically at the $\$$*else-part* placeholder of an if-statement. In fact, the AC heuristic did learn for two users to omit the else-statement. However, this customization had a low success rate (78% for one user, 67% for the other), and was automatically disabled after nine attempts in each case. Most of the more successful patterns learned by AC would require significant explanation of the ARL language in order for them to be sensible. However, one other

[2] As opposed to an editing command, such as *delete*, a cursor motion command such as *next-line*, etc.

customization that turned out to be very useful was that the AC heuristic learned to omit the exception handler that can be attached to certain statements. This customization was learned for four of the five users, with a success rate of 99% after 342 attempts.

### 4.4.2 Intermediate node construction

One of the biggest difficulties users have with structure editors is that the editors require each piece of structure to be built explicitly. But users naturally tend to think in terms of the text they wish to see on the screen, rather than in terms of the structures that must be built to reach that state. In many cases there is a one-to-one mapping between structure and text, and in those cases the structure commands are typically chosen as a prefix of the desired text. For instance, a procedure is both a structural component and corresponds to some textual entity (in most languages, including ARL). The command to construct the procedure structure is *procedure*, which can be abbreviated to any prefix, such as *p* or *proc*. However, some structures do not have any textual representation of their own, but rather are displayed simply as the textual representations of their components. For instance, a variable declaration structure in ARL does not have any textual representation of its own, but is displayed simply as the concatenation of the textual representations of its type and variable components, as shown in Figure 4.1. In such situations, users often forget to build the declaration structure and instead try to build a type, which the editor prohibits, because it is structurally incorrect. This is the first example given in Section 4.2.4.

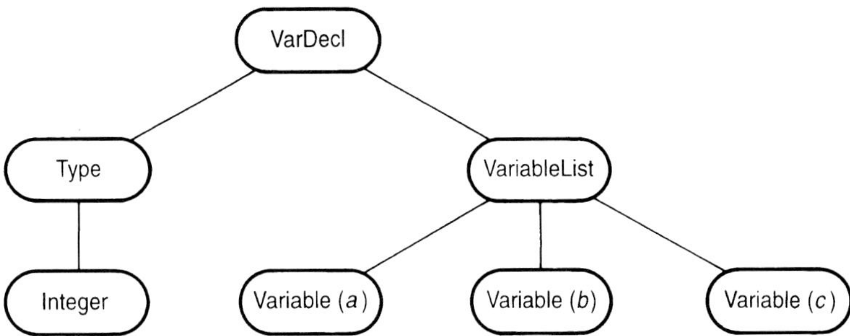


**Figure 4.1** A variable declaration in ARL.

The purpose of the *intermediate node construction* (INC) heuristic is to overcome this difficulty with structure editing. It allows constructive commands to be applied at placeholders at which they are not legal, if it can make a good guess about which structure the user neglected to build explicitly. It uses environment knowledge to determine if there is a unique constructive command legal at the selected placeholder. If there is, it will automatically apply that command and try to re-apply the user's command at the first placeholder of the new structure. If there is more than one legal constructive command, it will look at the history of user interactions to determine which constructive command the user used the most in the past, automatically apply that command and then try to re-apply the user's command. In either case, if the user's command still cannot be applied successfully, the heuristic will delete the structure that it has constructed.

This heuristic is fired whenever the "unknown command" error is reported by Lantern's kernel. This occurs whenever the command interpreter cannot find a legal command that matches the user's input. The INC heuristic uses the same logger as the AC heuristic to collect data concerning which command the user normally applies at each placeholder of each structure. It might seem, then, that the INC heuristic is redundant since the AC heuristic would have automated the construction of the structure if the user had exhibited a preference for some command. However, this is not the case since the INC heuristic is more ambitious about performing constructions than the AC heuristic, because it has more reasons to explain why the construction is appropriate:

- *Environment knowledge* indicates that the construction is legal.
- *Environment and user knowledge* indicate that the construction is likely, because it is either the only legal constructive command, or the one preferred by the user.
- The heuristic can *justify* its customization, because it maps the current state, in which the command results in an error, into one in which the command can be legally applied.

In contrast the AC heuristic has only the first two reasons to perform constructions, namely, legality and the user's preference.

The INC heuristic learned to create the variable declaration structure when the user tried to construct a type, as described above, for four of the five users. For one of the users, this customization became disabled because it had a success rate of only 50% after six attempts. For the other three users, it had a combined success rate of 92% after 12 attempts. This is an example of a situation where a customization was found that is only appropriate for some of the users of an environment. The remaining customizations again deal with structures that are very ARL-specific and would be difficult to describe concisely in this chapter.

### 4.4.3 Destination help

One of the most confusing aspects of ARL, particularly for novices, is the move-statement. The move-statement is an ARL construct, similar to an assignment-statement, that a programmer uses to move a cursor[3] to point to a node in a tree. A simple example is:

```
move thecursor to right;
```

This moves the cursor referenced by the variable *thecursor* to point to the right sibling of its original value. The difficulty ARL users have is in finding the correct structures to build to describe the desired destination node, since there are many choices to select among.

The *destination help* (DH) heuristic provides active help on the construction of destinatiom expressions. It looks for a sequence beginning with a construction at a *destination* placeholder, followed by zero or more commands that do not modify the tree, and ending with a deletion of the constructed destination. This pattern of creation and immediate deletion is interpreted as confusion on the user's part. When such confusion is observed, the heuristic presents a screenful of help text that describes each legal structure briefly, and provides pointers to the appropriate pages of the ARL manual.

The DH heuristic differs from the previous heuristic in two major respects. It is both environment-specific and user-independent. As a result, it can only be used in the ARL environment, and it cannot adapt to individual users. It therefore provides the same help message for all users.

This heuristic was rarely used in experimentation, but that is to be

[3] A cursor in ARL is a special type of pointer. It points to abstract syntax tree nodes. Cursors cannot be directly assigned to. Instead, the **move** statement is used to perform the assignment.

expected of active help heuristics, since users will quickly learn how to perform the desired actions, and will no longer require the help.

### 4.4.4 Summary

Lantern is an extension of an existing Gandalf environment with the addition of five heuristics. These consisted of two automation heuristics, two tolerant command interpretation heuristics, and one active help heuristic. As with all Gandalf environments, the Lantern programming environment is a generated environment. It is produced by linking an environment description with the reusable Gandalf kernel. Since all of the heuristics but the active help heuristic are environment independent, they can be incorporated directly into the kernel and reused in any Gandalf environment. In fact other Gandalf environments have been constructed that incorporate these heuristics, including the Gandalf tools Aloegen, DBGen (Staudt *et al.*, 1986), LexGen (Green, 1989) and HeurGen, as well as an environment for the specification language Larch (Guttag *et al.*, 1985). In the next section the results of experimentation with Lantern are discussed.

## 4.5 Experimental results

Two sets of experiments were performed with Lantern. The first set, referred to as the "canned experiment", involved the collection of a large number of scripts of users using the standard ARL environment, which has no heuristics. These scripts were then used as a guide in performing the same operations using Lantern. An experimental script matched the corresponding original script, until heuristics performed some customization.[4] At that point the experimental script differs to compensate for the customizations performed by the heuristics. For

[4] In some cases there were some additional minor deviations in the two scripts besides the changes caused by customizations. For instance, some changes and bug fixes unrelated to heuristics were made to the ARL environment and the Gandalf kernel between the time the scripts were collected and the time the scripts were used in the experiment. In general, these modifications did not have a significant effect on the number of commands or errors. When there was a significant difference, the command and error counts of the original script were modified to appear as if the original script had been run with the current version of the environment without the heuristics.

instance, if a customization automatically constructs a structure correctly, then it is unnecessary for the experimental script to include the commands to perform the construction. Also, when the active help heuristic fired, the help was assumed to be sufficient to resolve the user's confusion, and therefore the command that the user ultimately executed was executed immediately after receiving help. Data was collected to compare the number of commands used in the original and experimental scripts, the number of error messages that occurred in each, the success rates of each heuristic's customizations, and the customizations learned for each user.

The second set of experiments, called the "live experiment", involved the use of Lantern by three users in their daily work. The initial history file for each user was the one obtained at the end of the canned experiment for that user. Therefore, the user interface for the live experiment was already performing customizations when the users began using it. The live experiment was primarily used to evaluate the usefulness of the interface as viewed by human users. In addition, measurements were taken to determine the success rate of each heuristic's customization, and the customizations learned for individual users.

Table 4.2 summarizes the experience level of each user, as well as the amount and type of data provided by each. Data was collected from

**Table 4.2** Summary of data sets of experiment

| | 1 | 2 | 3 | 4 | 5 |
|---|---|---|---|---|---|
| Canned or live | Canned | Canned | Both | Both | Both |
| Experience | Moderate | Novice | Novice | Expert | Novice |
| Type of work | New code | Tutorial<br>New code | Tutorial | Old code | Browsing |
| Number of: | | | | | |
| Scripts | | | | | |
| Canned | 142 | 146 | 49 | 23 | 16 |
| Live | | | 32 | 80 | 30 |
| Commands | | | | | |
| Canned | 5683 | 6115 | 3777 | 358 | 67 |
| Live | | | 2209 | 3907 | 238 |
| Errors | | | | | |
| Canned | 244 | 236 | 143 | 14 | 10 |
| Live | | | 35 | 44 | 15 |

five users. Of these, there were three novice users, one intermediate user and one expert. Two users were involved only in the canned experiment, while three users were involved in both the canned and live experiments. In the latter case, the numbers for both the canned and live scripts are given separately. The data shown in the table consists of the number of scripts, commands used, and error messages reported.[5] The numbers given for commands and errors are for the scripts entered by the user. For canned scripts, these numbers come from the original scripts, and therefore are without heuristics being active. For the live scripts, the numbers reflect the use of heuristics.

Users used Lantern primarily for four different types of tasks: writing new code, maintaining existing code, following an extended example in a tutorial, or browsing. The new code task requires a great deal of construction of new ARL code, as well as a fair amount of editing and some browsing. The maintenance task concentrates more on the editing of existing ARL code, but also involves some browsing and construction. The tutorial task concentrates almost exclusively on construction. The browsing task is one in which the user concentrates on browsing existing code, performing only occasional edits or constructions.

A number of hypotheses were developed concerning the expected behavior of Lantern. They are the following:

- *Effectiveness hypothesis.* Heuristics reduce the number of commands required to complete a task, and the number of error messages reported while performing a task.
- *Accuracy hypothesis.* Simple success/failure criteria evaluate the correctness of a customization accurately.
- *Learning hypothesis.* Heuristic performance improves over time.
- *Adaptiveness hypothesis.* Heuristics learn different customizations for different users.
- *Acceptance hypothesis.* Users find the heuristic-enhanced interface to be an improvement over the standard Gandalf interface.

[5] Counting error messages omits a collection of errors, namely those where the user executes a legal command, but the command does not have the effect the user desires. Nevertheless, counting error messages provides an objective measure of errors, as opposed to the subjective measure that would be necessary to account for the latter class of user errors.

This section describes how the experiments were used to evaluate these hypotheses. Only data related to the heuristics described in Section 4.4 are presented here. A description of all implemented heuristics as well as a complete list of customizations learned for all users and all heuristics can be found in Lerner (1989).

### 4.5.1 Effectiveness hypothesis

The effectiveness hypothesis claims that heuristics reduce the number of commands required to complete a task, and the number of error messages reported while performing a task. If this hypothesis holds, then it is possible for heuristics to reduce the amount of work expected of a user to complete a task.

This hypothesis was tested solely through use of the canned data. For each canned script, a count was made of the number of commands and the number of errors within the original scripts and their corresponding experimental scripts.

For these comparisons only the three users who have more than 500 commands in their canned data (1, 2 and 3) were considered. Figure 4.2 shows the ratio of commands in the experimental scripts compared to the original scripts. Each datapoint represents the end of a script. Thus, the low point for user 3 is only the second datapoint, even though it

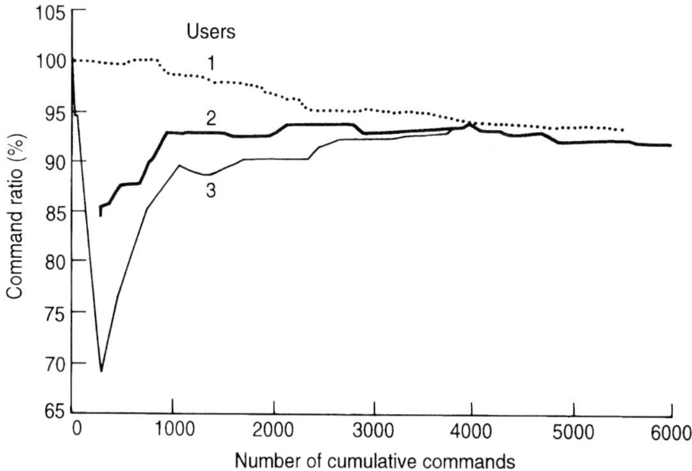


**Figure 4.2** Comparison of number of commands used with and without heuristics.

represents approximately 300 commands. This indicates that automation was quite successful in the second script for this user, but reached a more moderate degree of automation in the long run. It is interesting to note that while the command ratio differs greatly for each user initially, after about 2500 commands the three curves converge at approximately 7% command reduction. Overall this data suggests that heuristics do reduce the number of commands required to achieve a task.

Figure 4.3 shows the ratio of errors reported in the experimental scripts compared to the original scripts. For users 2 and 3 the error ratio converges after about 75 errors, while the error ratio for user 1 is significantly higher. The reason for this disparity can be traced to two causes. First, the amount of error prevention that occurs depends upon how well the tolerant command interpretation heuristics match the types of errors the user encounters. Since the matching was better for the novices, more error reduction occurred for them. Second, construction is a major source of errors for novices. Since the AC heuristic automates construction, the users need to do fewer constructions themselves. By reducing the number of constructions the novices needed to perform, the number of errors was also reduced.

It is interesting to consider how the 7% command reduction and up

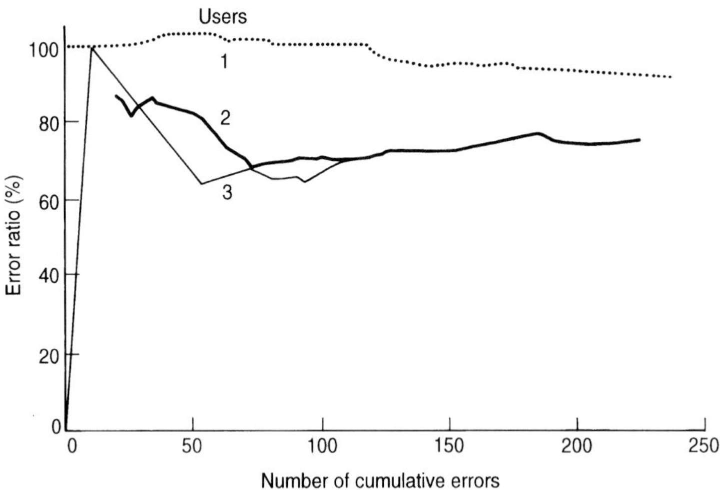


**Figure 4.3** Comparison of number of errors encountered with and without heuristics.

to 25% error reduction affect the user interface in a qualitative sense. Does this significantly reduce the user's effort?

There are three situations which lead directly to noticeable reduction in effort. First is the handling of deviant input. Without tolerant command interpretation heuristics to perform these actions, the user would be required to read and understand an error message, and determine how to reformulate the input so that it is acceptable. With heuristics the original input can be interpreted directly. Obviously, this will result in significant savings, particularly for novice users who experience the most error reduction.

Second, patterns that represent long-term user preferences will account for significant reduction in effort because the user will incorporate those actions into his model of the system. He will therefore expect those actions to be performed, and will not be surprised by their occurrence. (An example of this actually occurring during experimentation is presented in Section 4.5.3.)

Finally, in some situations automated actions will be chained together. That is, the user will enter a command that will cause a heuristic to take an action. This heuristic action may itself trigger other heuristics, resulting in a chain of automated activity. These chained actions will result in significant savings, regardless of whether these are long-term preferences or short-term preferences that may only be applicable to the current editing session. This is simply because of the number of commands that are automated through the invocation of a single command.

There are also two situations that lead to an increased user effort. First, failed heuristics obviously lead to increased effort, because the user must undo the inappropriate actions.

Second, patterns that are just being learned might result in increased effort temporarily since the user may be surprised by their automation initially, and be expecting to perform those actions manually. In such cases, the user must recognize that the action has already occurred, which requires more cognitive processing than if only the normal, expected action had occurred. This surprise effect should be much less serious for novice users, since the user interface should learn these routine tasks nearly as quickly as the user incorporates them into his normal behavior. In any case, if this develops into a long-term pattern, the long-term benefits of automation should outweigh the short-term costs of the surprise.

One final point should be made about the amount of command and error reduction observed in Lantern. These reductions were observed in an environment that contained only five heuristics. It is reasonable to expect that a more complete system with even more heuristics would be able to improve upon these numbers.

In conclusion, the effectiveness hypothesis is supported by the data. Heuristics do, in fact, reduce the effort required to achieve a task.

### 4.5.2 Accuracy hypothesis

The accuracy hypothesis claims that simple success/failure criteria evaluate the correctness of a customization accurately. This hypothesis was evaluated by analysing each customization in the scripts from both experiments. For each application, the result of the success/failure criteria was noted, as well as what the true result was based upon human analysis of future user actions. This resulted in each applied customization being recorded as one of the following types: true success, true failure, false success, false failure, or irrelevant. A *true success* occurs when both the success/failure criteria and the script analysis indicate success. Similarly, a *true failure* occurs when both indicate failure. A *false success* occurs when the success/failure criteria indicate success, while the script analysis shows the application to be a failure. A *false failure* occurs when the success/failure criteria identify the application as a failure, but script analysis reveals it was a success. An *irrelevant* application is one in which the customization neither helped nor hindered the user. For instance, if the AC heuristic constructs a structure, but the user later deletes an enclosing structure, the construction neither provided a useful service nor interfered with the user's actions. Lantern did not identify any customization as irrelevant. All irrelevant customizations were actually classified either as successes or failures. A summary of the results of this analysis is presented in Table 4.3. Total successes are the number of successes as determined by script analysis; total failures are the number of failures as determined by script analysis.

Accurate success/failure criteria are essential components for the optimal performance of the heuristic. When the success/failure criteria work properly, successful customizations will gain confidence. Unsuccessful customizations will be disabled. In contrast, if false successes occur, heuristics will gain confidence in undesirable customizations and will commit these undesirable customizations more often. False failures may cause customizations that are actually providing useful

**Table 4.3** Accuracy of success/failure criteria

| | User 1 | User 2 | User 3 | User 4 | User 5 | Overall |
|---|---|---|---|---|---|---|
| Attempts | 515 | 817 | 689 | 212 | 9 | 2242 |
| Total successes | 444(86%) | 622(76%) | 590(86%) | 115(54%) | 7(78%) | 1778(79%) |
| True successes | 433(84%) | 610(75%) | 581(84%) | 113(53%) | 7(78%) | 1744(77%) |
| False failures | 11(2.1%) | 12(1.4%) | 9(1.3%) | 2(0.9%) | 0 | 34(1.5%) |
| Total failures | 46(8.9%) | 128(16%) | 74(11%) | 67(32%) | 0 | 315(14%) |
| True failures | 41(8%) | 106(13%) | 61(9%) | 42(20%) | 0 | 250(11%) |
| False successes | 5(1%) | 22(2.7%) | 13(1.9%) | 25(12%) | 0 | 65(2.9%) |
| Irrelevant | 25(4.9%) | 67(8.2%) | 25(3.6%) | 30(14%) | 2(22%) | 149(6.6%) |

behavior to be disabled. This phenomenon explains why 66% of all incorrectly identified customizations are failures, while only 13% of the correctly identified customizations are failures.

Overall we see that 98% of the successful customizations are correctly identified, while only 79% of the failed customizations are correctly identified. The reason for this disparity is because the false failures result in those customizations being disabled, thereby limiting the number of times the same mistake is made. However, when a false success occurs, it becomes even more likely that the same mistake will occur again since the heuristic gains confidence in the incorrect customization. The causes for misidentifications can be classified into one of three categories: weaknesses of Lantern, weaknesses of the model, or unavoidable. The problems caused by Lantern's weaknesses involve bugs in the criteria, missing timeouts, and customizations attempting illegal actions. Fixing them would improve the accuracy of recognizing failed customizations to 89%. The primary weakness in the model is the assumption of success when a timeout occurs. By allowing heuristics to define evaluation functions to be called when a timeout occurs, instead of just assuming success, the accuracy of recognizing successful customizations would be 99%, while failed customizations would be correctly identified 94% of the time.

The remaining misidentifications seem much harder to correct. For

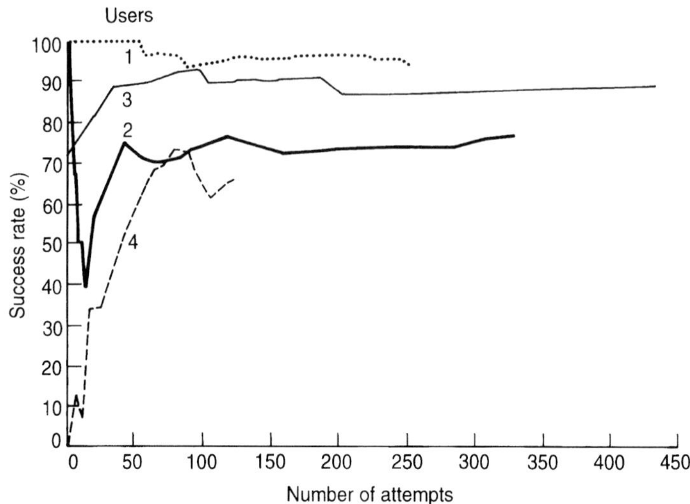


**Figure 4.4** Success rate of the automatic construction heuristic.

instance, several false failures were identified by comparing original scripts to those used in the canned experiment. In some situations, the original user script contained sequences of commands in which later commands undid earlier commands. There were five situations in the canned experiment in which customizations performed these early commands. When the later user commands undid these customizations, they were identified as failures. However, it is hard to consider the customizations at fault, since they did faithfully mimic the user's behavior.

### 4.5.3 Learning hypothesis

The learning hypothesis claims that heuristic performance improves over time. This hypothesis was evaluated by recording for each heuristic how many customizations it attempted, how many attempts resulted in success, and how many resulted in failure. The success rate is the percentage of successful customizations compared to the total of successful and failed customizations.[6] The data presented here is for those users who have over 500 total commands. Figures 4.4 and

[6] Irrelevant customizations are not figured into the success rate.

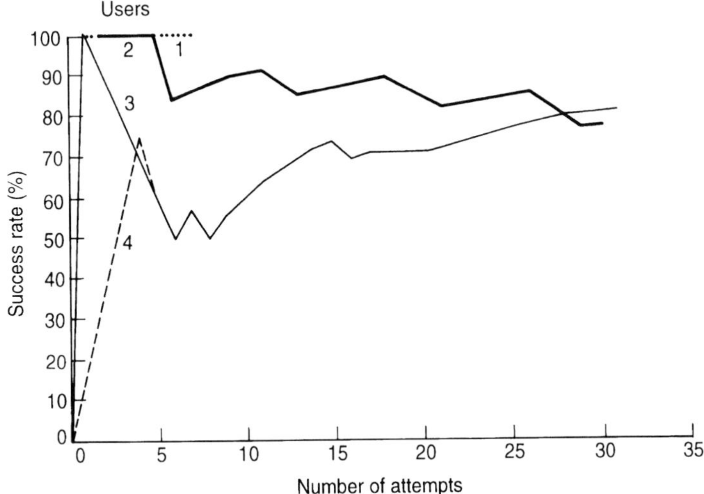


**Figure 4.5** Success rate of the intermediate node construction heuristic.

4.5 show the success rates for the individual heuristics. (The DH heuristic is not shown because it was rarely triggered during experimentation.) For the AC heuristic, we see that heuristic performance is erratic initially, but generally reaches a stable state by the 100th attempt, while the INC heuristic never becomes stable.

Overall, the performance of the INC heuristic is the worst of all the implemented heuristics. There are two reasons for this. First the confidence threshold used by the INC heuristic is lower than for the other two. Since the INC heuristic applies customizations that it has a comparatively low confidence in, it is not surprising that the success rate is also lower. The second reason for the lower success rates is simply because the INC heuristic has been applied many fewer times than the two heuristics that automate routine tasks, since it is triggered by a much rarer event than those. AC can be activated by any constructive command, but INC can only be activated by a specific error. As a result the INC heuristic may require many more commands to occur before it can reach its steady state.

One of the more interesting results of the experimentation was the observation of a user incorporating a heuristic's actions into his model of how the system worked. This phenomenon can be observed by comparing the data for user 3 in Figures 4.5 and 4.6. Figure 4.6 graphs

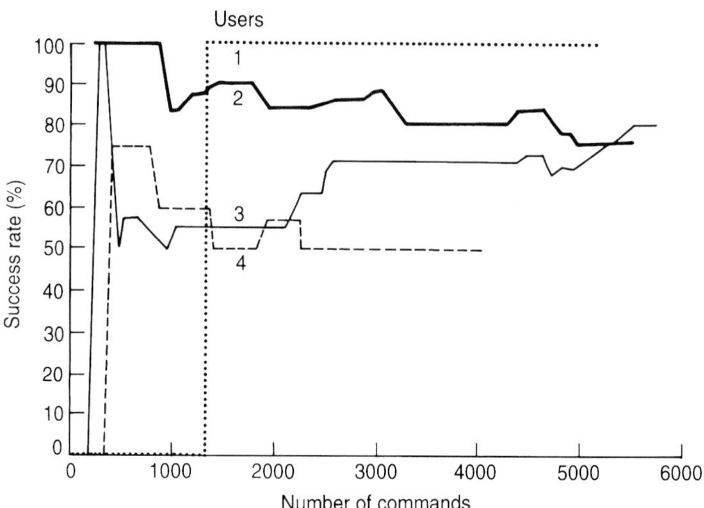


**Figure 4.6** Success rate of the intermediate node construction heuristic.

the success rate for the INC graph as a function of the number of commands, rather than the number of attempts. User 3's data consist of canned data up to command 3777, which corresponds to 15 attempts. After that point the data represents live data. For some time, the INC heuristic is unused by user 3. However, once he begins using it, he uses it heavily, resulting in the sharp increase in the success rate when graphed against the number of commands. This indicates that the user has found this particular tolerant command interpretation heuristic to be very useful, and has incorporated the customization into his model of how the command interpreter works. Insteasd of learning that certain commands cause errors in certain contexts, he is now able to apply those commands successfully. In effect, the heuristic has learned what user 3 means when he applies certain commands in a context in which they are not legal, and is therefore able to interpret his input even though it is not correct.

In summary, the learning hypothesis seems to hold for the heuristics and users for which there was a large quantity of data. Learning occurred during early interactions with the system, and then performance levelled off at a steady state. For those heuristics and data for which there was little data, the results are inconclusive.

### 4.5.4 Adaptiveness hypothesis

The adaptiveness hypothesis claims that heuristics learn different customizations for different users. In order to test this hypothesis a record was kept of the customizations that the various heuristics attempted for each user. Figures 4.7 and 4.8 show the distribution of the customizations per number of users for which the customization was applied. Since the DH heuristic is user-independent by nature, it is not expected to develop different customizations for different users, and so is omitted from the discussion here. This data shows that most of the customizations are single-user customizations, thereby demon-

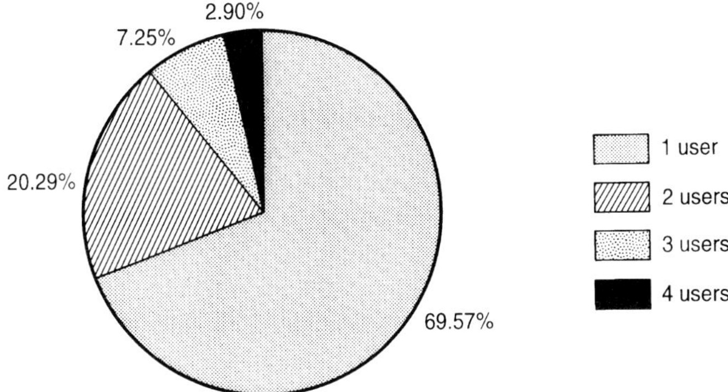


**Figure 4.7** Distribution of customizations for automatic construction.

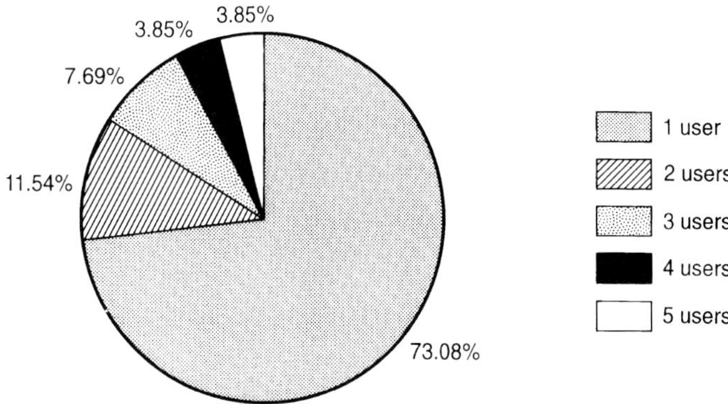


**Figure 4.8** Distribution of customizations for intermediate node construction.

strating that the heuristics do learn the habits of individual users. For instance, 70% of the customizations found by the automatic construction heuristics were only used for a single user, 20% of the customizations were common to two users, 7% were common to three users, 3% were common to four users and no customizations were shared by all five users. The other heuristics exhibit similar behavior.

### 4.5.5 Acceptance hypothesis

The acceptance hypothesis claims that users will find the heuristic-enhanced interface to be an improvement over the standard Gandalf interface. This hypothesis is aimed at making some qualitative judgement of the user interface. In order to evaluate the hypothesis, it is necessary to look at the reactions of the users from the live experiment.

Of the three users, only user five found the performance to be a serious drawback and withdrew from the experiment as a result. He had not used Lantern long enough for it to have learned any interesting customizations or been beneficial to him. This user was working on a loaded VAX 11/780, while the others were working on single-user Sun 3s.

User 3 preferred the experimental version to the standard version after becoming accustomed to it, because the automation it provided simplified his job. However, he did encounter some difficulties with Lantern. The most notable problem was that he was unable to type commands ahead, because he was unsure of how they would be interpreted.

User 4 also reported that he found Lantern easier to use. Initially, he noticed a great deal of automation, which did not occur in later sessions, and expressed disappointment over this change. From looking at scripts of his sessions it is clear that the automation stopped simply because his patterns changed. He believed that the initial automation was being done primarily by the AC heuristic and therefore did not understand why the constructions did not occur in later editing sessions. In fact most of the automation was caused by a different heuristic, which learned to paste in cut buffers. When he stopped using this cut-and-paste paradigm, the heuristic was unable to perform the same automation.

Another comment made by user 4 was that he wanted to have more control over what the heuristics did. In particular, he wanted to be able to identify the customizations he liked explicitly so that they would gain more confidence. Part of this desire most likely grew out of his misunderstanding of what the heuristics were actually doing. In particular, he wanted to boost the confidence of the AC heuristic in certain contexts, when in fact the automation he observed was not being performed by that heuristic at all. His comments do raise an interesting question about what the user's involvement should be in the use of heuristics. This topic is discussed in Section 4.6.2.

Perhaps the most telling evidence that the heuristic-enhanced interface is preferable to the standard interface is the fact that the two users who completed the experiment continued to use Lantern instead of the standard ARL environment for several months beyond the completion of the experiment. They discontinued using Lantern only when an upgraded version of the standard ARL environment, which did not include heuristics, was released. It therefore seems that the heuristics do, in fact, improve the user interface in some qualitative sense.

## 4.6 Future work

Experimentation has shown that the ideas presented in this chapter can be used effectively to automate customization of a user interface. Of course, this is a young research area with a great deal of room for improvement. This section describes some topics that are worthy of further research.

### 4.6.1 Analogical reasoning

Currently the patterns recognized by a heuristic are available only for use by that heuristic, and are generally further restricted to a particular context. One interesting extension of this work would be to allow patterns to be shared in different contexts by the same heuristic, or even different heuristics. The major advantage of sharing is that heuristics might be able to perform history-based customizations even when they encounter a situation for the first time. Of course, heuristics will not benefit from randomly sharing patterns, but are most likely to benefit if the patterns being shared are similar to the patterns they expect to observe themselves. Thus, sharing will be most beneficial among heuristics that are intended to customize analogous actions in analogous contexts.

How can such analogous heuristics and contexts be identified? First, a heuristic is obviously intended to customize analogous actions in different contexts, since it executes the same code, simply on different data. Thus, a heuristic should be able to share patterns developed in different contexts if those contexts are somehow related. Second, heuristics that automate similar actions may wish to share patterns, even if they are triggered by different events. For instance, two heuristics of Lantern both automate construction, the former (AC) as the result of another construction event, the other (INC) by an error. It is reasonable to believe that the INC heuristic would have benefitted greatly if it could have used the patterns identified by the AC heuristic for the same context.

Analogous contexts can be identified in a number of ways. First, the contexts are generally associated with a particular type of structure in the environment. These structure types can be organized into a type hierarchy, which can be used to make contextual analogies. Siblings in the type hierarchy are analogous in many respects. They probably share many methods, and hence have similar behavior. It is therefore reasonable to expect that a user might develop habits that could be the same among sibling types. For instance, the type hierarchy for a typical imperative programming language might have a type *Looping-Statement* with subtypes *While*, *For* and *Repeat*. It seems reasonable that a user might develop patterns that could be shared by these three types, such as the automatic construction of *begin . . . end* blocks for their bodies. Heuristics should be able to use the type hierarchy to support sharing.

A second kind of analogous situation is the use of a kernel-defined type across several domains. While some types are kernel-defined, an individual environment does not distinguish between kernel-defined and environment-specific types, and therefore collects data about the type within that environment. It is quite likely that the user will develop similar habits when interacting with these kernel-defined types regardless of which environment is currently being used. However, unless the patterns from one environment are shared with another environment, each environment will have to learn these patterns individually. Sharing patterns across environments should provide the user with a more consistent user interface for the different environments.

Finally, analogies can be drawn across users. If a collection of users all develop the same pattern when in the same situation, it is quite likely that a new user will develop the same pattern. This is particularly true if some other knowledge about the users can be applied. For instance, some set of users may typically perform the same type of tasks, they might all be novices, etc. By comparing the customizations made for several users, we can develop stereotypes that could be useful for new users.

The three types of analogies discussed above seem useful in three different ways. Analogies about users seem most useful when developing the initial patterns to be used for a new user. Analogies across environments could be used to create the initial patterns for a particular user in a new environment (in addition to using user analogies for the new environment). Analogies across types could be used when the user encounters a new context. Thus contextual analogies should be drawn dynamically, while the former two analogies are done before the user begins using the new environment.

It seems that contextual analogies should be performed only upon a user's request. For instance, a *suggest* command could be used to indicate that the user is unsure of what to do and would like help. At this point it would be reasonable for heuristics to use whatever knowledge they can to aid the user, and also to lower their confidence thresholds so that they are more likely to find a customization to apply. Since the heuristic is considering customizations that it has less confidence in, the options that heuristics find should be presented to the user so that he may decide which customization to perform, rather than be performed without confirmation.

### 4.6.2 User interaction

Lantern was designed to require minimal user intervention in the customization process. The user is only asked to intervene when multiple proposals are made that would affect the same object, and the confidence in each is nearly equivalent. This was useful for the experiment to exercise the limits of complete automation of the customization process. However, it would be more realistic to expect users to provide more direct feedback during the customization process. This direct feedback should be voluntary. That is, automated customization should be able to proceed as described in this chapter. However, if the user chooses to provide direct interaction with the heuristics, the heuristics should be able to benefit from it.

User interaction could occur in any of the three phases of customization: pattern identification, pattern application, and evaluation. A user can intervene in pattern identification by providing explicit examples of the actions he wishes to have automated, in the style of keyboard macros. In this way, heuristics should be able to learn those patterns more quickly. In addition, they should use techniques from research in macros by example (such as those found in Lieberman, 1982; Maulsby and Witten, 1989; Olsen and Dance, 1988) and therefore be able to learn and automate more complex patterns.

Users could intervene in the pattern application process by explicitly invoking a customization. This would have two effects. First, automation could occur even in those situations in which the heuristic does not have enough confidence in a customization to perform it automatically. Second, it would allow heuristics to learn patterns that directly incorporated other customizations, rather than just the implicit incorporation achieved via the chaining of customizations.

Finally, users should intervene in the evaluation phase of customization. In particular one can imagine an *applause* command intended to reinforce a customization and a *hiss* command intended to show contempt for the customization. These commands can be used to modify the confidence threshold that must be achieved before reconsidering the customization. An applause command would lower the threshold, while a hiss command would raise it, and perhaps even completely disable the customization.

### 4.6.3 Customized plan

Plan recognition and automation have been used to provide intelligent user interfaces to various systems, including software development environments (for example, see Bisiani *et al.*, 1988; Huff and Lesser, 1988). For example, planning can be used to perform minimal recompilation, ensure that all the necessary actions are taken when releasing a new system etc. In general, planning is intended to aid the user in solving high-level tasks. In contrast, automated customization is directed at the low-level operations the user performs.

Because of the high level for which plans are intended, they frequently need assistance from the user in order to provide some low-level commands or information needed to complete the plan. The automated customization techniques developed in this research might be able to extend the automation capabilities of planners by allowing them to learn customized plans useful for individual users. In this way, planning can be used to guide users through complex tasks, while automated customization can further reduce the burden on the user by automatically performing the routine tasks normally carried out by the individual user in completing the plan.

## 4.7 Related work

In this section some related work in intelligent user interfaces is discussed to clarify how the work presented in this chapter fits into this larger body of research.

### 4.7.1 Error recovery

One of the most popular programming environments to incorporate error recovery was Interlisp through a facility called DWIM (Do What I Mean) (Teitelman, 1972). DWIM applied some error recovery techniques, most notably spelling correction, to input that could not be interpreted directly by Interlisp. Another programming environment, COPE (Archer, 1981), provides more extensive error recovery through an incremental parser combined with sophisticated error repair algorithms.

While tolerant command interpretation heuristics are similar to error recovery rules in these two systems, there are several significant differences. First, the same mechanisms are used to support not only command interpretation, but also automation, and active help. Second, the heuristics used are not fixed by the system as they are in COPE, but can be supplemented by the developer of a programming environment. Since DWIM is part of an open Lisp programming environment, it is possible for users of Interlisp to extend the functionality of DWIM to handle more errors. However, there is little support that would allow new user-defined DWIM heuristics to be able to reason about past user actions, while the mechanisms presented here directly support the collecting, abstraction, and saving of data collected from the user's interactions with the system.

In addition, heuristics support more flexible customization than either of these systems. In Interlisp, the error recovery facilities can only be enabled by the user in an all-or-nothing manner. In COPE, error repair is "automatic, inescapable, applicable throughout the system" (Archer *et al.*, 1984: p. 14). In contrast, individual heuristics can be turned off for particular users and particular situations. Furthermore, the heuristics themselves decide when they should be turned off based on their performance, rather than requiring the user to disable those heuristics that are not helpful.

### 4.7.2 Self-adaptive parsing

Self-adaptive parsing (Lehman and Carbonell, 1989) is used to adapt a natural language interface so that it better matches the styles of individual users. Like DWIM, the adaptations undertaken are driven by the desire to understand input that deviates from the form expected by the system. Self-adaptive parsing is accomplished by starting with a small grammar and modifying it so that the resulting parser can accept all of the input from a particular user. The modifications are driven by failures in the parser. When an input cannot be parsed, the parser weakens its grammatical rules and applies knowledge of the domain and some linguistic rules to find a way to parse the input. If it can find a parse using these weaker rules, it presents its interpretation to the user to see if it is correct. If the user confirms the interpretation, the parser modifies the grammar so that if the same input is repeated, it will be accepted simply using the grammatical rules in the parser. The goal is to learn a grammar that is both easy to parse (since it is small) and easy for a particular user to use.

Self-adaptive parsing differs from the thesis research in several respects. First, it is only used for command interpretation, while automated customization is also used for automation and active help. Second, it requires user confirmation of the adaptations, whereas success/failure criteria evaluate the customizations without such explicit user interaction.

### 4.7.3 Macros by example

Some research has been done in the area of learning macros by example. These systems require the user to provide one or more examples of the actions that should be made into a macro. The system then translates these actions into the corresponding macro. Extensive user interaction may be required to create powerful macros. For instance, Olsen and Dance (1988) use a single example, but require the user to edit the macro to identify parameters and to encode control flow statements explicitly. In Tinker (Lieberman, 1982) the user must also explicitly identify the parameters of the macros. However, the control flow portions of the macro are derived through the use of multiple examples.

In a similar line of research, Nix (1985) uses input/output pairs as examples of desired behavior in a text-editing application. He then derives a pattern/replacement pair such that the pattern matches all the input components and application of the replacement results in the corresponding output components.

In the current incarnation of automated customization, macros by example is really a complementary approach. With the use of explicit examples, the system can be expected to expend more resources in the quest for automation than is feasible with just automated customization, since the system can be fairly certain that the effort will result in a useful macro. As a result it should be possible to develop more complex macros when the user provides the examples explicitly. On the other hand, automated customization can provide automation for more simple, routine tasks without requiring the user to expend any extra effort, or even recognize that automation could be useful. It can therefore benefit all users, not just those who know how to use a specific feature of the interface.

### 4.7.4 User modeling

Much research has been done in the area of user modeling. User modeling involves the representation of the user's model of the system in a knowledge base. The user model encodes what the user knows or believes about the application and/or the application's domain. This knowledge of the user is used to tailor help, error messages, and advice for individual users (for examples, see Finin, 1983; Kass, 1987; Matthews and Biswas, 1985; Reiser *et al.*, 1985; Rich, 1983; Shrager and Finin, 1982; Wilensky *et al.*, 1988; Wolz and Kaiser, 1988; Zissos and Witten, 1985).

There are a number of differences between this work in user modeling and the research presented here. Most obviously, the goals are different. User models are used to provide customized messages to the user. The contents of the messages vary depending upon what the system believes the user knows. In this way, messages can be tailored so that they can be understood by the user without being overly verbose. Typically, messages are not presented to the user until the user either asks for help or encounters an error. User modeling and automatic customization are therefore complementary approaches. Automatic customization provides help automatically when the user appears confused. User modeling can be applied to tailor the help messages given. Automatic customization also attempts to resolve errors without the user's assistance, but when it is unable, it would be useful to present the user with error messages that are tailored to his ability and experiences.

## 4.8 Conclusions

Experimental results with Lantern support the three major claims of this chapter. First, customization can be performed without direct user intervention. Second, the history of interactions between the user and the system is a rich source of knowledge to drive such customizations. Third, customizations can be evaluated automatically using success/failure criteria.

The basic mechanisms developed in the research are loggers, heuristics, an arbiter and success/failure criteria. These four mechanisms provide the functionality of history abstraction, automation, conflict resolution and evaluation respectively. Separating the functionality in this way works well with the object-oriented architecture of the underlying system. The system-defined success/failure criteria show how evaluation can be automated not only from the user's perspective, but also from the user interface designer's perspective.

One of the most powerful techniques a user interface designer can employ to create simple, yet useful heuristics is reuse. Environment-independent heuristics and loggers can be used by a large collection of structures across a wide range of environments, providing a very powerful reuse mechanism. Customization is then achieved by relying heavily upon knowledge of the user derived from history. By collecting user data for each environment and structure separately, environment-independent heuristics are able to perform environment-dependent, context-dependent and user-dependent customizations.

Lantern was developed to allow experimentation with the mechanisms investigated in this research. One of the most important lessons learned from Lantern is the power of just a few heuristics. The two environment-independent heuristics that automated routine tasks provided a great deal of customization since they were applicable in a large number of situations. It seems that a few such general purpose heuristics should be sufficient to provide a reasonable amount of automation. Tolerant command interpretation heuristics should be developed to address the most commonly encountered errors, so that users can benefit the most from a few heuristics. Finally, active help heuristics should be used liberally. In addition to providing help actively, the active help heuristics should select messages based upon models of the user's knowledge of the system.

While further research will undoubtedly improve upon the mechanisms and techniques investigated in this research, the feasibility of automated customization without user intervention has been demonstrated.

## Acknowledgments

This research was carried out at the School of Computer Science, Carnegie Mellon University, Pittsburgh, PA 15213, USA.

I am indebted to my thesis advisor Nico Habermann for his guidance during my thesis research. I also extend my thanks to Brad Myers, Charlie Krueger, Richard Lerner, David Miller, and an anonymous referee for their comments on earlier drafts of this chapter.

Support for research on Gandalf is provided in part by ZFE F2 KOM of Siemens Corporation, Munich, Germany and in part by the Defense Advanced Research Projects Agency (DOD), ARPA Order n4976, under contract F33615–87–C–1499 and monitored by the Avionics Laboratory, Air Force Wright Aeronautical Laboratories, Aeronautical Systems Division (AFSC), Wright-Patterson AFB, Ohio, 45433–6543.

The views and conclusions contained in this chapter are those of the author and should not be interpreted as representing the official policies, either expressed or implied, of the Defense Advanced Research Projects Agency, the US government, or other supporting institutions.

## References

Archer, J. Jr (1981). *The design and implementation of a cooperative program development environment.* Ph.D. thesis, Cornell University, Ithaca, NY, USA.

Archer, J. Jr, Conway, R. and Schneider, F. (1984) User recovery and reversal in interactive systems. *ACM Transactions on Programming Languages and Systems*, **6**: 1–19.

Bisiani, R., Lecouat, F. and Ambriola, V. (1988). A planner for the automation of programming environment tasks. In *Twenty-first International Hawaii Conference on System Sciences*, Kailva-Koni, HI, USA, January.

Broverman, C. A., Huff, K. E. and Lesser, V. R. (1986). The role of plan recognition in design of an intelligent user interface. In *Proceedings of 1986 International Conference on Systems, Man, and Cybernetics*, pp. 863–868, Atlanta, GA, USA, August.

Card, S. K., Moran, T. P. and Newell, A. (1983). *The Psychology of Human–Computer Interaction.* Lawrence Erlbaum Associates, Hillsdale, NJ.

Finin, T. W. (1983). Providing help and advice in task oriented systems. In *Proceedings of the International Joint Conference on Artificial Intelligence 1983*, pp. 176–178, Karlsruhe, Germany, August.

Green, R. M. (1989). *LexGen User's Manual and Tutorial.* Carnegie Mellon University, Pittsburgh, PA, USA.

Guttag, J. V., Horning, J. J. and Wing, J. M. (1985). The Larch family of specification languages, *IEEE Software*, **2**: 24–36.

Huff, K. E. and Lesser, V. R. (1988). A plan-based intelligent assistant that supports the software development process. In *Proceedings of the ACM SIGSOFT/SIGPLAN Software Engineering Symposium on Practical Software Development Environments*, Boston, MA, November.

Kass, R. (1987). *Implicit acquisition of user models in cooperative advisory systems.* Technical Report MS–CIS–87–05, Department of Computer and Information Science, University of Pennsylvania, Philadelphia, PA, USA.

Lehman, J. F. and Carbonell, J. G. (1989). Learning the user's language, a step towards automated creation of user models. In Wahlster, W. and Kosba, A. (editors), *User Modeling in Dialog Systems.* Springer-Verlag, Berlin.

Lerner, B. S. (1989). *Automated customization of user interfaces.* Ph.D. thesis, Carnegie Mellon University, Pittsburgh, PA, USA.

Lieberman, H. (1982). Constructing graphical user interfaces by example. *Proceedings of Graphics Interface '82.*

Matthews, M. and Biswas, G. (1985). Raising user proficiency through active assistance: an intelligent editor. In *Proceedings of the Second Conference on Artificial Intelligence Applications*, pp. 358–363, Miami Beach, FL, USA, December.

Maulsby, D. L. and Witten, I. H. (1989). Inducing programs in a direct-manipulation environment. In *CHI '89 Conference Proceedings*, pp. 57–62, Austin, TX, USA, May.

Nix, R. P. (1985). Editing by example. *ACM Transactions on Programming Languages and Systems*, **7**: 600–621.

Notkin, D. (1985). The GANDALF Project. *The Journal of Systems and Software*, **5**: 91–105.

Olsen, D. R. Jr and Dance, J. R. (1988). Macros by example in a graphical UIMS. *IEEE Computer Graphics and Applications*, **8**: 68–78.

Reiser, B. J., Andersen, J. R. and Farrell, R. G. (1985). Dynamic student modeling in an intelligent tutor for Lisp programming. In *Proceedings of the International Joint Conference on Artificial Intelligence 1985*, pp. 8–14, Los Angeles, CA, USA, August.

Rich, E. (1983). Users are individuals: individualizing user models. *International Journal of Man–Machine Studies*, **18**: 199–214.

Shrager, J. and Finin, T. (1982). An expert system that volunteers advice. In *Proceedings of the National Conference on Artificial Intelligence*, pp. 339–340, Pittsburgh, PA, USA, August.

Staudt, B. J., Krueger, C. W., Habermann, A. N. and Ambriola, V. (1986). The GANDALF system reference manuals. Technical Report CMU–CS–86–130, Computer Science Department, Carnegie Mellon University, Pittsburgh, PA, USA.

Teitelman, W. (1972). Automated programmer: the programmer's assistant. In *Fall Joint Computer Conference Proceedings.*

Wilensky, R., Chin, D. N., Luria, M., Martin, J., Mayfield, J. and Wu, D. (1988). The Berkeley UNIX consultant project. *Computational Linguistics*, **14**: 35–84.

Wolz, U. and Kaiser, G. E. (1988). A discourse-based consultant for interactive environments. In *Fourth IEEE Conference on Artificial Intelligence Applications*, pp. 28–33, San Diego, CA, USA, March.

Zissos, A. Y. and Witten, I. H. (1985). User modeling for a computer coach: a case study. *International Journal of Man–Machine Studies*, **23**: 729–750.

# CHAPTER 5

# Support for software design, development and reuse through an example-based environment

*Lisa Neal*

## Abstract

Example-based programming is a methodology which supports software design, development and reuse. The integration of examples into a software development environment results in a tool which is especially effective for the support of experienced software developers who are working in new domains. Example programs are used as instances of language constructs, thus providing syntactic information through instantiations of templates, or as examples of algorithms or programs. Examples selected from a library can be viewed, totally or partially copied, or run. The initial example-based programming environment was implemented for Pascal on the Macintosh computer. The system was successful in addressing problems with the use of structure editors and in facilitating the reuse of software. In its initial conception, example-based programming was designed to provide passive assistance in the software development process; however, it is a paradigm which is extensible to provide more active and intelligent support. The initial system and results from an empirical study are presented, and current and future developments are discussed, with a focus on extensions which increase the amount and types of support provided by the environment.

Structure-based Editors and Environments
ISBN 0-12-681890-8

## 5.1 Introduction

We define an example-based environment as one which incorporates examples to facilitate the design, development and reuse of software. The concept of integrating examples into an environment arose from empirical studies of structure editors (Neal, 1987a,b). The editors under study, which were oriented primarily toward novice programmers, were designed to support the construction of Pascal programs through a structured and restrictive approach to input, modification and navigation. Templates were available for all language constructs, and were the means by which programs were entered.

The highly structured approach to program construction and editing was limited in its success because users lack thorough knowledge of the structure of their program and its underlying representation. More experienced users, in particular, found the restrictiveness of the editors frustrating, especially when they were familiar with and comfortable with the use of non-language-based text editors (Neal and Szwillus, 1990). The template approach to input theoretically had advantages over textual input due to factors such as less focus on syntactic details and a greater emphasis on semantics and algorithms. In fact, template insertion was often cumbersome or awkward, and, more importantly, did not effectively prompt users.

In examining alternative approaches to structure editing which both aid users more effectively and take a less restrictive approach, it was considered likely that program fragments, as, in essence, instantiations of templates, would prompt users more effectively than templates. However, program fragments lack the contextual information, such as declarations, necessary to understand and effectively utilize a program fragment. Hence, it was determined that actual examples of programs were likely to provide more useful information. The use of preexisting code is common in many instances: novices use code in textbooks and their previously written programs, experienced programmers use their own and other's code, and teams of programmers share code. However, access to preexisting code is not actively supported by most programming environments.

While it seemed likely that this approach would be beneficial for development, the benefits for design and reuse were only discovered following the implementation of the example-based environment. Likewise, empirical evidence showed that this approach was especially beneficial for programmers who are experienced but are working in a new domain. Since novices do not remain novices for very long, but experts are always becoming novices in a new domain or have domains in which they work infrequently, we felt it was important to provide support within a programming environment for this population (Neal and Szwillus, 1990).

## 5.2 Initial implementation

A syntax-directed editor for Pascal on the Macintosh was augmented with an example window and access to an example library. The editor uses a palette to present the available templates and has a command completion capability which allows the enter key to be used after typing a keyword in order to invoke a template. In addition, textual input is incrementally parsed upon the use of a semicolon or the enter key. The editor hence provides templates for input, but allows free textual input as well. Modification and navigation can likewise be either structural or textual.

Examples in a separate example window can be viewed, totally or partially copied, or run. The only restriction on use of the example window was that edited examples could not be saved to the same program name, in order to preserve the quality of the library. For the initial implementation, examples were accessed only through descriptive program names. Examples are selected through a dialog window which has unlimited size.

## 5.3 Empirical study

An empirical study was performed with the initial version of the system in order to determine how the examples were used and their effectiveness (Neal, 1989). It became clear through the study that the example-based system had the potential to aid in comprehension, design and reuse as well as program construction.

The primary results of the study follow. The examples were more heavily used than any of the structure-based capabilities of the syntax-directed editor and provided capabilities lacking in the syntax-directed editor. The examples were used in a variety of ways: to aid in design, to aid in writing code, to aid in comprehension of syntax and semantics, and for reuse. Specifically, examples were used: to determine the use of Macintosh-specific language features and procedures; for I/O, both syntactically and semantically; to discover approaches to solving problems; to be executed in order to better understand or to test an understanding of the semantics of a program; for inspiration and guidance towards a solution; and as guidelines for formatting and standardizing code.

While the participants in the empirical study were given a constrained task rather than given the environment to use in solving a problem on which they were already working, the results were strong enough and the feedback enthusiastic enough that we were encouraged to pursue the approach further based on a generalization of the results. We were especially encouraged by the response of experienced programmers who were novice or infrequent users of Pascal. They found that they were easily able to construct Pascal programs which they understood and had confidence in the correctness of, rather than the process being a struggle which resulted in poorly written and semantically incorrect code.

## 5.4 Use of an example-based system

From the empirical study and from observation of more casual use of the system, we found that the example-based approach is effective in a number of ways. Examples facilitate the learning process (Lieberman, 1993) and are essential to processes such as case-based reasoning, in which known solutions are adapted to solve problems. It is very common for textbooks and manuals to make heavy use of examples and for system developers to use examples from these sources, as well as their own or other's code. When an experienced programmer is a new or infrequent user of a programming language, examples are used to learn or refresh knowledge of a language. Likewise, examples are effective for starting in new domains, such as the use of windowing systems or interface builders, or for programming on a parallel architecture on which the programming languages may be new or hybrid languages. Once a programmer has developed expertise in one or more languages, learning another language by looking at existing programs and extracting the relevant information is not especially difficult and is less time-consuming than using a tutorial or reading a textbook, which is likely to give too much information at a greater level of detail than is needed.

The example-based environment has been successful in helping users in the design process. The availability of examples within the editing environment means that a user can easily scan examples, studying approaches to solving similar problems. Examples in the example library are well-written and well-documented, aiding in this process. Examples can be reused in part or in full, through the cut-and-paste facility or through procedure and function calls. Reuse is encouraged because of the accessibility of examples and because examples can be viewed and run, which allows a programmer to feel that the routine is much better understood than if the only access is to a routine name and parameter list. Additional support for design and reuse is included in enhancements to the system.

## 5.5 Current and future directions

Current development of the example-based programming system is in two directions: new domains and system enhancements. The former includes support for programming on the Connection Machine for non-parallel programmers and support for E-L. The latter includes additional capabilities and support; for instance, the system is being enhanced to allow multiple access mechanisms for the example library and to annotate examples with information related to the design and use of the artifact.

The initial mechanism for access to examples, through descriptive program names, proved to be effective but too simplistic. When it became clear through observations of use that users often wanted an example of a particular language construct without having to scan multiple examples, access to examples through a hypertext-like capability was added, where a language construct could be selected in the main editing window and an example would be displayed in the example window which included the use of the construct. Examples were linked so that a number of examples would be made available, with simple examples first, followed by more complex instances of the construct. The linking of the examples had to be done explicitly as additions were made to the example library.

In considering more sophisticated access mechanisms, we rejected approaches which overly constrained the user in the process of coding or in the use of the system. We chose the use of an embedded design language, which is natural language in comments within the code being written. Guidance is provided through a help window for the optimal use of comments, and keywords are suggested. Also, examples provide additional guidance. For example, a program to sort would include in its comment the word "sort". In the same manner that selection of syntactic elements invoke examples including those elements, the use of "sort" would invoke examples including that within their comments. This approach has the advantages of supporting the inclusion of design information without forcing the user to code in a particular way or use an artificial language, and of extending the retrieval capabilities of the system. The better access to examples the user has, the more likely it is that they will be taken advantage of as code to be examined or reused.

Additional enhancements include annotations to examples. Examples can become large, and the annotations provide additional, selectable information without increasing the example size. Annotations include design information, explanations of syntax, snapshots of memory, and instances of inputs and outputs. While the embedded design language can be used to provide well-documented routines, we wanted to include more extensive information which recorded the design history and alternatives (MacLean *et al.*, 1990). This provides information which allows code to be more effectively understood and reused, and is especially desirable for the maintenance and enhancement of code. In addition, we are exploring more graphical approaches for specifying design (Szwillus, 1989). The other annotations are primarily to allow easy access to information which may prove useful and can increase a user's understanding of a routine or its components with as little effort as possible. For instance, even though examples can be run, the availability of sample inputs and outputs means that the user need not switch contexts in order to find out more about what a program does or what its output looks like.

Future directions for the example-based environment include further work on system enhancements. The inclusion of a more knowledge-based approach to the use of examples is being explored. We would like to incorporate access to relevant examples through an intelligent assistant that would monitor the user's input and example use in order to determine what the user's goals and plans are and provide appropriate examples to aid the user in achieving that goal. Such approaches have been successful in systems dealing exclusively with semantic knowledge (Johnson and Soloway, 1985). More semantics-based retrieval could aid in one of the significant problems with software reuse: the identification of appropriate and relevant reusable elements (Tracz, 1987). Attempts to automate assistance have met with limited success, hence we are examining approaches in which the user is in control but with intelligent support.

While the syntax-directed editor integrated into the example-based environment has not been especially successful in aiding in design or development, we would like to further consider how to effectively utilize the knowledge about a language that structure editors have. One approach is to think of a structure editor as a knowledgeable assistant, which monitors a user's actions, and guides, rather than forces, a user. The structure editor could give experts advice when requested and give novices continual feedback, acting more like an unobtrusive intelligent tutoring system for designing and programming. Alternatively, the structure editor can be used to aid more in understanding code than in writing code; views of the code which reveal its structure or provide condensed views can aid in a higher level understanding.

## 5.6 Conclusions

Example-based programming was originally conceived as a supplement to structure editors. Empirical evidence showed the effectiveness of the example-based approach for the design, construction and reuse of code, and its superiority over some of the capabilities of structure editors. Current and future enhancements to the system are increasing the amount of information included in examples and the accessibility of relevant examples. The example-based approach is showing effectiveness, particularly for experienced programmers working in new or infrequently used domains, since the information provided by the examples allows a programmer to more easily get started through the extraction of useful and relevant information. The examples provide support for the design process and encourage the reuse of existing code, in addition to supporting the development of code.

## Acknowledgments

This research was supported, in part, by US Army Research Office Grant #DAAG29–83–G0008 and, in part, by the Defense Advanced Projects Research Agency under contract N00014–85–C–0710. I would especially like to thank Tom Cheatham and Ugo Gagliardi for their support of this work.

## References

Johnson, W. L. and Soloway, E. (1985). PROUST: Knowledge-based program understanding, *IEEE Transactions on Software Engineering*, SE–11 (March): 267–275.

Lieberman, H. (1993). Tinker: a programming by demonstration system for beginning programmers. In Cypher, A. (editor), *Watch What I Do: Programming By Demonstration*. MIT Press, Cambridge, USA.

MacLean, A., Bellotti, V. and Young, R. (1990). What rationale is there in design? In *Proceedings of INTERACT'90 3rd IFIP Conference on Human–Computer Interaction*, Cambridge, England, 27–31 August.

Neal, L. R. (1987a). Cognition-sensitive design and user modeling for syntax-directed editors. In *Proceedings of CHI+GI'87 Conference on Human Factors in Computing Systems and Graphics Interface*, Toronto, 5–9 April. ACM, New York.

Neal, L. R. (1987b). User modeling for syntax-directed editors. In Bullinger, H.-J. and Shackel, B. (editors), *Human–Computer Interaction–INTERACT'87*. North-Holland, Amsterdam.

Neal, L. R. (1989). A system for example-based programming. In *Proceedings of CHI'89 Conference on Human Factors in Computing Systems*, Austin, 30 April–4 May. ACM, New York.

Neal, L. R. and Szwillus, G. (1990). Report on the CHI'90 Workshop on structure editors. *SIGCHI Bulletin*, **22**(2): 49–53.

Szwillus, F. (1989). Editing graphical structures. In Smith, M. J. and Salvendy, G. (editors), *Work with computers: organizational, management, stress and health aspects*. Elsevier Science Publishers B.V., Amsterdam.

Tracz, W. (1987). Software reuse: motivators and inhibitors. In *Proceedings of Compcon '87 (Spring), Thirty-second IEEE Computer Society International Conference*, 23–27 February, Cathedral Hill Hotel, San Francisco, CA.

# CHAPTER 6

# Experimental data on the usefulness of a structured editor

*Pierre-N. Robillard, Mario Simoneau, Jean Mayrand and Daniel Coupal*

## Abstract

This chapter presents a method to evaluate the characteristic of the use of a structured editor and of the outlining facilities. The method uses volume and effort metrics to quantify measurements. Data were obtained from a commercial project developed using the structured editor SCHEMACODE. The results show that the effort required at the syntactic level is reduced and that the outliner provided an efficient mechanism to improve program readability. The tool presented in this chapter can be obtained at the following address: http://www.rgl.polymtl.ca/schema/schema.htm.

## 6.1 Introduction

The benefits in terms of productivity to be derived from the use of a structure editor are intuitively obvious to the programmer using it; however, it is difficult to demonstrate productivity gain on the basis of experimental evidence. It is costly to run two identical reasonably sized projects, one with a structure editor and the other without. In addition, team formation, and the background and motivation of the team members are important factors that are difficult to control.

To overcome this difficulty, we used the same method of measuring from two different viewpoints. First, we applied the metric measurements to the project based on the source instructions, as if the project were being carried out without any special editor. Next, we applied the

Structure-based Editors and Environments
ISBN 0-12-681890-8

same measurements to the same project based on the schematic instructions provided by the structure editor. The method is applicable to any high-level representation.

The structure editor used has two distinct and specific features. One is that it is based on schematic pseudocode (Robillard, 1986) and the other is that the outliner is based on the operational comment or scope comment approach. The software tool automatically translates the schematic instructions into programming language instructions. From the programmers' standpoint, the source code generated by the tool is not significantly different from that which is usually generated.

The automated outliner implements and documents the stepwise refinement approach. New statements, called operational comments, have a certain scope within which they expand. These comments define refinements (Kaebling, 1988; Robillard, 1989).

The project is a commercial software product developed during the past three years by a team of three programmers. Everyone on the team was familiar with the tool before the project began, since they had used it for many years in the development of commercial software in various programming languages. The project, which was entirely implemented with this tool, was selected as a case study after it was completed.

We are looking for objective and automated measures. Halstead's measures of psychological effort (Halstead, 1977) are used to compare the two approaches. We do not intend to validate this measure but rather consider it as an indicator of change. Other measures, based on levels of abstraction, are also used to evaluate the impact of the outlining feature.

## 6.2 Tool description

This section describes the structure editor, SCHEMACODE, in order to help the reader to understand the details of the analysis. However, the approach presented is not specific to this tool.

The SCHEMACODE software tool allows the programmer to write a program with a schematic structural editor. It draws a visual representation of the control structure on the screen, and outlines the documentation. The underline method is called Schematic Pseudocode (SPC) (Robillard, 1986).

The documentation approach identifies two types of comment, and specifies when and how each is used. The narrative comment describes the meaning of a statement (*What*). The operational comment abstracts a group of statements (*How*). The implementation of an operational comment is a refinement. Operational comments are the realization of the stepwise refinement approach.

Schemas represent control statements (conditions, loops, etc.), and are applicable to any procedural programming language. There are only three types of schema: one for sequential constructs, one for conditional constructs and one for repetitive constructs. Sequential constructs take three forms: sequential statements, narrative comments and operational comments.

Figure 6.1 shows a sample program developed with SCHEMACODE. It is a program to update inventory from a transactions file (Bergland, 1981). Refinement zero (R0) shows the first step. Narrative comments, preceded by a dash, identify the author and describe the function. The narrative comment is a comment used to document locally. The next two statements, preceded by a two-digit number, are operational comments. They define the include files and the variables. The operational comment is a one-line summary of a refinement. An operational comment expresses an action and usually begins with a verb. The two-digit number facilitates referencing.

Three other operational comments represent the main steps: 03 OPEN FILE, 04 UPDATE INVENTORY and 05 CLOSE FILE. Refinement zero (R0) contains five operational comments. Once implemented, operational comment 03 becomes refinement R3. This refinement contains two operational comments. The operational comment, 06 OPEN TRANSACTION FILE, becomes refinement R6. Refinement R6 contains only two programming language statements.

The operational comment 04, UPDATE INVENTORY, becomes refinement R4. It includes a repetitive structure represented by the schematic parallel lines, some executable statements and seven operational comments. Repetitive constructs are made up of sequential statements, operational comments and one or more exit conditions. The loop construct executes repeatedly until one of the exit conditions becomes true. An underlined boolean expression and a star within the parallel lines identify the exit condition.

Operational comment 12 UPDATE INVENTORY FILE becomes a refinement composed of one conditional structure and many sequential programming language statements. Conditional constructs divide into

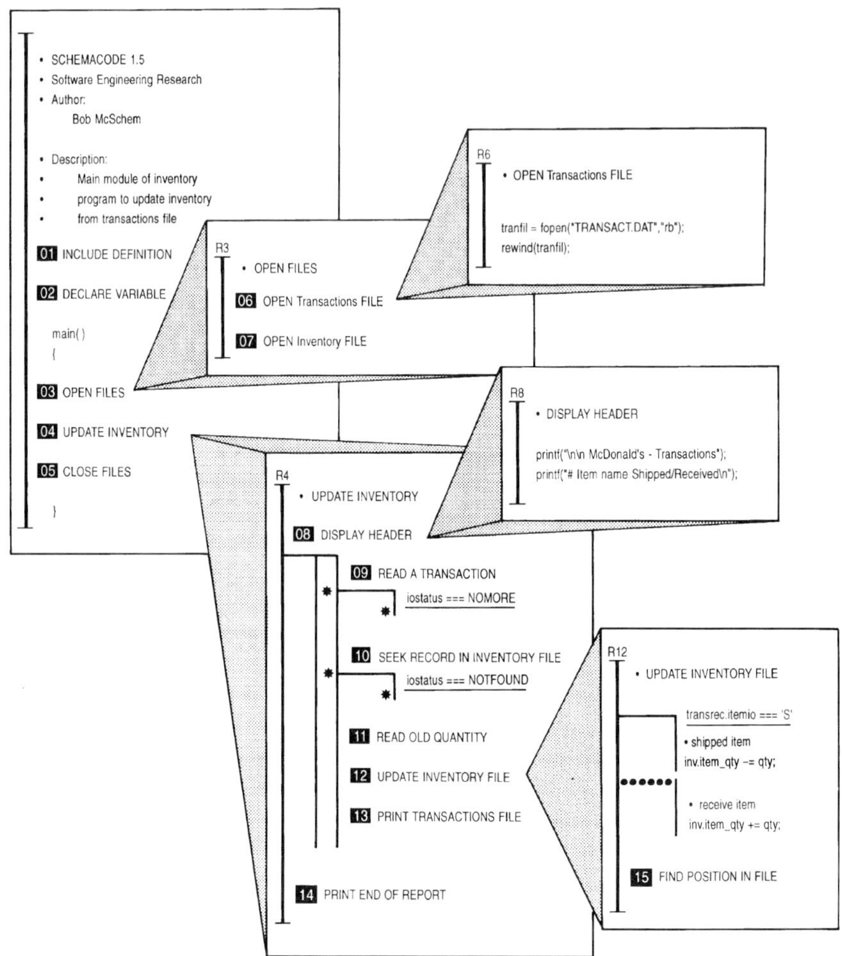


**Figure 6.1** Sample program developed with SCHEMACODE.

branches. The IF branch is the basic element of the conditional construct. It contains text expressing the condition in the programming language. Sequential constructs make up the body of the branch. The ELSE IF branch is optional, and there can be more than one of these (there are none here). The ELSE branch is also optional.

A zoom feature enables the programmer to directly access any refinement. Operational comments automatically become narrative comments once they implement a refinement. The programming task ends when there are no more operational comments left.

SCHEMACODE automatically integrates all the documentation, and also generates program source code in the target programming language. The schematic structures become programming statements according to pre-defined rules optimized for particular languages. The computer languages now operational are BASIC, COBOL, C, C++, dBASE IV, FORTRAN and PASCAL.

## 6.3 Lexical token versus schemas

The SPC writes control structures in terms of schemas instead of conventional lexical tokens. Lexical tokens that describe programming language syntax are grouped into BNF rules (Kernighan and Ritchie, 1988). Schemas replace underlined tokens. These increase the abstraction level since fewer symbols or characters are needed to represent the same information. The token ratios are measured.

Table 6.1 shows examples of lexical equivalence of the schemas and the computed ratios. For example, in the C language the conditional construct, with a boolean expression $b_1$ and a group of statements $s_l$, is represented by the expression **if** $(b_1)$ $\{s_l\}$. The schema replaces five tokens. They are **if**, (, ), { and }. The conditional structure with an ELSE part has eight lexical tokens replaced by two schemas. A schema also has the advantage of defining the structure and its scope. To achieve the same result, a programming language requires the linking together of many tokens. The measures proposed are then a lower limit.

A repetitive structure with an exit at the beginning (WHILE) has a ratio of 5/2. A repetitive structure with an exit at the end (DO WHILE) has a ratio of 7/2. There is one schema for the loop construct and one schema for the exit. This can be generalized to a loop structure with $n$ exits.

These various ratios are inherent in the schematic generation mechanism. However, programs are made up of various statements of which semantic constructs are only part. For example, a fully sequential function with no control construct does not benefit from the use of a schematic structured editor. An experimental study was

**Table 6.1** Examples of ratios

| Schemas | Lexical equivalence | Ratio |
|---|---|---|
| $b_1$ $s_1$ | **if** ($b_1$) {$s_1$} | 5/1 |
| $b_1$ $s_1$ $s_2$ | **if** ($b_1$) {$s_1$} **else** {$s_2$} | 8/2 |
| * $b_1$ $s_1$ | **while** (!$b_1$) {$s_1$} | 5/2 |
| $s_1$ * $b_1$ | **do** {$s_1$} **while** (!$b_1$): | 7/2 |

conducted to determine the benefits of using a schematic structured editor averaged over a complete project.

The aspect measured is related to the physical nature of the structured editor which provides a more abstract representation of the control constructs. This results in fewer tokens or keys to manipulate. Productivity should then be affected in terms of the effort required to implement a function.

## 6.4 Effort comparisons

We present here a method to quantify the differences between SPC and programming language statements. The basic elements needed to compute Halstead's Effort ($E$) are:

$$V = (N_1 + N_2)\log_2(n_1 + n_2)$$
$$E = n_1 N_2 V/(2n_2)$$

where

- $n_1$ is the number of distinct operators;
- $n_2$ is the number of distinct operands;
- $N_1$ is the total number of operators;
- $N_2$ is the total number of operands.

Halstead's Effort is used as an indicator of change and not as a measure of the effort. Interpretation of this metric as a validated effort measure is questioned.

The first step is to define operators and operands for both the source code and SPC. Operands are tokens used to represent user-defined names (such as variables and constants). Operators are tokens such as {, }, →, reserved words and schemas (for SPC).

The results come from a project of 284 functions. The computation of $E$ involved both the schematic pseudocode and the generated source code. Haldstead's effort measures the improvement in programming with SCHEMACODE. The computation of effort shows that the mean ratio is:

$$\frac{\text{Total effort (SPC)}}{\text{Total effort (code)}} = 73.9\%$$

Figure 6.2 shows the distribution of Halstead's effort for the 50 largest functions. Black columns represent the effort needed to write a function in programming language (code). Overlapping grey columns represent the measures based on SCHEMACODE. This metric indicates a net gain of 35.3%.

## 6.5 Operational comment efficiency measurement

The operational comment described in Section 6.2 is a mechanism to abstract information (Robillard and Coupal, 1988). An operational comment outlines a task. This task is implemented with programming statements and other operational comments. This can be viewed as macro instruction. This mechanism reduces the amount of information seen by the programmer. Operational comments make the program

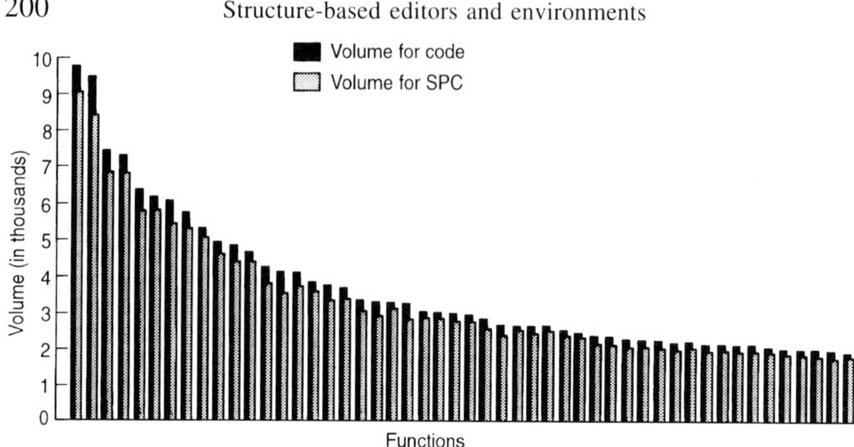


**Figure 6.2** Distribution of function effort.

more abstract. The task is described by fewer words. This section measures the effects of this approach on resulting programs.

Figure 6.3 based on the example of Figure 6.1, shows the full development of refinements R3, R6 and R7. Successive refinements build a tree structure where the leaves contain only programming statements.

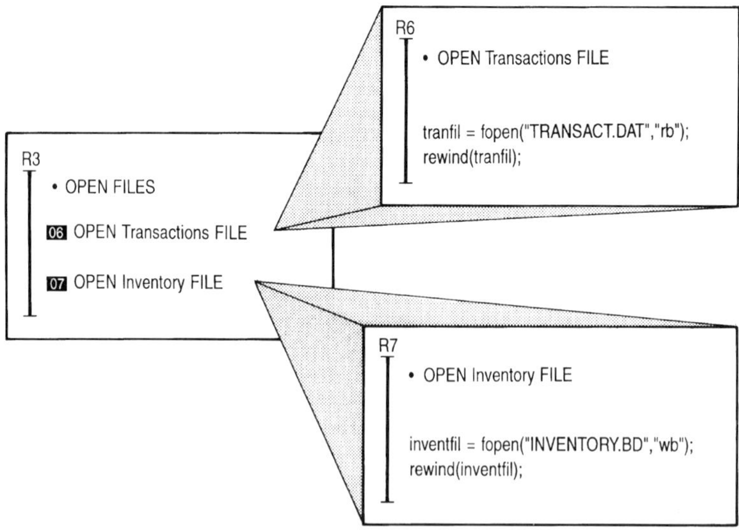


**Figure 6.3** Full development of refinement R3.

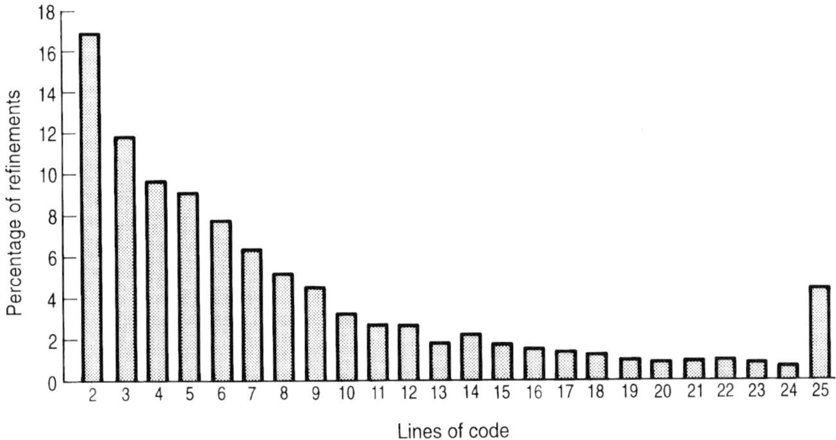


**Figure 6.4** Distribution of lines of code within refinements.

Figure 6.4 shows the distribution of the number of lines in refinements. For example, the first column shows that 17% of the refinements have only two lines. A line is a statement or an operational comment. Refinement R3 in Figure 6.3 has two lines which are two operational comments. The real number of statements represented by refinement R3 is in fact much larger. It includes the content of refinements R6 and R7.

Figure 6.5 shows the distribution of the total numbers of lines embraced by a refinement. For example, the first column shows that 36% of the refinements embrace between one and five lines (median is three). Refinements with more than 125 lines cumulate.

Figure 6.6 shows the distribution of the ratio actual lines to embraced lines. For example, 4% of the refinements have a ratio between 0% and 4% (median is two). This means that for those refinements, 96% to 100% of the content is embraced by operational comments. At the other end of the scale, only a few refinements (under 0.2%) contain 98% of the statements. Sixty-three percent of the refinements (not shown) have a ratio of 1.0. These refinements have no operational comment. They are leaves, like R6 and R7 (see Figure 6.3).

The project has 3881 refinements. Thirty-seven percent of these (1418) have at least one operational comment (like R3 and R4). The average abstraction ratio of these non-leaf refinements is 0.33.

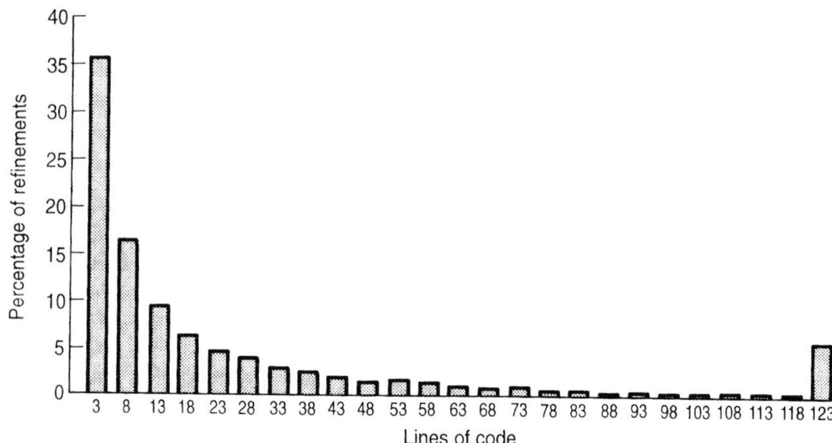


**Figure 6.5** Distribution of lines of code embraced in refinements.

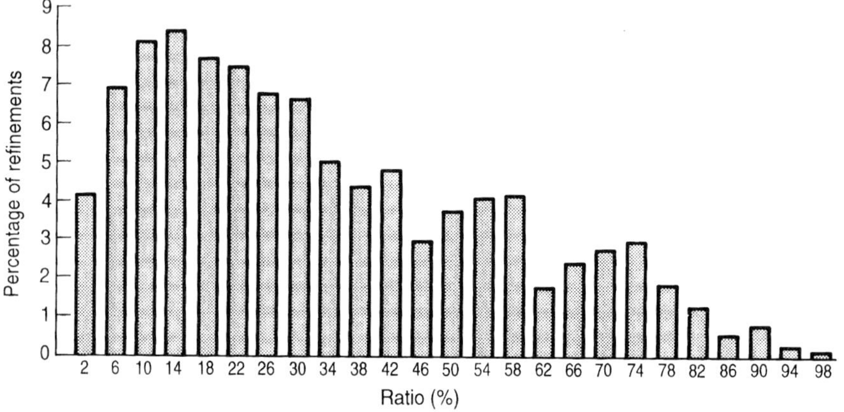


**Figure 6.6** Distribution of the ratio (%) of lines within refinements/lines embraced in refinements.

On average, these refinements show only the third of the lines. For example, R3 contains only two lines. These lines are an abstraction of two programming language statements each.

These data can also be seen as a measure of the use of operational comments by individual programmers. Some programmers may use fewer operational comments than others. However, studies based on the projects of various groups of programmers did not show significant variations in this respect.

## 6.6 Conclusion

This chapter presents a tool that supports the schematic description of control flow structure. This tool automatically translates schemas into programming language statements. It also has an outliner that enables the abstraction of programming language statements. Operational comments implement the abstraction mechanism.

The outliner feature of the tool organizes the program into a hierarchic tree. Refinements form the nodes of the tree, and 63% of the nodes are leaves. Leaves contain only programming statements. Thirty-seven percent of the refinements contain operational comments that make abstractions of programming statements. On the average, these refinements abstract 67% of the information.

The net benefit of the increased program organization provided by the operational comments is that roughly one-third of the program text (including comments) is enough to look at the whole program. However, to obtain the maximum benefit from this feature, we need meaningful operational comments. This approach makes comment quality control most rewarding.

A study has been performed to measure the effect on program development of SPC and outlining, using Halstead's Effort metrics. The results of this study are as follows. The schematic approach reduces Halstead's Effort metric for the program by 26%. This result is solely based on the use of abstract tokens to describe the control flow. Visual aspects provided by the schematic control flow have not been measured and are likely to add to the benefits.

This analysis, based solely on the program's objectively measurable parameters, provided a lower limit indicator on the potential benefit of the use of a structure editor and an outliner. The analysis looked only at the implementation phase. Structured editors produce programs that are written with less effort and that are also easier to read and understand. The increased program readability has a positive impact on code inspection, and the test and maintenance phases.

## Acknowledgements

This work was in part supported by Bell Canada and the National Research Council of Canada under grant # A0141.

## References

Bergland, G. D. (1981). A guided tour of program design methodologies. *IEEE Computer* (Oct.) 13–37.

Halstead, M. H. (1977). *Elements of Software Engineering*, Elsevier, North-Holland.

Kaelbling, M. J. (1988). Programming languages should NOT have comment statements. *SIGPLAN Notices*, **23**(10): 59–68.

Kernighan, B. W. and Ritchie, D. M. (1988). *The C Programming Language*. Prentice Hall Series, Englewood Cliffs, New Jersey.

Robillard, P. N. (1986). Schematic pseudocode for program constructs and its computer automation by SCHEMACODE. *Communications of the ACM*, **29**(11): 1072–1089.

Robillard, P. N. (1989). Automating comments. *SIGPLAN Notices*, **24**(5): 66–70.

Robillard, P. N. and Coupal, D. (1988). Data on the automation of program's comments. In *Proceedings of the 6th Annual Pacific Northwest Software Quality Conference*, Portland, Oregon, 19–20 September 405–417.

Tripp, L. L. (1989) Bibliography on graphical program notations. *ACM SIGSOFT*, **14**(6): 56–57.

# Part III Graphical structure editors



Parts I and II of this book represent the classical field of structure editing and its current development with innovative extensions. The systems discussed so far support the process of editing structured texts, mostly programs. The concepts and techniques learned from these systems, however, can be exploited to deal with graphics. We introduce here the term "graphical structure editor" in two different senses.

First, a graphical structure editor can be a conventional syntax-based editor equipped with a graphical user interface to display all or parts of the structure worked on by the system graphically. By conventional, we mean that the structure under consideration is typically a tree-like structure, i.e. a hierarchy. An example for such a system is GRIPSE, discussed in Chapter 7, by Halewood and Woodward. The GRIPSE system stays within the programming field, but supports a visual programming technique based on Nassi–Shneiderman diagrams. The system gives interactive access to the graphical representation of a program's control and data structures. The system's features include automatic layout of diagrams, easy selection, refinement and editing features, automatic zooming, intelligent cursor movement and graphically animated execution. Apart from the graphical outfit, the system is basically a "classical" structure editor system, which renders particularly interesting how the system evaluates in comparison to a text-based editor.

Secondly, we can think of a graphical structure editor as a system leaving the "classical" tree-oriented basis, but generalizing this to the editing of graph structures, which are graphically depicted to the user. This generalization adds greatly to the complexity of structure editors in terms of specification of syntax rules, user interface implementation and administration of the structured document to be edited.

The ideas presented by Szwillus in Chapter 8 are concerned with the specification problem. Whereas there are well-established techniques for defining tree structures deriving from textual languages, the same does not apply to arbitrary graph structures. This chapter shows where the problems are when generalizing from trees to graphs, as it incorporates a subtree structure as a basis for graph definition. The specification of the syntax rules of graphical languages is based on techniques known from formal language theory. Emphasis is on the integrated definition of connection rules, outlook, semantical attributes and overall conditions of the graph, with the path specifications as one single unifying construct.

In Chapter 9 by Bos, a graph editor is presented which generalizes the principles of structure editing to non-tree structures. Its main purpose is to allow easy implementation of graph manipulation tools. This is done by defining rules about the graphs to be edited, which are specific to a given domain. These rules specify the graphical outlook of nodes and arcs, general graph layout properties, restrict the way connections are introduced and give general graph properties such as connectedness. The system is universal as it can be modified in its rules to adapt to new application domains. This individualization has to be done by adding or modifying LISP expressions.

All three chapters show that the concepts of structure editing, as originating from very early syntax-based work, can be highly useful for today's graphical-oriented applications, editors and user interfaces.

# CHAPTER 7

# A uniform graphical view of the program construction process: GRIPSE

*K. Halewood and M. R. Woodward*

## Abstract

This chapter describes the ideas leading to the development and construction of a Graphical Integrated Programming Support Environment (GRIPSE) for a modified form of the Pascal language. In order to represent all procedural and declarative aspects of a Pascal program, the original Nassi–Shneiderman diagrams forming the graphical basis of GRIPSE, have been augmented into a multi-level, three-dimensional system capable of supporting all static and most dynamic aspects of a program in a uniform manner. The dynamic view exists by virtue of an interpreter that has been incorporated along with debugging aids. A planned extension to the environment is an interactive form of mutation testing called "firm" mutation which exploits the middle-ground between "strong" and "weak" mutation, both of which have traditionally been applied in a non-conversational, non-interactive mode of use.

## 7.1 Introduction

GRIPSE is an integrated programming support environment that supports the construction, debugging and testing of medium-sized programs written in a variant of the Pascal programming language. GRIPSE is based on an earlier project, NSEDIT (Halewood and Woodward, 1988), a syntax-directed, structure-mode Pascal program editor using Nassi–Shneiderman charts. NSEDIT's diagnostic capabilities were limited because it could only check gross syntax; a semi-

Structure-based Editors and Environments
ISBN 0-12-681890-8

automatic interface between the editor and the standard VAX/VMS Pascal compiler was used to perform full syntax checking and compilation. Errors were reported back to the editor via operating system interprocess communication facilities.

GRIPSE offers considerable improvements over NSEDIT. Its syntax-directed editor is much more comprehensive in that it performs full syntax and static semantics checking, providing immediate error feedback. Program execution facilities are built into the environment in the form of an interpreter that is also capable of continued execution even after minor program modifications. The interpreter is also capable of handling debugging events such as code and data breakpointing and variable watchpointing. Finally, a test tool based on a variant of mutation testing has been especially devised in such a way that it becomes an integral part of GRIPSE.

An "integrated" programming environment implies a degree of uniformity across all facets of GRIPSE. This is achieved in a number of ways: in the traditional program construction and testing loop, the programmer has to switch between separate tools such as editors, compilers and debuggers each with their own command sets and output formats. GRIPSE attempts to overcome this by making all debugging and testing facilities functions of the syntax-directed editor. Further functionality is obtained by incorporating *pragmatic remarks* or directives into the program under construction. Such pragmatic remarks are checked in the same way as actual program code is syntax checked. Real execution, debugging and testing commands take the form of switches, stating to the interpreter that various types of directive are to be acted upon or ignored.

Section 7.2 describes how integration is further achieved using a common graphical representation. Subsequent sections describe the syntax-directed editor, interpreter and firm mutation testing components. A limited experiment with a small number of users is described and finally, some comparisons with similar systems are made, along with future directions.

## 7.2 Graphical representations

Another way of achieving uniformity across all of GRIPSE's facilities is by the adoption of a common output format. Whether editing or testing, GRIPSE attempts to convey information to the user using the same display method. Naturally, normal program output, dependent upon the code being executed, is exempt. A graphical approach has been chosen because in all cases GRIPSE is working with structures, be they procedural or declarative in nature, and these are best represented pictorially. The design of software is often graphically based, so code generation should reflect design origins. Suitable graphical methods are also amenable to information control, limiting and browsing. As GRIPSE's target language is the procedural language Pascal, a graphical system capable of representing Pascal's procedural structures, declarative structures, data structures, values and execution flow must be used. It must also be capable of representing debugging and testing constructions.

Previous work with Nassi–Shneiderman structure charts (Halewood and Woodward, 1988; Woodward, 1987) led to their adoption in this project because it was felt that their simplicity offered scope for improvement and extension. In their original form, N–S charts (Nassi and Shneiderman, 1973) are simple nested box structures, procedural in nature, capable of expressing the three building blocks of structured programming: sequence, selection and repetition. With very minor additions, there is sufficient scope for representing the gross statement structure of a Pascal program. N–S charts in this form are still incapable of representing the structures of expressions, assignments, procedure and function calls. For this, modified Pagan charts (Pagan, 1987) were used and merged with N–S charts from which they were derived. Pagan charts were originally derived for representing expressions in functional and applicative languages. In simplified form, they are more than suitable for expressing Pascal assignments and expressions in functional form.

Wirth (1976) listed the main types of data structures found in Pascal along with the statement structure most likely to be used to process variables of such a type. For example, an array of data objects is most likely to be processed using the **for-do** loop structure. Similarly, record structures with variants are usually processed using conditional statements such as the **case** statement. If the N–S graphical form of loop statements and conditional statements are taken as a basis, the graphical form for most of Pascal's datatypes is similar.

The representation of data types may be extended still further to allow the display of values of specific types. In general, this is achieved by using the data type structure, replacing the simple type marks such as "char" and "integer" with actual values such as "X" and "15". Figure 7.1 shows how a type structure diagram and its associated value structure diagram coexist. The data type consists of an array definition whose five elements are characters and whose type name is STRING5.

A sequence of declarations, whether type structures or variables, is represented in the same way as a sequence of executable units; a vertically stacked set of boxes. Declarative and procedural parts of N–S program charts are segregated into separate parts of the N–S structure representing a Pascal block.

There are now sufficient N–S building blocks to represent a static Pascal program structure. Before considering the debugging and testing requirements of the representation, aspects of information hiding and management must be explained because it is on this platform that the debugging and testing constructions are based. At first sight, a method of limiting the amount of information on an N–S chart is desirable; although a chart may possess any area desired, display constraints such as terminal or window widths impose practical limits. Complicated N–S charts are composed of a main chart and a

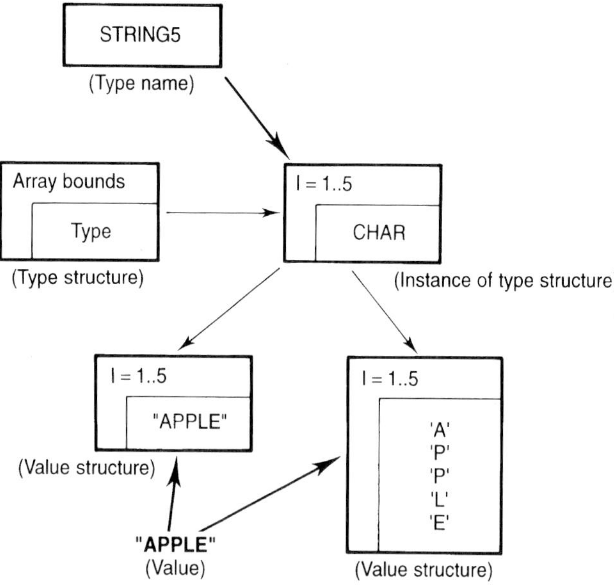


**Figure 7.1** N–S value representation: array objects.

number of subcharts referenced by numbered boxes in higher level charts. GRIPSE's interactive way of constructing N–S charts requires something more dynamic; subcharts must be created automatically where display resolution limits are being approached and there is a danger of crowding too much information on the terminal display. The information limiting technique employed is called zooming; this is discussed later as a tool for limiting congestion and as a means of overlaying or surrounding sections of structure for the purposes of debugging and testing.

## 7.3 The GRIPSE program editor

The GRIPSE program editor is a template-based structural editor for a Pascal-like programming language. The templates are in the form of Nassi–Shneiderman diagram elements which, combined with restriction and immediate error feedback, ensure correct program syntax. How the user interacts with this is explained in the following sections.

### 7.3.1 Implementation environment

The whole of GRIPSE is implemented in an environment without high-resolution, bitmap workstations or mouse-controlled pointing cursors; instead it relies on nothing more sophisticated than industry standard display terminals. Their limited facilities include character-cell text which may be output in a number of renditions including boldface and reverse-video. N–S chart display is made possible by the use of character-cell line drawing building blocks. Input facilities are similarly restricted to an alphanumeric keyboard with *compass points* cursor keys and a dual purpose auxiliary numeric keypad which can be programmed as a function-key array.

The underlying operating system and programming support libraries provide terminal screen management facilities which enable the construction and control of virtual displays (text and graphical windows) which are similar in effect to the window display capabilities of workstation displays, although severely restricted in terms of resolution and area. These are used to implement rudimentary but quite effective pop-up menus which may safely obscure the main work areas of the terminal display without permanent disruption. Multiple virtual displays are also used to act as destinations for the other support tools of GRIPSE, namely interpreter and test tool output.

The simple nature of graphical output facilities available on such a video terminal, augmented by screen management support libraries, offers surprisingly few obstacles when representing N–S charts which are essentially a series of nested rectangular boxes augmented with textual information. The few oblique lines used in *classical* N–S diagrams are not present in the variant that is employed by GRIPSE.

The syntax-directed editor – indeed the whole of GRIPSE – presents itself, upon execution, as a series of overlapped display areas which naturally assumes most of the display area of the video display. Apart from program output during execution, which may take any form dictated only by program logic, the sole *unit* of output is the Nassi–Shneiderman chart. Classical N–S charts may assume any horizontal and vertical dimensions. Clearly, the video terminal's capacity is quite limited and so, instead, must act as a scrollable window onto a portion of the N–S program chart. A text-editing analog is employed whereby the chart's width is constrained to be the same as the display window's width which then need only scroll vertically to accommodate the editing cursor's position. The concept of line-wrapping in a typical text editor to achieve the same restricted width effect does not directly apply to the N–S chart rectangles. However, textual items within those rectangles will *wrap around* onto lower display lines, causing an overall increase in chart length.

A few terminal display lines are reserved at the top and bottom; these are used to relay the global status of GRIPSE to the user as well as the program file name currently being edited, cursor position in terms of cartesian coordinates and structural terms (structure type and its context). The status of auxiliary editing buffers such as the delete buffer is also displayed. Sufficient space is also available for brief (single line) error reports; more verbose reports require the appearance of a pop-up display which may then be dismissed once the report has been understood. Figure 7.2 shows a typical screen display.

GRIPSE's command structure is simple and is based entirely on the use of control-shifted keys, function keys and pop-up menus. The lack of a direct access pointing device such as a mouse or puck, limits the utility of having most command options available on menus since selection may only be either via numeric choice, which has no mnemonic value, or by serial selection using up/down cursor movement. Figure 7.3 shows a typical menu that drops down from the status line as

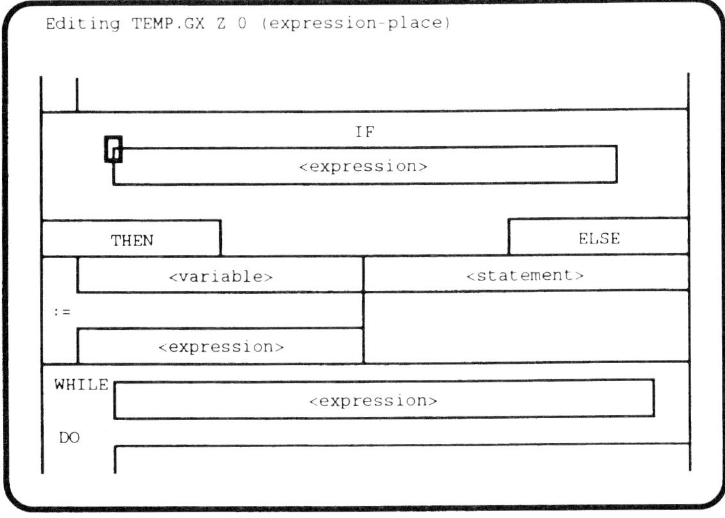


**Figure 7.2** Terminal display usage.

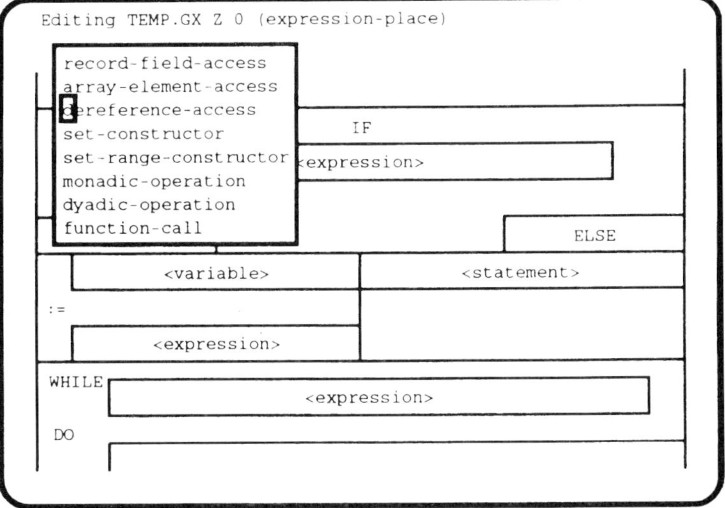


**Figure 7.3** Selection of menu items.

a result of a refinement operation (described later). Pop-up menus are therefore only used sparingly where, for one particular operation, there are a large number of choices and where the set of choices presented may alter radically, depending upon the context in which the command is executed.

Although the programming language employed by GRIPSE is graphical in nature, its underlying control and declarative structures are that of the Pascal programming language. The particular variant of the language conforms to the original user manual and report with three exceptions:

- GOTO statements and label declarations are not permitted. Their use is incompatible with the aims of structured programming and have been omitted from this version of the language.
- The type attribute PACKED has been omitted. In an interpreted system such as GRIPSE, packed types have no place because they represent a change in the way variables of a particular type are represented and allocated. The absence of packed types liberalizes certain type checking constraints present in standard Pascal's handling of input/output and subprogram parameter passing rules.
- Pointer types are not permitted. The implementation allows their use in program construction but at the present time the interpreter does not have the necessary routines to handle the creation and management of pointer objects and the heap.

### 7.3.2 Template-based structure editing

GRIPSE's syntax-directed editor provides an environment in which the user may construct and modify Pascal programs in the form of N–S diagrams. Program construction proceeds in top-down fashion by inserting templates or textual lexemes into predefined points at the current editing position. The templates are predefined N–S diagrams consisting of graphical *layout* such as rectangular boxes and keyword text, and placeholders. Placeholders are the non-terminal symbols of the template, into which other templates or text may be inserted. Figure 7.4 shows the format of the program template which forms the root of the program structure and is always displayed when GRIPSE is initialized. Environment templates for Pascal procedures (a) and functions (b) are shown in Figure 7.5. The reader will immediately note that there are instances of *canned* text at strategic points throughout the

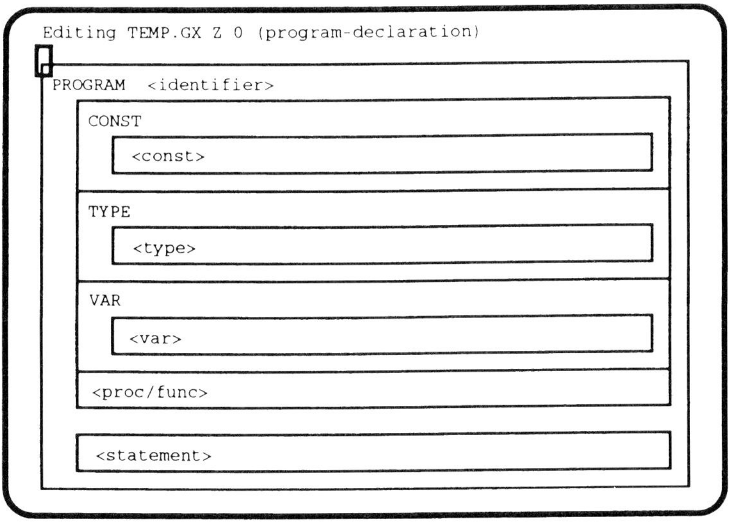

```
Editing TEMP.GX Z 0 (program-declaration)
PROGRAM <identifier>
CONST
<const>
TYPE
<type>
VAR
<var>
<proc/func>
<statement>
```

**Figure 7.4** The program template display.

diagrams that correspond almost exactly to Pascal keywords. N–S diagrams depict general structure (environmental in this case); keywords are necessary to display specific types of environment. The same is true for the procedural template diagrams of Figure 7.6 where, for example, FOR and WHILE loops are both instances of iteration structures where the looping condition is evaluated prior to any execution of the loop body. Keywords are necessary to distinguish between the two. The FOR loop control part is also expanded out into the form of an assignment. Unit (b) represents a procedure call. Pascal type structures are shown in Figure 7.7; units (b) and (c) represent the set-type and enumerated data type, respectively. Unit (h) is the subrange datatype structure and is represented as an operator.

The process of gradual placeholder replacement is called refinement and there exists a template for every production which may be applied to a placeholder except where the placeholder may be refined to a lexical unit. For example, in Figure 7.8(a), the placeholder <identifier> cannot be refined to a template because one does not exist for identifier text. In this case, refinement is effected by moving the cursor anywhere within the placeholder boundaries in Figure 7.8(a) then the text of the identifier is typed directly over. As the first character is typed, the

(a)

```
PROCEDURE <identifier>
  <parameter>
  CONST
    <const>
  TYPE
    <type>
  VAR
    <var>
  <proc/func>
  <statement>
```

(b)

```
FUNCTION <identifier>
  <parameter>
  : <identifier>
  CONST
    <const>
  TYPE
    <type>
  VAR
    <var>
  <proc/func>
  <statement>
```

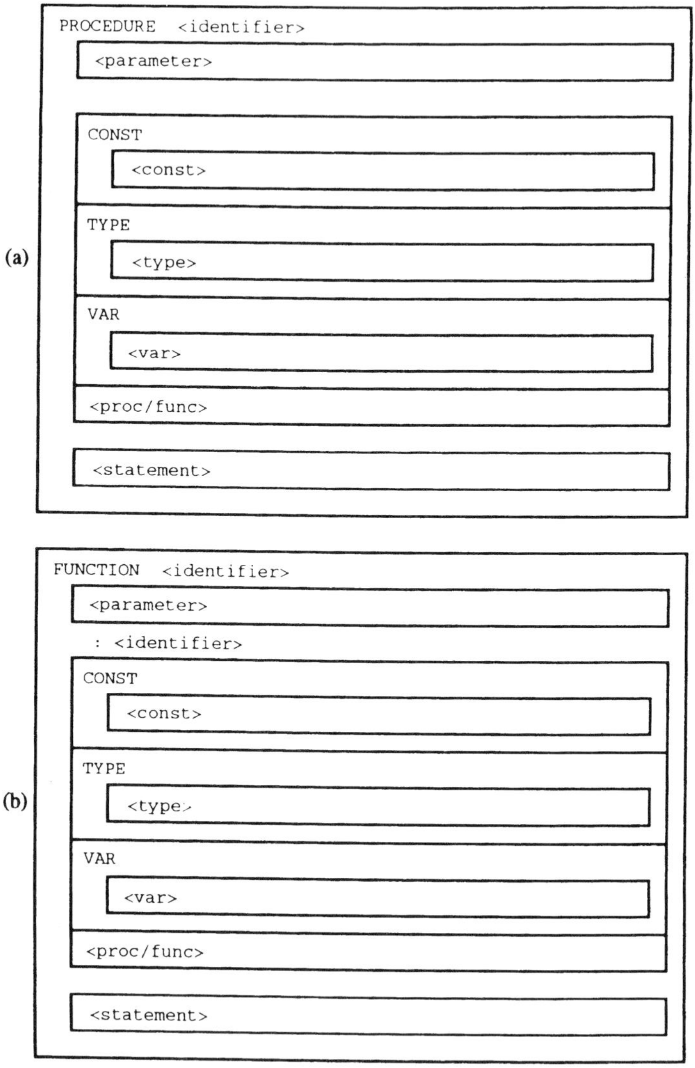

**Figure 7.5** Procedure and function environment templates.

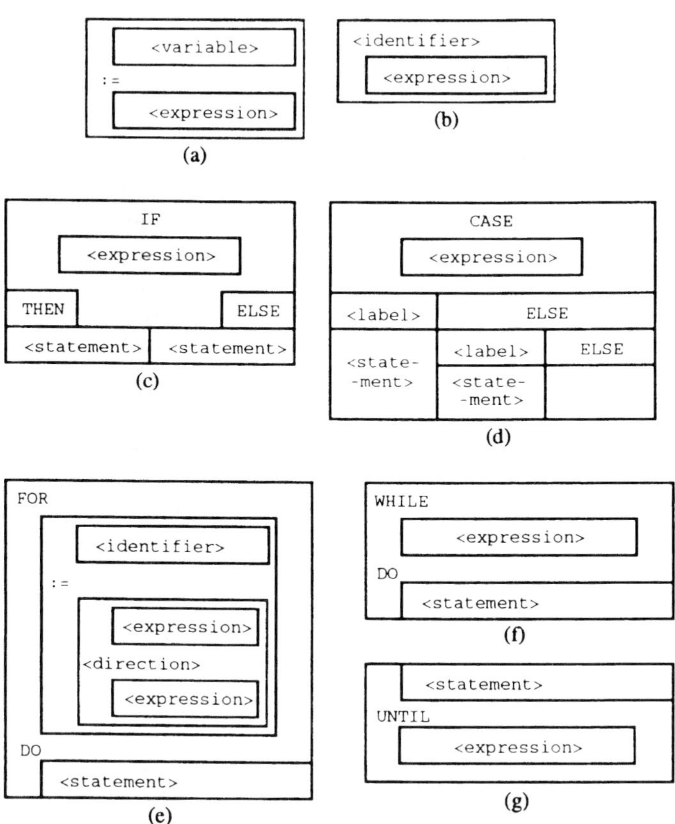


**Figure 7.6** Pascal statement diagrams.

placeholder vanishes as in Figure 7.8(b). When the cursor is moved away from the text area, lexical analysis takes place to ensure that a valid textual item was typed. An error will force the cursor back into the editing area until the fault is rectified or all of the text is removed.

Structural refinement is achieved by moving the cursor to a placeholder and issuing the refinement command (the TAB key). The list of possible template refinements is then presented as a pop-up menu from which the user may either select one item or abandon the refinement altogether. The refinement menu may be *navigated* in one of two ways: the selection cursor may be moved by the up/down cursor keys or by typing the first few characters of the menu items which will then cause

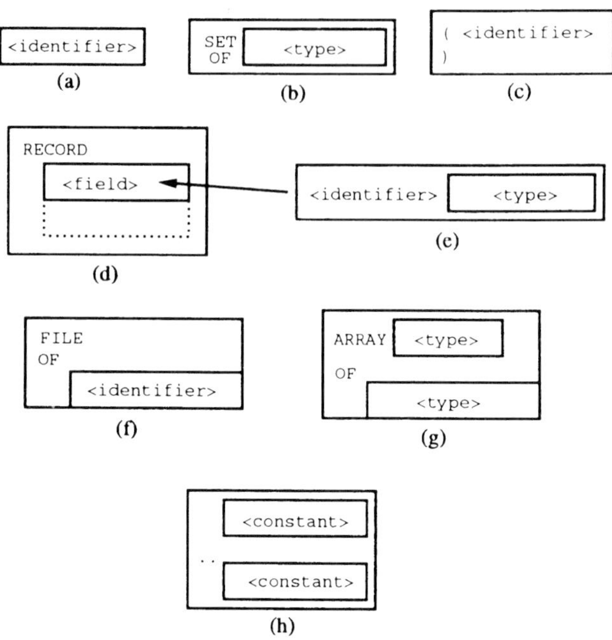


**Figure 7.7** Pascal type structure diagrams.

the selection cursor to jump to the nearest match. In either case, pressing the RETURN key causes the refinement to be obeyed; the SPACE bar abandons the refinement command. Figure 7.9 shows the example of refining a <statement> placeholder into a REPEAT-UNTIL loop template. When RETURN is pressed, the result is displayed as in Figure 7.10. The placeholder box is replaced with a suitably scaled REPEAT loop template, and the cursor is automatically positioned within the area of the loop body which is the next legally modifiable position in the template. This statement placeholder is subsequently refined into an IF-THEN-ELSE statement, the result of which is displayed in Figure 7.11.

Some placeholders are only capable of refining to one template; these are usually in places where the Pascal language grammar rules permit the optional occurrence of an item. In this case, the refinement command omits the redundant menu selection, selects the only possible template and performs the refinement immediately. Several placeholders exist that permit both structural refinement as well as lexical

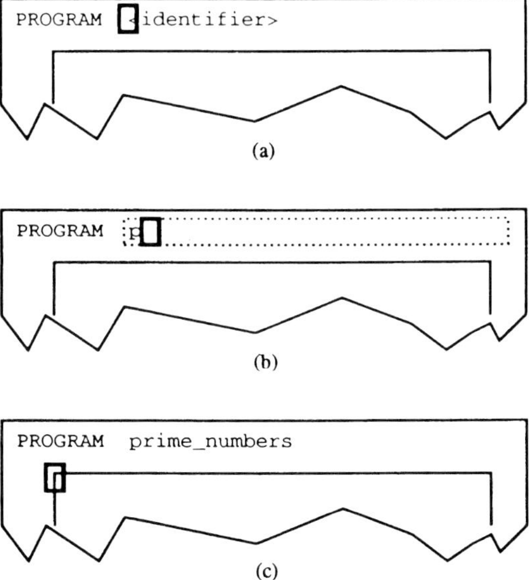


**Figure 7.8** Textual refinement of placeholders.

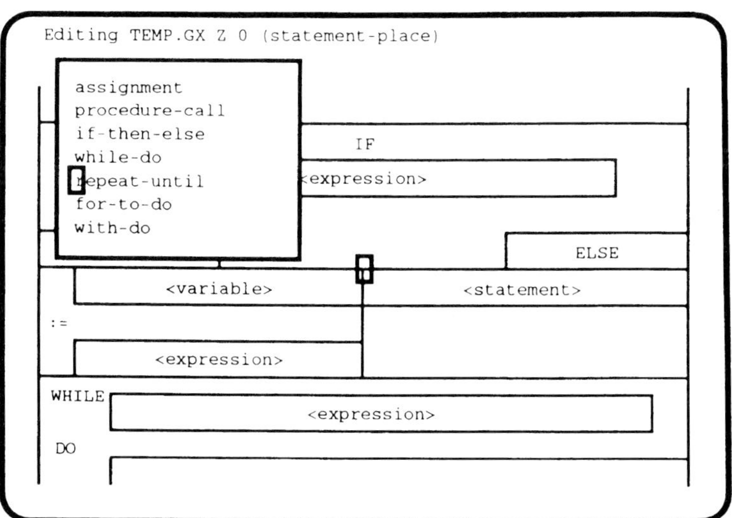


**Figure 7.9** Structural refinement of the < statement > placeholder.

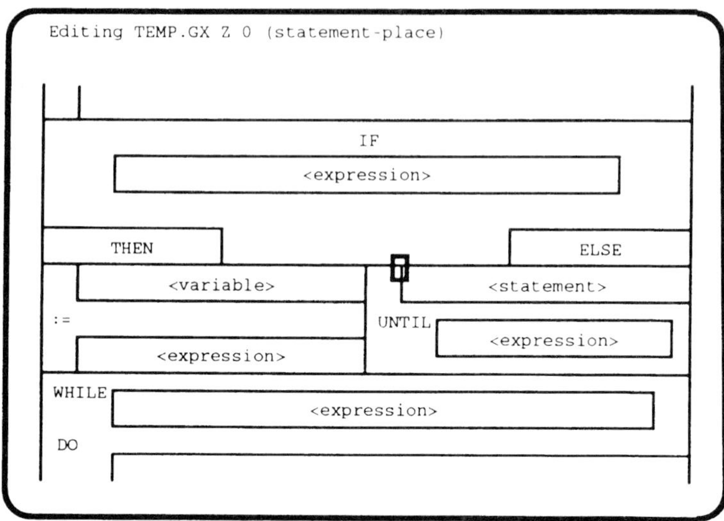


**Figure 7.10** After initial refinement.

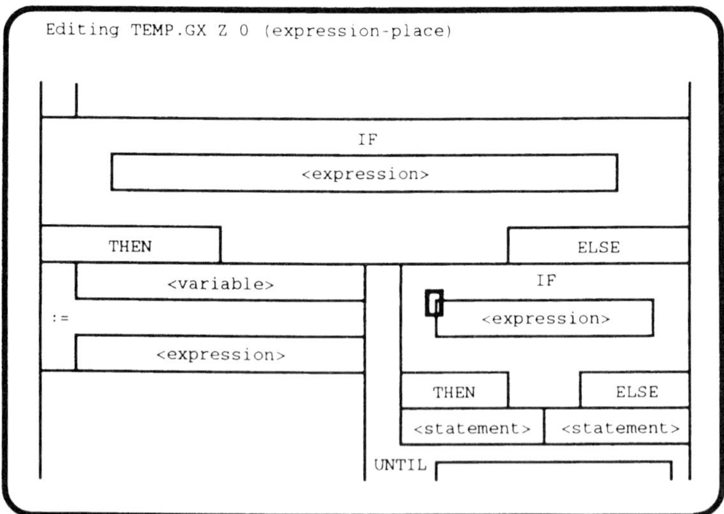


**Figure 7.11** After further refinement.

refinement. Text may be typed at these points indicating lexical refinement or the structural refinement command may be applied. Attempts to refine placeholders structurally which admit only lexemes are not permitted. Figure 7.12 shows the schematic refinement of the <expression> placeholder which admits both textual and structural refinement.

Lexical refinement is restricted to single lexemes. In the case of the previous example involving the <expression> placeholder, the only permissible refinements would consist of an identifier, a numeric literal value or a character string literal value. Any other lexeme or sequence of lexemes is flagged as erroneous, and attempts to move the cursor away from the textual refinement are not allowed until either the

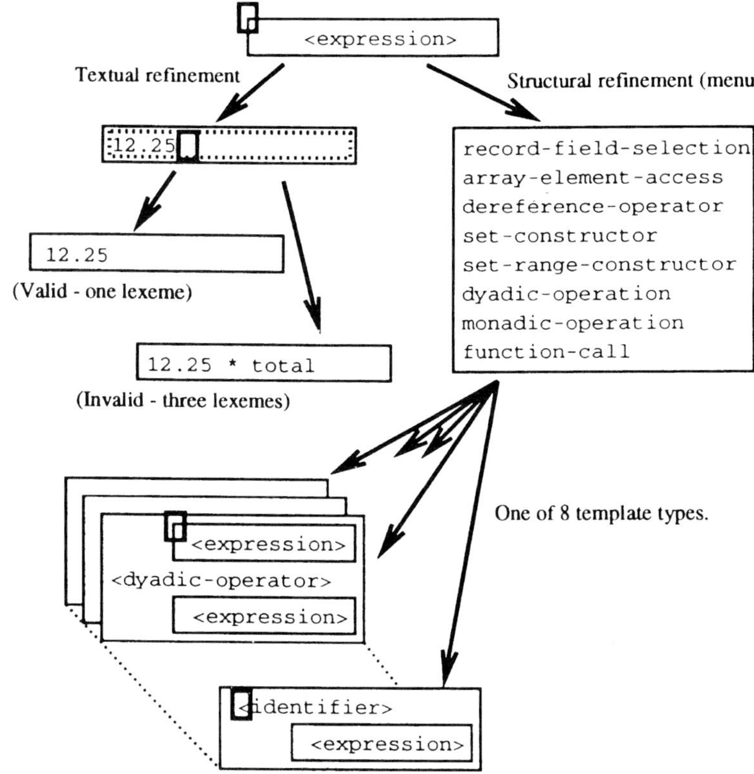


**Figure 7.12** Schema for textual and structural refinement of <expression>.

error is corrected or the refinement abandoned. With the exception of comments for which there is a specific placeholder, lexeme strings are forbidden as program structure is entirely expressed by N–S diagrams and is created using the structural refinement command. For example, to construct the expression structure for "$a + b$" would involve the structural refinement of <expression> to a *dyadic-operation* within which the lexemes "$a$", "+" and "$b$" may be used to refine the placeholders of the template (Figure7.13).

Any structural entity refined from a placeholder may be removed; this is the reversal of the refinement operation. The excised template (and its possibly refined substructure) or lexical unit is not immediately destroyed but is merely moved to the delete buffer; the original placeholder reappears at the point of removal. The contents of the delete buffer may be reinserted into the main program structure as a

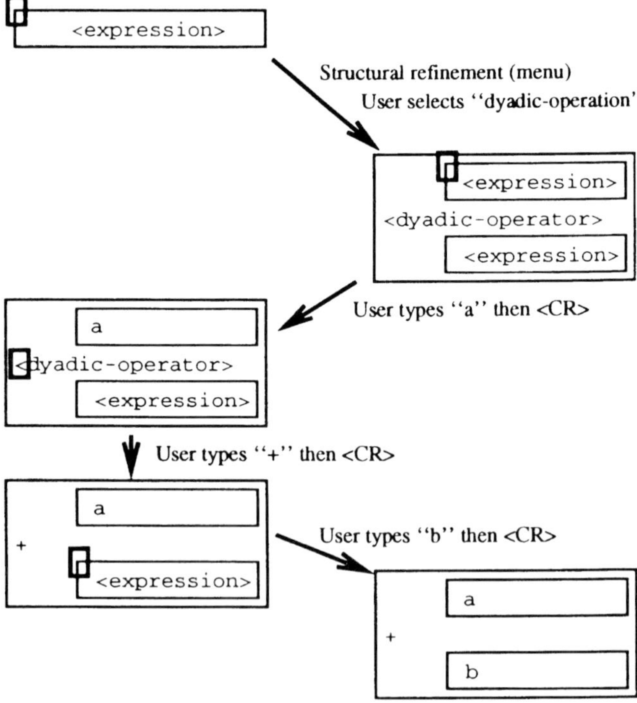


**Figure 7.13** Refinement of <expression> into "$a + b$".

special form of refinement. The refinement operation from the delete buffer needs to occur at the site of the original removal and is only permitted if the placeholder admits a template or lexical item of the same type as the unit being reinserted. During the removal operation, the structural unit or lexeme being deleted loses the context of its insertion. For example, a <variable> placeholder may be refined into an identifier lexeme which may subsequently be deleted and reinserted to refine an <expression> placeholder. The equivalent loss of contextual obstruction occurs with structural reinsertions.

The ability to remove and reinsert program elements is very convenient on three counts:

- Structures may be moved from one part of the program to another. This is especially useful as reinsertion of a deleted unit involves only a copy of the unit. Consequently, a deleted unit may be copied as many times as is required.
- Reinsertion also fulfils the role of an *undo* facility. Inadvertent removal of a structure of lexeme may be immediately corrected. The undo facility is particularly important because the editing cursor is capable of referring to quite large areas of structure for the purposes of removal.
- Direct higher-level syntactic transformations using graph transformation rules proposed by Arefi *et al.* (1990) and subsequently expanded by Dykes and Cameron (1990) are not provided. At the time it was felt that the range of such transformations was too great for GRIPSE's envisioned usage. Instead, the user must consider a transformation as a multi-stage editing task: the contents of a structure are moved to one of the delete buffers. The structure is also removed and discarded. A new structure is refined into its place. Finally, the saved contents are restored into the new structure.

### 7.3.3 N–S subchart generation

During the process of continual refinement of a template or series of templates, and as the N–S diagrams become more nested, a point will be approached where the available width for newly refined templates becomes so narrow that the contents of the template cannot be displayed. When this is likely to happen, GRIPSE creates a subchart into which the newly refined template is placed. The position in the main chart at which the template would normally be displayed is replaced

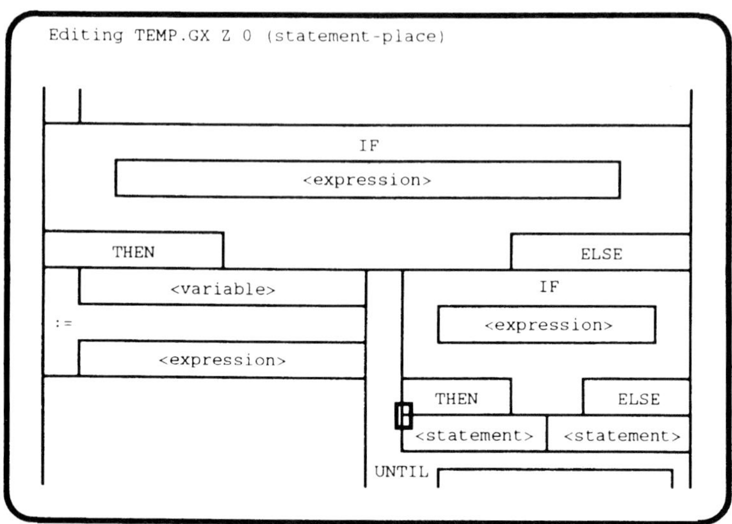


**Figure 7.14** Structural refinement in limited width.

with a *zoom box* containing a numerical reference to the subchart. The generation of zoom boxes and subcharts may also be effected manually in order to hide the detail of any desired structure. Subcharts may also be unzoomed and brought into the context of its parent chart. However, if the zooming thresholds are violated, unzooming cannot take place. Figure 7.14 presents the scenario in which structural refinement is about to take place at the indicated cursor position. An assignment statement template is to be inserted. GRIPSE judges that the available width is insufficient, so a subchart is created and the display is conceptually magnified to accommodate the new template (Figure 7.15). Again, the cursor is moved onto the next legally modifiable entity, in this case a variable placeholder. Without changing the magnified status of the subchart, the cursor may be zoomed out again so that the view of the parent chart is restored. This is shown in Figure 7.16. The cursor is positioned within the area of the zoom box, which also contains a reference to the number of the subchart, relative to its position in a sequence of possible subcharts whose parent is currently displayed. The position indicator in the status line shows the cursor pointing to an assignment and this is generally the case; structurally, zoom boxes do not really exist and an attempt to delete the structure at the cursor

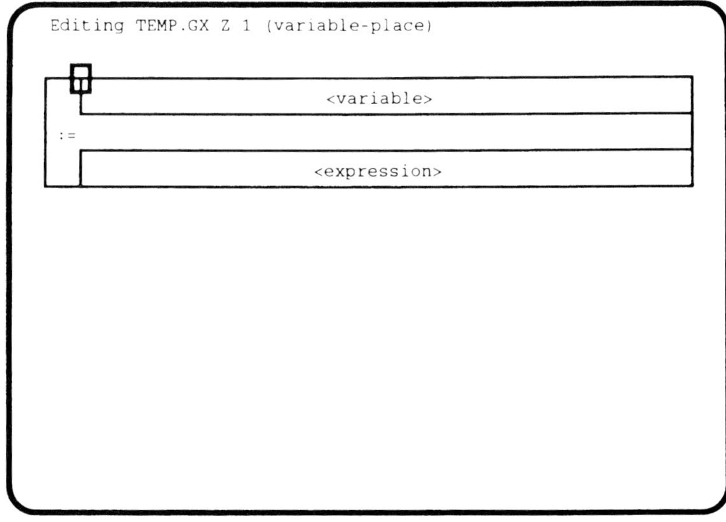


**Figure 7.15** Automatic zooming into subcharts during refinement.

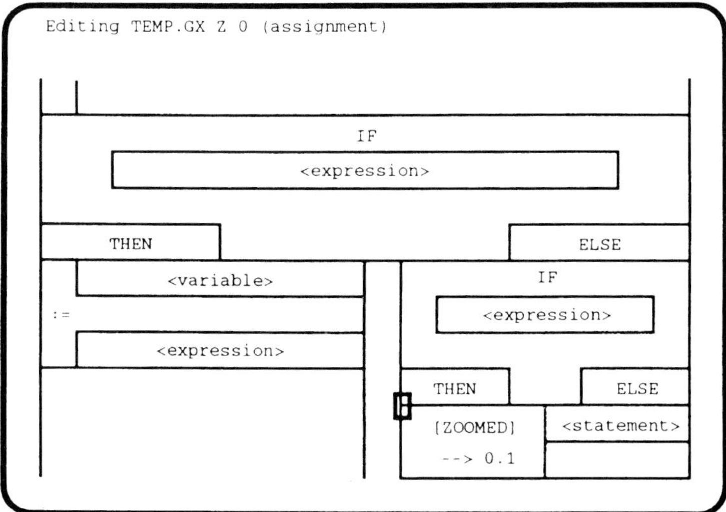


**Figure 7.16** Zooming out from the subchart.

position will result in the removal of the underlying assignment template.

Lexical units and placeholders may not be zoomed and moved to subcharts because their displayed structure is textual in nature and considered to be infinitely flexible (lines of words and characters may be wrapped to any constraints), certainly beyond the threshold at which templates are zoomed.

### 7.3.4 Cursor movement and context

The position of the editing cursor is indicated on the terminal display by the familiar character cell flashing block. All templates, placeholders and lexical units are associated with a notional bounding rectangle whose area exactly encloses that of the structural entity. The entity whose bounding rectangle contains the cursor is said to be *under the cursor* and is the unit to which all editing operations then apply. Naturally, the cursor is capable of existing within the bounding box of several structures simultaneously; the structure with the smallest area is chosen, i.e. has the highest reference priority.

There are two types of cursor movement operations. The cursor may be moved in increments derived from program structure in several step sizes. The smallest step size permits the cursor to be moved to the next previous structural or lexical entity nearest the current structural position. For example, with the cursor positioned on an IF-THEN-ELSE template, moving FORWARD causes the cursor to jump to the <expression> placeholder and then to each individual nested position within the THEN-branch of the conditional, followed by the same behavior down the ELSE-branch (Figure 7.17). The exact reverse operation is also possible.

In structural terms, the cursor may be moved so that it follows only gross structural sequence, ignoring the *internal* structural complexity of any template to which the cursor currently refers. This mode of cursor movement is shown in Figure 7.18. From the assignment template, the cursor is moved in sequence (it ignores the two placeholder positions internal to the template) to the REPEAT-UNTIL template. Both branches of the conditional structure are considered to be at the same sequential *level* so cursor movement may continue to follow the gross sequence of the ELSE-branch of the conditional structure. In either direction, when a sequence has been exhausted, the cursor jumps to the next unit in the immediately enclosing sequence.

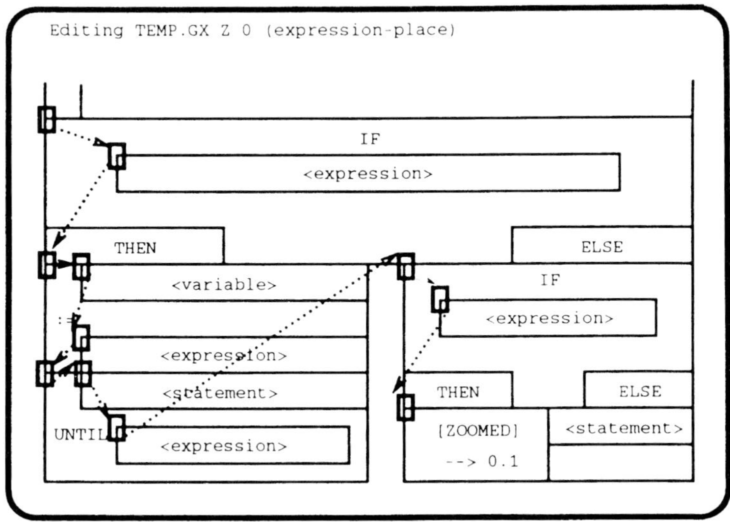


**Figure 7.17** Structural cursor movement in minimal increments.

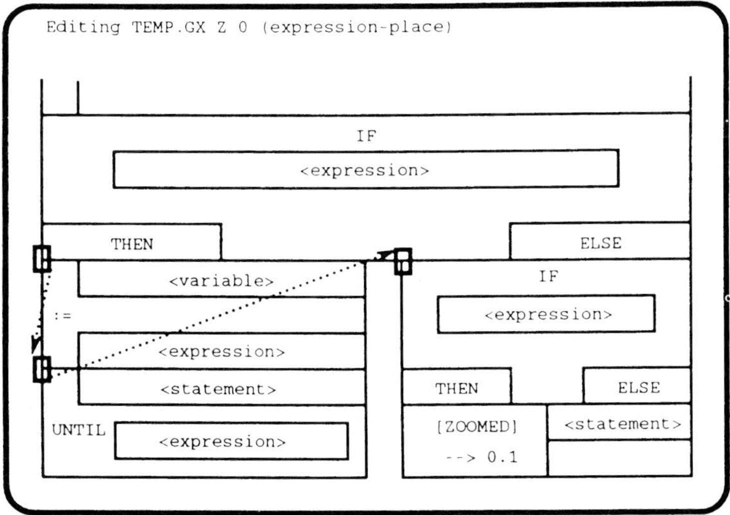


**Figure 7.18** Structural cursor movement in gross sequential increments.

The two step sizes have a rough textual analog, the terminology of which is carried over into the structural domain. Moving LEFT and RIGHT correspond to the structural movements with minimal stepping distance; in textual terms, it is similar to moving the cursor through the contents of a line, only moving to the next or previous line when the cursor is moved off the boundaries of the current line. Moving UP and DOWN correspond to the structural movements that follow only gross structural form; in textual terms, the cursor is moved between lines without having to visit every character of each line.

A further structural cursor movement permits the user to position the cursor at the template which directly contains the unit currently under the cursor. Both in N–S graphical terms and following the textual analog, this cursor movement is equivalent to a DIAGONAL cursor movement.

During structural cursor movement, the flashing block cursor is positioned at the *standard* reference point of a template or placeholder. This is always the upper left-hand corner of the notional bounding rectangle for the template or placeholder and has been the indicated cursor position throughout all of the previous illustrations. Structural cursor movement ignores the presence of zoom boxes and subcharts; the process of zooming between subcharts is handled automatically.

It may appear that this limits the addressability of lexical units to the extent that only the lexeme in its entirety may be manipulated without regard to its composition by characters. Indeed, this would be the case, except that the cursor may be moved in character cell increments (by character and by line) whilst within the bounding box of the lexical unit, thus giving access to the individual characters composing the lexeme. Modification of the lexical unit may now proceed in character increments. In structural terms, the cursor is always referring to the whole lexical unit so a structural removal would still be possible. After modification, the cursor will be prevented from leaving the bounding box of the lexeme if the modification results in something that is malformed or invalid for the context.

The movement of the cursor by character cell increments is liberalized still further, resulting in completely unrestrained cartesian cursor movements. The cursor may be moved anywhere within the bounding box of the entire program N–S chart (Figure 7.19). Consequently, there is a large area over which the cursor continues to refer to any particular template so care must be exercised when positioning the cursor prior to a removal operation, hence the availability of an undo facility. Cursor

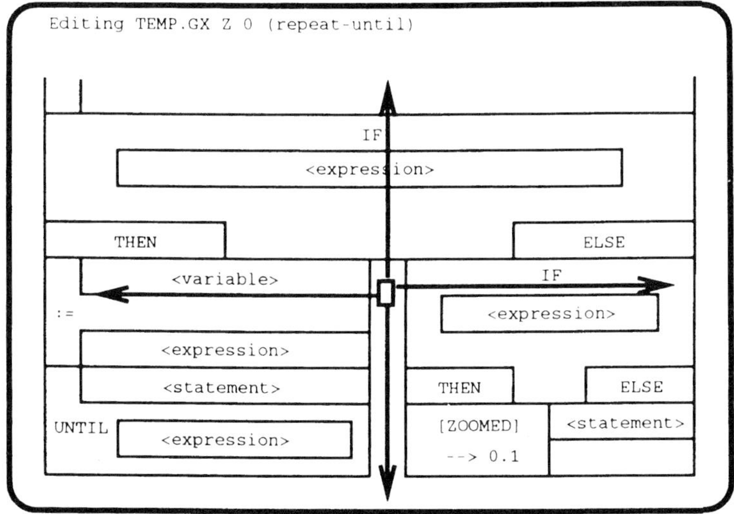


**Figure 7.19** Cursor movement in cartesian increments.

movement in cartesian terms is again available in several step sizes: by individual character cells and lines, by the limits of the window and by overall chart limits. For both cartesian and structural cursor movements, the editing window scrolls automatically to accommodate the new cursor position.

The cartesian cursor movement commands described above, work in a two-dimensional plane only. Zoom boxes are treated as valid templates although structurally, the cursor refers to the template that forms the root of the subchart under the zoom box. Separate cursor movement commands allow the cursor to be moved into the third dimension between subcharts and their parents. From any position within a subchart, the cursor may be moved into its parent chart.

### 7.3.5 List placeholders

In the previous discussion regarding templates, placeholders and their refinement, no mention was made of the class of placeholders that indicate the position of lists, i.e. the position where optional or iterated refinement may take place. This class of placeholder is displayed textually as a normal placeholder except that braces are used instead of angle brackets. In order to prevent the unnecessary display of list placeholders which are notionally present at the beginning, end and in between every element of a list sequence, these placeholders are not usually displayed except to mark the position of a completely empty list.

Instead, a separate structural cursor movement is used to open up list sequences to allow the insertion of more list members. In effect, a roving list placeholder follows the cursor position and breaks through to be displayed where it is possible to insert a new list member. Figure 7.20(a) sets the scenario: the cursor position is about to be moved into the loop body sequence of statement templates. In place of the RIGHT cursor motion command, the RETURN command is used. Because it is possible to insert a statement template prior to the assignment template, a statement placeholder is created automatically when the cursor is moved as in Figure 7.20(b). If the placeholder is not refined, the redundant placeholder is removed when the cursor is moved away from it towards the assignment template in Figure 7.20(c). Using RETURN to move past the assignment template (through both internal placeholders) will result in another statement placeholder being created (Figure 7.20(d)) which will again disappear when the cursor is moved away from it. This cursor movement command forms the normal mode

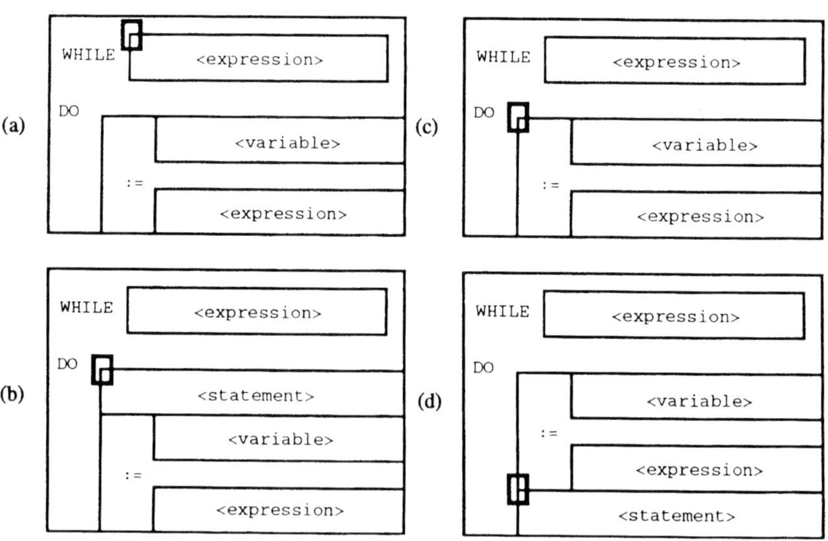


**Figure 7.20** The "list-oriented" cursor movement.

of program entry because it always opens up sequences of lists for subsequent refinement and yet acts as a normal *move to next structural unit* when not in the context of a list.

### 7.3.6 Grouping operations

All list contexts have implicit grouping capabilities which enable the user to refer to more than one structural entity at a time for the purposes of removal and possible reinsertion. A grouping is created by marking the start and end of a sublist then issuing the GROUP-CREATE command. Structures within the group are still available for individual manipulation but the difference between a list and a grouping is that the cursor may refer to a grouping as a single entity but a list context is essentially anonymous, i.e. its bounding box is always obscured by the list contents. Lists consisting entirely of lexical items (identifier lists, for example) are right-shifted by one character. Lists of structural templates are reduced in width to accommodate a double-line down the left- and right-hand sides of the sequence. An example of grouping is shown in Figure 7.21 which is concerned with surrounding structures and overlays.

### 7.3.7 Syntax and semantics checking

It has been stated that the syntax-directed editor can guarantee, through restrictions enforced during the application of the refinement operation, the syntactic correctness of the program under construction. However, the process of refinement is powerless to prevent errors arising from the violation of Pascal language static semantic rules.

Semantic checks are not performed immediately after refinement of a placeholder. Instead, certain classes of templates have *phrase* status which works in a similar way to lexical refinement. The user is allowed to refine the internal structure of a template that has phrase status without impunity until an attempt is made to move the cursor away from the bounding box of that template. Full semantic checking then takes place with respect to the entire program and any errors are reported.

### 7.3.8 Surrounding structures and overlays

As mentioned in a previous section, the zooming operations and the creation of zoom boxes that refer to a hidden subchart may be put to

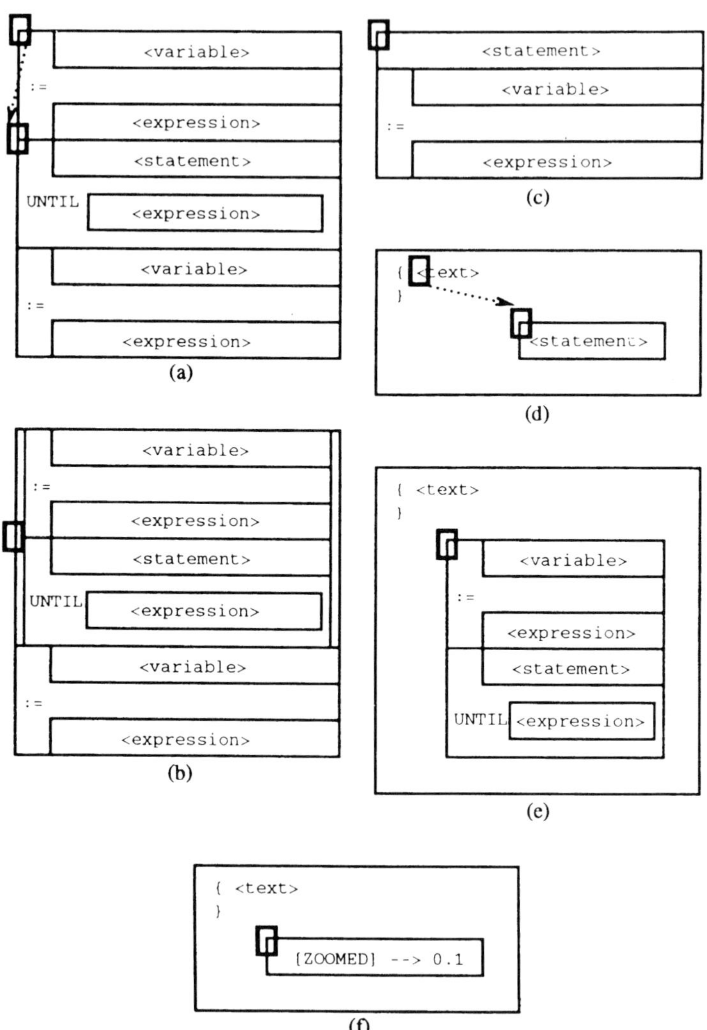


**Figure 7.21** Grouping and surrounding/overlaying.

other, less cosmetic uses. Whole sequences of structures may be zoomed into subcharts and the zoom window surrounded or overlayed by a comment abstraction documenting the action of the hidden structures. At the current display level, the kind of detail represented by the structures themselves may not be important.

A template or structural placeholder in any context may be surrounded or overlaid by a general purpose surrounding template which may be put to several uses. This section demonstrates its use as a documentation aid in which comment text may be used to hide underlying program structure. The interpreter and testing components use the same mechanism to specify tracing and test control instructions.

It will have been noticed in previous sections on refinement, particularly lexical refinement, that there is little scope for the positioning of comment text. Most textual programming languages permit properly delimited, arbitrary text to be placed wherever whitespace may be found. GRIPSE's singular handling of lexical units prevents informal placement of comment text. However, the surround facility is ideally suited to the task of comment creation.

Comment structures may be attached to any structure or structure grouping without interfering with basic syntax. The example of Figure 7.21 illustrates: (a) the selection of the endpoints of a sequence of statement templates which, (b) are then grouped to be removed (c) into temporary storage (the delete buffer). In (d), the statement placeholder resulting from the removal is surrounded by a comment overlay into which the contents of the delete buffer are then restored (e) which also demonstrates that comment text and its associated structural realisation may coexist at the same chart zoom level. Unit (f) is probably more demonstrative of comment template usage; the structure is obscured (by zooming it into a subchart) by its documentation. The zoom box may be considered equally redundant so in this case, GRIPSE permits the box to be suppressed completely.

### 7.3.9 Auxiliary facilities

In addition to the construction and editing of programs and providing an interface to other GRIPSE subsystems, the syntax directed editor provides hardcopy and secondary storage operations.

The program under construction within the text editor will, as its complexity increases, spread to an increasing number of separate subcharts, connected by zoom boxes. Similarly, the use of explicit zooming and overlays will further increase this number. GRIPSE therefore allows the dumping of program charts with user-specifiable limits:

- The zoom level at which the cursor is positioned determines the *root* of the printed chart.
- A depth value may be specified by the user to limit the nesting depth of subcharts output to the printer. The absence of a depth value implies that all subcharts below the one selected are to be printed. A depth value of zero means that only the currently selected chart is to be printed without its subcharts.

Charts and subcharts are printed separated by a form feed and each labeled with its zoom box label. Subcharts are output in depth first order because this is the natural reading order.

Programs may be saved to secondary storage to be reloaded at a later date. The editor stores programs in a linear form of the internal abstract syntax tree. Only syntax structure information is stored in such files. During reloading, the static semantic attributes and symbol tables are reconstructed. This has the advantage that the user may save the program under development and leave GRIPSE's environment without having to correct errors that have been flagged by the editor. The errors are rediscovered during the program load operation when environmental information is reconstructed and the cursor will be placed at the location of any uncorrected syntax or static semantics violations.

Programs may also be converted into their textual representation as a pretty-printed source file but this is of limited use since the dialect of Pascal used by GRIPSE is not sufficiently standard to be accepted by most compilers. Specifically, unpacked arrays may be used in operations where standard Pascal would only permit the use of packed arrays.

## 7.4 Program execution

At any stage of program construction, the user may invoke the interpreter to begin program execution or to continue from the last point of execution. Single stepping through the program structure is also possible and this is performed with respect to the editing window; the cursor denotes the structure which is about to be executed. At any stage of execution, the user may return to editing. The user may move between program editing and execution without restriction provided that none of the active declarative structures (those participating in execution) are modified. Initially, an attempt to modify such structures is prevented; they are locked by the interpreter. The user may override this but will then be prevented from resuming execution which must be restarted from the first statement structure in the program.

The N–S chart display of program structure has a more dynamic aspect during execution and may be used to animate the evaluation of

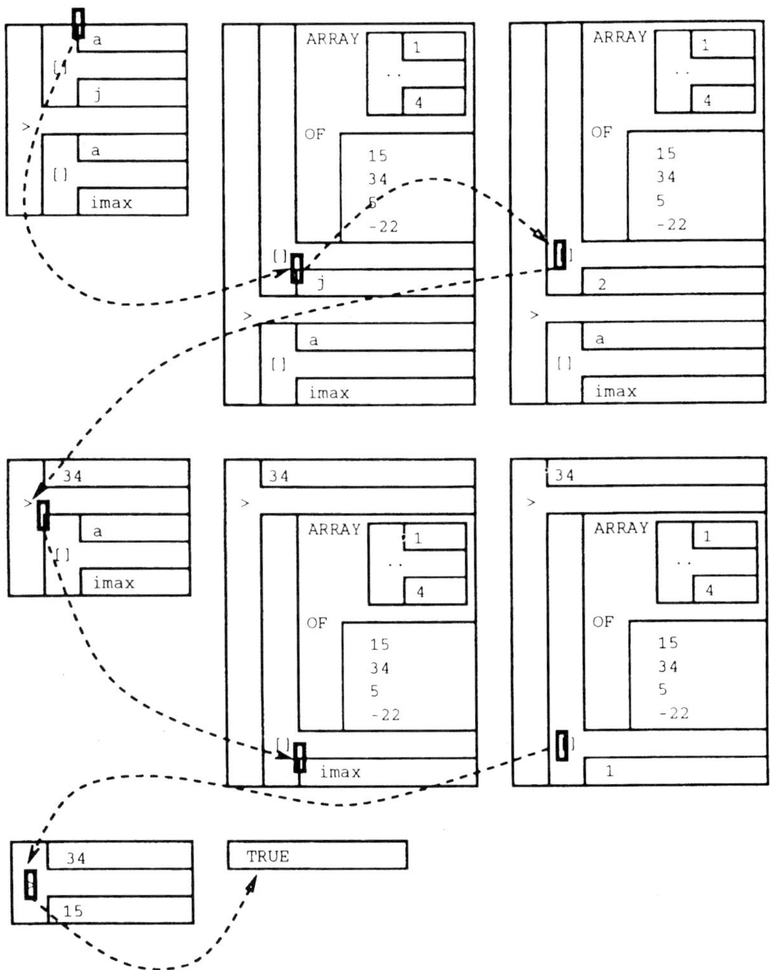


**Figure 7.22** Expression animation.

expressions. Figure 7.22 shows the result of requesting an animated evaluation of the expression `a[j] > a[imax]`. Successive evaluation of each expression element causes its replacement by actual value structures. This is gradually reduced to the result of evaluating the relational operator.

A dynamic view of an active variable declaration structure for any procedure, function or the main program is available. This is an editable N–S diagram which permits the user to modify the values of variables.

As well as executing Pascal program structures, the interpreter acts upon the pragmatic overlays that pertain to debugging (and to testing, described later). The user may have overlaid or surrounded a sequence of statements during the execution of which, watchpoints or data breakpoints are to be set on selected variables. Debugging events are signaled in the same way as error messages, and the editing cursor is positioned at the executable structure that caused the event.

In addition to debugging facilities, the interpreter provides execution and output trace data structures, and a simple save and restore interpreter state mechanism (a simple reverse execution system). These are intended for use by the mutation testing facility of GRIPSE.

## 7.5 Mutation testing

It was stated in an earlier section that the aim of GRIPSE was to provide an environment within which programs could be constructed, executed, debugged and tested. The rôle of tester would inevitably fall upon the user of GRIPSE and it was felt that no matter what the degree of sophistication of any testing system, the overall support of the environment and the testing system's integration with the rest of it might prove more important.

A program testing technique called mutation testing was considered as the candidate, because, at the time it was felt that current mutation testing systems were *report-orientated.* This means that the testing systems merely accepted a program and associated control data, analysed it and, without involving the tester in any other way, would produce as output a detailed report. Another term which could be employed for such a mode might be *batch-orientated* which although more applicable to the practice of non-interactive data processing or the submission of program compilation jobs to a batch processor, has an important analog.

GRIPSE provides an almost ideal environment in which mutation testing may be supported: mutation testing is a form of automatic, syntax-directed program editing and execution. It may therefore be considered as another *user* of GRIPSE, one that has greater powers of control than the human user, having direct access to GRIPSE's abstract syntax tree, environment tables and executable structures.

Mutation testing (Budd *et al.*, 1979; Budd, 1981), unlike most other forms of software testing, attempts to aid the tester in *improving* already supplied test data. The idea is based on the assumption that the software was constructed by able programmers, the "*competent programmer hypothesis*". Such programs are likely to be *almost* correct, their differences from the ideal, correct program are almost certainly very small. Mutation analysis, at its simplest, attempts to determine whether supplied test data would be capable of distinguishing between a program and its ideal.

### 7.5.1 Strong mutation testing

The original test program ($P$ in Figure 7.23) is *mutated* into a series of programs ($M_1$ to $M_n$ in Figure 7.23) by the single application of a *mutant operator* whose only constraints are in terms of size and that the application must result in a program that still obeys the rules of the programming language in terms of syntax and static-semantics; put simply, it must compile. Superficially, each mutant is almost identical

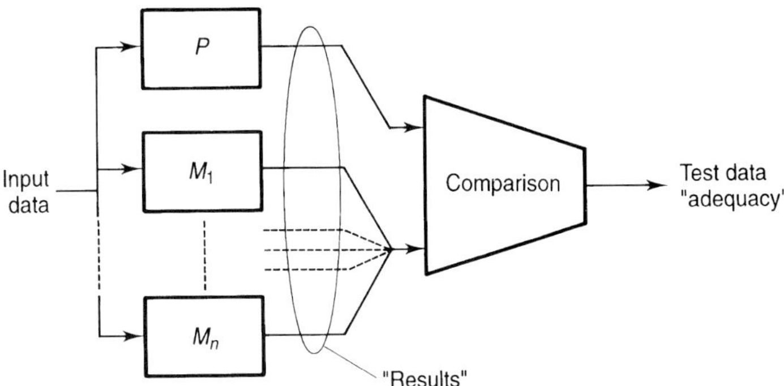


**Figure 7.23** Strong mutation testing.

to the original unmutated program which has now, in effect become the ideal version of the program.

Each mutant program is executed using the same set of test data. The outcome from each execution falls into two possibilities:

- The resulting output is different from that obtained by executing the original test program with the test data. In this case, the test data *kill* the mutant and are considered sufficiently adequate to distinguish between the mutant and the original. Note that the failure of the mutant due to a software exception, an arithmetic overflow for example, would also be a valid example of test data killing a mutant.
- The resulting output from the mutant is the same as that obtained by executing the original program. This may arise for two reasons:

  - The test data are not sufficiently well constructed, i.e. are inadequate, to distinguish the mutant from the original program. This may be due to any number of reasons, for example, the mutated part of the program may not have been executed.
  - The mutated program is equivalent to the original program and no matter what test data are used, would still produce the same output. It is usual to find 3–5% of live mutants fall into this class (DeMillo, 1989).

The full spectrum of mutation operators is obviously language dependent to a certain degree, but will certainly include the following classes of operator:

- comparable operator replacement;
- unary operator insertion or removal;
- comparable constant/variable replacement;
- off-by-one constant replacement/addition/removal;
- statement removal;
- *break* statement insertion;
- statement label replacement.

An analysis of the possible mutant operators applicable to the C programming language has been performed in Agrawal *et al.* (1989) which also includes mutant operators applicable to declarative units.

Within these classes of mutation operator, with respect to a particular data set, its adequacy can be represented as a ratio:

$$\text{Test data adequacy} = \frac{\text{number of dead mutants}}{\text{total number of non-equivalent mutants tested}}$$

One of the problems with this mode of mutation testing, termed strong mutation, is that it consumes a large amount of computing resources mainly because the number of mutants that can be generated from a program tends to approach $O(n^2)$ where $n$ is the number of lines in a program (Howden, 1982). Not only will there be problems in storing each mutant but also in executing every mutant. Each mutant is still a complete program and if the original takes significant time and computing resources to execute, then so will the mutants. A number of solutions to this problem have been proposed including resource shifting (DeMillo *et al.*, 1988), computationally intensive analysis away from the user's workstation onto faster vector processing hardware. Patch-orientated mutant generation (DeMillo *et al.*, 1990) works in close consort with a special compiler which generates object code for the original program and a series of patches for each mutant. Thus, compilation need only take place once.

Strong mutation has a poor definition of program outcome as being the *output* generated by the execution of the program. Similar weak phrases have described live mutants as "giving the same results" as the original (DeMillo *et al.*, 1978). Program output may take many forms; in the data processing world, these will almost always take the form of files, although for one piece of software, the output may be split across several different files. In the world of high-performance graphical workstations and their application software, program output will amount to a series of high-resolution rectangular areas of pixels or windows.

If the range of possible output representations is so great, then the idea of comparing original outcome with that of mutant outcome might prove to be too wide in scope for a general purpose testing system. In the case of comparing high-resolution bitmaps (or any large or distributed output data set), this operation might prove too time consuming or even too sensitive to distinguish outcome. For textual output, Lipton and Sayward (1978) propose that whitespace be filtered out, prior to comparison.

A way is therefore needed of presenting program outcome to the testing system that is independent of its *presentation*. The sensitivity of outcome comparison should also be modifiable, providing a filtering or amplifying service for comparison. This will also have an effect of presentation independence.

### 7.5.2 Weak mutation testing

Weak mutation analysis (Howden, 1982) has only one fundamental difference to that of strong mutation: the program under test is never physically mutated. Instead, at significant points during program execution, elementary components are considered, along with possible errors within them with respect to the current program state. Test data is then constructed such that the component would generate a different outcome if re-execution were to take place.

Weak mutation has certain advantages over strong mutation testing. Explicit mutants need never be constructed because the conditions required for test data to kill mutants can be decided in advance. Because weak mutation analysis is passive with respect to program state changes, within the course of one program execution, for each component under analysis, several mutated components can be considered. Program outcome is also less of an issue than with strong mutation: test data need only produce a *difference* in the program component being examined even though the overall outcome or output may be the same as that produced by the original, unmutated program.

Weak mutation also has several disadvantages. Its name correctly implies that test data which are declared adequate under weak mutation testing criteria might fail under the more stringent conditions required of strong mutation testing. Also, the range of applicable mutation operators available in weak mutation testing is a small subset of those available under strong mutation testing because of the difficulties in determining the *health* of the resulting notional mutant. Mutation operators available under weak mutation analysis are therefore concerned with:

- variable definition and reference;
- arithmetic expressions;
- relational expressions.

Although weak mutation is effectively *passive* with regard to the current program state during analysis, it is more *invasive* than strong mutation because weak mutation requires more information about intermediate program state whereas strong mutation analysis concerns itself entirely with program outcome or output. Consequently, prior to weak mutation analysis, the program under test must be analysed and instrumented with software *probes* (Girgis and Woodward, 1985) to record aspects of the program state pertinent to mutation analysis. Such probes will be responsible for recording:

- In the case of variable definition, whether that variable was given a new value. A new value would *kill* a mutant based on this result.
- In the case of variable reference, its value being unique out of the set of all comparable variable reference replacements would also kill a mutant based on this result.
- For arithmetic expressions, a non-zero result would constitute effective test data.
- For any relational expression of the form LHS $\circ$ RHS, values such that abs(RHS − LHS) $< \varepsilon$ where $\varepsilon$ represents machine precision in floating point expressions and unity in integer expressions.

Note that if the program is being executed in a specially constructed interpreted environment, probe insertion would not be necessary; the entire program state will always be available.

### 7.5.3 Firm mutation

Mutation in either of the previously described forms is concerned with the introduction of a small change in the original program under test. The output from the original program execution and its mutants is then compared. Strong mutation and weak mutation are basically the same except for:

1. The point in execution time at which a mutation operator is applied, say $t_{change}$.
2. The point during program execution at which the mutation operator is canceled, say $t_{undo}$. The process of canceling an applied mutant operator also results in the internal program state (this does not include the state of the mutation testing system) reverting to its condition immediately prior to $t_{change}$.
3. How and when comparison of *outcome* is performed and the definition of program outcome.

4. How sensitive the judgment is when pronouncing a mutant either live or dead.

Within the strong mutation testing environment, $t_{change}$ and $t_{undo}$ are prior to program execution and following program termination, respectively. The definition of program outcome is the (possibly filtered) stream of output whether from the *standard* output channels or the output to data files. The degree of output filtering determines the sensitivity of the live/dead judgment, for example, the output may be subject to the removal of all typographical layout (blanks, tab characters, newlines, etc.) prior to comparison. Comparison is performed well beyond $t_{undo}$.

Within the weak mutation environment, $t_{change}$ and $t_{undo}$ are notionally before and after each single program component under test. Comparison of outcome is performed immediately after $t_{undo}$ and the definition of outcome in this case is generally one of uniqueness from the state of all other comparable components.

For *firm mutation* (Woodward and Halewood, 1988), the values of $t_{change}$ and $t_{undo}$ become controllable by the tester, and need not be prior to and following the entire program under test. Consequently, mutants created by the application of a mutation operator persist for only part of a program's execution history, therefore, within one program execution, several mutants of the same program can be considered. Since under firm mutation, the program is physically mutated, the full range of strong mutation operators is available. The resulting program state changes due to mutation are of no consequence to continued execution, because at time $t_{undo}$, program state is restored. So firm mutation combines several aspects of strong mutation analysis, weak mutation and to a limited extent, reverse execution to offer a spectrum of controls to the tester which are ideally suited to their integration into an interactive programming support environment.

The application of firm mutation operators at $t_{change}$ and their reversal at $t_{undo}$ are not strictly related to linear time, and are instead tied to program structure, i.e. they form a `begin . . . end` bracketing around a section of program code that can now be considered *under test*. The bracketing can be further qualified as to its *activeness* based on some aspect of program state in the same way that the monitoring surrounding structures are guarded. In the previous paragraphs, it was implied that the time at which program outcome is compared follows $t_{undo}$, but this need not be the case. Continuous monitoring of the program state or serial outcome would enable the firm mutation testing system to kill a mutant immediately it begins to deviate. Such continuous monitoring would also ensure that, in the case of $t_{\text{undo}}$ never being reached due to the introduction of an infinite loop, the mutation testing system would still be able to kill the mutant. The `begin . . . end` bracketing implies the creation of an environment, and this is the case. This further solves the problem of $t_{\text{undo}}$ never being reached, perhaps due to the execution of a non-structured control statement or, in languages which support their use, the raising of an exception.

Once a program region has been selected for mutation analysis and surrounded by the testing structure, an interface for the program region can be constructed which identifies those data objects that *participate* in the program state for that region. Participation includes those data objects that are *referenced* (corresponding to input parameters), *defined* (corresponding to output parameters) and those which are both referenced and defined (both input and output parameters).

The surrounded program code therefore becomes a *headerless* procedure and could be transformed into a real subprogram whose only environment and outcome would be accessible via its parameters. Figure 7.24 shows the three-dimensional conceptual view of the headerless procedure with respect to the whole program. A firm mutation addition to the range of possible outcome comparisons would be on the set of constructed output parameters from such a headerless procedure to determine whether a mutant is alive or dead. It is envisaged that although initially, the construction of the headerless procedure para-

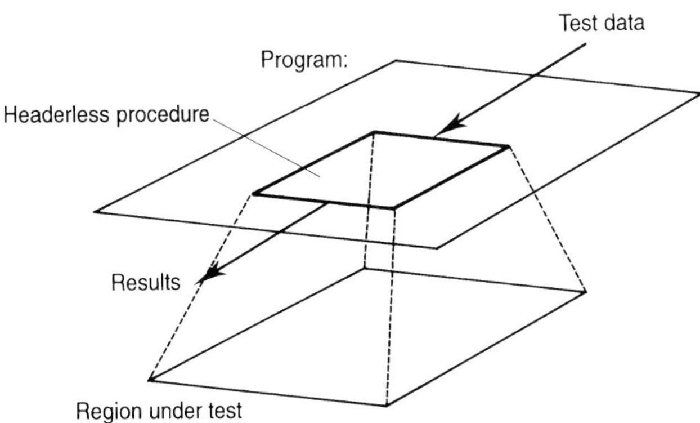


**Figure 7.24** Conceptual view of the headerless procedure.

meters will be constructed automatically by static data-flow analysis, the tester will be able to alter the procedure's interface to select a subset of output parameters for outcome comparison.

In addition to constructed outcome indicated above, since firm mutation needs to be as invasive as weak mutation testing in order to provide state reversal at time $t_{undo}$, the additional information which could be gathered during program execution and analysis would be sufficient to provide outcome comparisons based upon:

- *Data flow trace*. Aspects of weak mutation not amenable to inclusion into the spectrum of mutation analysis *strengths* can be provided.
- *Control flow trace*. This may be continuously monitored and filtered or simply compared after $t_{undo}$, depending upon the sensitivity of the outcome comparison required by the tester.
- *Actual physical output*. Subject to filtering, again, output may either be compared in batch after $t_{undo}$ or continuously monitored.
- *Exceptions*. The raising of exceptions, such as an arithmetic overflow or range constraint is considered a valid form of outcome. Its effect is obvious; the mutant is killed.

Because the tester will be controlling the mutation testing system in a very interactive environment, it will be possible to provide greater tester participation in the range and selection of mutant operators. Feedback from the system could guide the tester in providing a sample of mutant operators of any desired size. Because the application of a mutant operator will be within the constructed headerless procedure, the tester will have additional controls over the range of mutant operators based upon the frequency of applicability or some measure of the possible influence of a mutant operator. Further controls on mutant operators dealing with data objects manifest themselves due to scoping rules; comparable replacements for any particular data object may be restricted, at the tester's command to those within the selected region, an enclosing range, restricted to particular imported modules when the programming language permits it, or the global range.

## 7.6 User experience

In addition to extensive testing and debugging by the authors, the GRIPSE editor component has been subject to an experiment involving a group of 11 postgraduate students, each taking a one-year M.Sc. conversion course, either in Software Engineering or in Information Systems. All were experienced users of at least one conventional text editor, but only three had any experience at all of syntax-directed editors and even that experience was at an elementary level. The remaining eight had no previous knowledge of, or exposure to, such systems. The users were given a brief demonstration of GRIPSE and a summary sheet of the demonstrated features. They were then asked to enter a complete Pascal program. The program was the "FindNearestPrimes" program, taken from a standard Pascal textbook (Welsh and Elder, 1979) and was deliberately chosen for its use of a broad spectrum of syntactic features of Pascal. All users had six months experience of Ada programming, but only one claimed to have more than moderate knowledge of Pascal. In order to aid understanding, the program was presented in two forms, text form and Nassi–Shneiderman chart form, a notation familiar to all the students. On termination of the experiment each user completed a questionnaire.

Table 7.1 shows a list of significant facilities provided by the GRIPSE editor, together with an average measure of how much the users liked or disliked each feature, as determined from the questionnaire responses. A five-point scale was used with +2 indicating a strong liking for some feature, +1 moderate liking, 0 neutrality, −1 moderate dislike and −2 strong dislike. The third column in the table is determined from the same questionnaire given to six other students from the

**Table 7.1** Average user likes and dislikes on a scale +2 to −2

| GRIPSE feature | Actual | Anticipated |
|---|---|---|
| No need to enter syntactic keywords | 0.64 | 0.83 |
| Correct syntax ensured | 1.18 | 1.00 |
| Selection using menus | 0.82 | 1.17 |
| Automatic program layout | 0.73 | 1.33 |
| N–S chart display of syntactic units | 0.27 | 1.00 |
| N–S chart display of expressions | 0.00 | 0.83 |
| Automatic removal of unused placeholders | 0.27 | 0.67 |
| Zooming at deep nesting | 0.27 | 0.17 |
| Cursor movement in syntactic units | 0.27 | 0.50 |
| Cursor movement in cartesian units | 0.00 | 0.33 |
| Block cut and paste | 0.18 | – |
| Overall impressions | 0.18 | – |

same population, who were asked to enter the identical program using a conventional text editor with which they were unfamiliar. Since most of the features listed were not present in the editor they used, the students were asked to indicate how much they *would have liked* each feature to be present. It is the average of these values that is presented in the final column of the table.

From the table it appears that users most liked the way in which program entry was guaranteed to be free from syntactic and environmental errors. Menu selection of structural entities (based on context), automatic program layout in Nassi–Shneiderman chart form and the redundancy of much Pascal syntactic sugar (keywords and other punctuation) also rated quite highly. Non-GRIPSE users expressed a desire for all the same features as well as the structural display of expressions. Actual GRIPSE users, however, remained neutral on this latter aspect, mainly because the structural view demands a form of expression construction that is not as natural as merely typing the text. Although not listed in the table, since it is not a feature of GRIPSE, cursor movement using the mouse is something both GRIPSE and non-GRIPSE users would have liked. Even though cursor movement in syntactic units gave a greater speed of access to particular areas of the chart compared with cartesian cursor movement commands, the ability to point directly and select any particular construct was considered to have been a feature that would have been more useful.

The questionnaire also asked users to rank on a five-point scale, how easy or difficult they found certain specific tasks with the editor. As suggested previously, users had most difficulty in entering and correcting expressions in Nassi–Shneiderman chart form, even though most considered that they were quite easy to read in this fashion. This aspect is highlighted by the difference in responses between the non-GRIPSE users and those who actually experienced it. An enthusiastic response to a hypothetical feature loses some of its appeal when presented in reality. On the whole, however, users had no difficulty with other activities such as getting started, entering variable names or navigating using the cursor movement keys.

All six users of the conventional, but unfamiliar, text editor completed the task ahead of any of the GRIPSE users. The explanation for this outcome may be the fact that all participants in the experiment were experienced users of conventional text editors. Using an unfamiliar editor of the same style presents a far less dramatic change than using an unfamiliar editor of a radically different style. With any new editor a period of learning and readjustment is inevitable. An experiment with complete novices, having no particular editing style "ingrained", might well produce different results.

Another reason for the failure of the experiment to demonstrate any performance improvement is that it concentrated on use of the editor component, rather than comparing GRIPSE usage against the entire edit–compile–test cycle. Although such an experiment would be more difficult to set up and administer, it is hoped that this can be done in the future. It is still the authors' belief that there will be performance benefits, such as fewer errors, with use of systems like GRIPSE. However, it has to be admitted that this rather limited experiment has not yet demonstrated this.

## 7.7 Concluding remarks

GRIPSE, as originally envisaged, was based upon NSEDIT which only provided a syntax-directed editing tool. Full syntax checking, compilation and execution facilities were provided by the underlying operating system and tools. NSEDIT was also only capable of ensuring correct gross structural syntax; there was still much scope for error within the micro-syntax of textual refinement. GRIPSE has corrected most of NSEDIT's shortcomings in this area by ensuring totally correct syntax in programs under construction. This has been achieved by forcing the structural construction of a program as far down as individual lexemes such as identifiers. GRIPSE's syntax-directed editor also enforces consistent static-semantics such as type checking of expressions and in many respects is very similar to the Cornell Program Synthesizer (Teitelbaum and Reps, 1981) where the programmer is not allowed to proceed with program construction until erroneous code has been corrected.

In parallel with the editor's requirements that programs be constructed in structural terms, the graphical representation of such programs is also at a much finer level of structural resolution than was originally presented with NSEDIT. Augmented Nassi–Shneiderman charts capable of expressing the fine details of program structure further reinforce structural construction by templates. The uniform way in which both program structures (declarative and procedural aspects) and value structures, combined with the *pragmatic* way of introducing debugging and testing constructions, aids understanding of the program construction process as a whole; there is no need to understand wholly different representation schemes but only minor changes in structural semantics.

The use of N–S charts within programming environments is not new; the PIGS system (Pong and Ng, 1983) and Programming Support System (Frei *et al.*, 1978) are notable examples of N–S diagram usage but are limited in the amount of detail that can be expressed diagrammatically. FLOW (Dooley and Schach, 1985) employs generalized N–S diagrams to represent program structures and data structures for the FLOW-DL language which is then post-processed into real programming languages but again resorts to textual detail for expressions. PECAN (Reiss, 1985) provides a display-only N–S view of a program under construction and this is only one of many program views.

The direct use of GRIPSE's interpretive nature is merely a means to an end, providing in-environment execution facilities on which the debugger is layered. However, it enables the programmer to make code changes during program execution without having to re-execute the program under development right from the beginning. This mode of interactive editing and continued execution directly supports the notion of firm mutation. The mutation process will be completely interactive and, unlike strong mutation analysis, lends itself to application on partially complete programs or program fragments. Thus, mutation testing's mode of use becomes akin to that of debugging and can be used much earlier in the program development process.

It is intended that GRIPSE be employed in a teaching environment where novice programmers can experience program construction and testing without the need to learn the syntactic punctuation of the Pascal language. Many of the facilities provided in GRIPSE are a direct result of feedback from NSEDIT's use (its predecessor) in the same teaching environment. Increasing the number of structural units available to the user, thereby removing the need for textually represented syntax, appears at first to make program entry more cumbersome. However, anticipatory cursor movement in NSEDIT has been incorporated into GRIPSE with the result that only menu selection now takes significant keystroke activity.

The menu system used for selecting structures is aimed at novice users who then only need to remember one particular command, rather than several sets of commands for individual options. Currently, no provision is made, in user interface terms, for more experienced programmers. This could take the form of short-circuit key commands which bypass the menu display altogether. It might also appear that the provision for a mutation testing system within a system targetted for novices is perhaps too powerful a tool. However, it is considered that convenient access to something more sophisticated than execution tracing and debugging will enable students to experience the full power of the software development lifecycle *in situ*.

## Acknowledgments

Keith Halewood would like to acknowledge the Science and Engineering Research Council of Great Britain whose award of a research studentship made this work possible. Thanks are also due to Phil Jimmieson who helped to administer the user experiment.

## References

Agrawal, H., DeMillo, R. A., Hathaway, R., Hsu, W. M., Hsu, W., Krauser, E., Martin, R. J., Mathur, A. and Spafford, E. (1989). Design of mutant operators for the C programming language, SERC–TR–41–P, Software Engineering Research Centre, Purdue University and University of Florida, March.

Arefi, F., Hughes, C. E. and Workman, D. A. (1990). Automatically generating visual syntax-directed editors. *Communications of the ACM*, **33**(3): 349–360.

Budd, T. A. (1981). Mutation analysis: ideas, examples, problems and prospects. In *Proceedings of the Summer School on Computer Program Testing*, SOGESTA, Urbino, Italy, June, 129–48.

Budd, T. A., Lipton, R. J., DeMillo, R. A. and Sayward, F. G. (1979). Mutation analysis, Technical Report #155, Department of Computer Science, Yale University, April.

DeMillo, R. A. (1989). Test data adequacy. In *IEEE Proceedings of the 11th International Conference on Software Engineering*, May, Pittsburgh, 355–356.

DeMillo, R. A., Lipton, R. J. and Sayward, F. G. (1978). Hints on test data selection: help for the practicing programmer. *IEEE Computer*, **11**(4): 34–41.

DeMillo, R. A., Guindi, D. S., McCracken, W. M., Offutt, A. J. and King, K. N. (1988). An extended overview of the Mothra software testing environment. In *IEEE Proceedings of the Second Workshop on Software Testing, Verification and Analysis*, Banff, Alberta, July, 142–151.

DeMillo, R. A., Krauser, E. W. and Mathur, A. P. (1990). An approach to compiler-integrated software testing, SERC–TR–71–P, Software Engineering Research Centre, Purdue University and University of Florida, April.

Dooley, J. W. M. and Schach, S. R. (1985). FLOW: A software development environment using diagrams, *The Journal of Systems and Software*, **5**(3): 203–219.

Dykes, L. R. and Cameron, R. D. (1990). Towards high-level editing in syntax-based editors, *Software Engineering Journal*, **5**(4): July, 237–244.

Frei, H. P., Weller, D. L. and Williams, R. (1978). A graphics-based programming support system, *ACM SIGGRAPH*, **12**, 43–49.

Girgis, M. R. and Woodward, M. R. (1985). An integrated system for program testing using weak mutation and data flow analysis. In *IEEE Proceedings of the 8th International Conference on Software Engineering*, London, August, 313–319.

Halewood, K. and Woodward, M. R. (1988). NSEDIT: A syntax-directed editor and testing tool based on Nassi–Shneiderman charts. *Software–Practice and Experience*, **18**(10): 987–998.

Howden, W. E. (1982). Weak mutation testing and completeness of test sets. *IEEE Transactions on Software Engineering*, **SE–8**(4), July, 371–379.

Lipton, R. J. and Sayward, F. G. (1978). The status of research on program mutation. In *Digest for the Workshop on Software Testing and Test Documentation*, Fort Lauderdale, Florida, December, 355–373.

Nassi, I. and Shneiderman, B. (1973). Flowchart techniques for structured programming, *ACM Sigplan Notices*, **8**, 12–26.

Pagan, F. G. (1987). Program structure charts for applicative languages, *IEEE Transactions on Software Engineering*, **SE–13**(4): April, 490–493.

Pong, M. C. and Ng, N. (1983). PIGS – A system for programming with interactive graphical support, *Software – Practice and Experience*, **13**(9): 847–855.

Reiss, S. P. (1985) PECAN: Program development systems that support multiple views, *IEEE Transactions on Software Engineering*, **SE–11**(3): 276–285.

Teitelbaum, T. and Reps, T. (1981). The Cornell program synthesizer: A syntax-directed programming environment. *Communications of the ACM*, **24**(9): 563–573.

Welsh, J. and Elder, J. (1979). *Introduction to Pascal*, Prentice-Hall.

Wirth, N. (1976). *Algorithms + Data Structures = Programs*, Prentice Hall Series in Automatic Computation.

Woodward, M. R. (1987). The use of Nassi–Shneiderman charts and supporting tools in software engineering education, *Journal of Computers and Education*, **11**(4): 267–279.

Woodward, M. R. and Halewood, K. (1988). From weak to strong, dead or alive? An analysis of some mutation testing issues. In *IEEE Proceedings of the Second Workshop on Software Testing, Verification and Analysis*, Banff, Alberta, July, 152–158.

# CHAPTER 8
# User interface definition based on graphical structure editors

*Gerd Szwillus*

## Abstract

Human users and computer programs communicate via user interfaces. This process can be described in terms of editing activities: both the user and the application edit the screen and exchange information in this way.

Research efforts on user interface management systems aim at extracting as much user interface information as possible from the application and transferring it to application-independent instances. We claim structure editors to be well-suited for this task. Graphical structure editors can serve as presentation and dialog components in terms of the Seeheim model.

In this chapter we present a concept called GEGS (Generation of Editors for Graphical Structures). With the GEGS system the user can generate a graphical structure editor from an appropriate specification. The input to the generator contains the definition of a graphical language. It comprises a specification of the graphics to be edited, as well as information about the editor's interface.

The meta language for specifying graphical languages contains various features necessary for providing state-of-the-art, object-oriented user interfaces for the generated editors. It will become obvious that there is no clear distinction between the presentation of the editing object and the graphical user interface elements. Both aspects are closely linked and will be treated with one common approach.

User interfaces can be described very naturally in terms of graphical languages. The generated structure editors can be considered as comfortable user interfaces without the presence of a "real" application program. There are features treating the lower-level aspects usually

Structure-based Editors and Environments
ISBN 0–12–681890–8

adressed as interaction techniques, and others dealing with higher-level aspects like dialog architecture.

Although the GEGS system is in operation only partially, we think that the concepts involved, especially the specification mechanisms, are in themselves interesting enough for presentation.

## 8.1 Graphical structure editors

Structure editors are well-known as editors for structured texts, mostly programs, from systems like Gandalf (Haberman, 1978), Mentor (Donzeau-Gogue *et al.*, 1980), or the Cornell Program Synthesizer (Teitelbaum and Reps, 1981). In these systems structure editing is based on context-free syntax enriched by attributation. The concept of outlining (Hershey, 1985) in state-of-the-art text processor systems is a tree structure editing facility; whereas hypertext environments (Conklin, 1987) are based on directed graphs.

A common property of all these systems is the presence of some internal editing object which is represented graphically, pseudo-graphically ( "pretty-printed" text (Rubin, 1983)), or textually on the screen. The user seemingly operates on the representation, the external editing object, but implicitly modifies the internal structure as well.

The advantages are numerous, and include the following.

1. The structure editor can offer high-level structural operations to the user, as these structures are internally available.
2. The user can be freed from learning syntactical or technical details of the supported language as the representation is not "typed in", but derived from the internal structure.
3. The editor can perform syntactic or semantic checks on the user's input "on the spot", thus reducing the trouble of later error-detection.

With GEGS we want to generalize the ideas of textual structure editing to graphical languages. Graphical structure editors exist as special purpose graphical tools, like SPECIF-X (Lissandre and de Vaulx, 1987), a graphical editor for SADT diagrams, or GreatSPN (Chiola, 1987), a system for interactively editing and simulating Petri-Nets. But as in the textual case, we want to generate structure editors from an appropriate language specification. The analogous step has been taken for textual structure editing systems, e.g. in Gandalf's ALOE-GEN, as well.

Approaches for generating graphical structure editors, in most cases restricted to customizable graph editors, have been followed by several groups. Examples are the LOGGIE system (Backlund *et al.*, 1988), the universal graph editors Kb-edit (Tichy and Newbery, 1987) and EDGE (Newbery, 1988), the GRANOT system (Hekmatpour *et al.*, 1987), GDL within the ECLIPSE project (Beer *et al.*, 1987), and Göttler's graph grammar approach (Göttler, 1988).

The main problem is the design of the generator input language. This language has to cope with the specification of the graphics to be edited and with the specification of the editor's user interface. We call the input to the GEGS generator, containing information on both aspects, the *specification* of a graphical language.

As editing of graphics necessitates much more user interaction than text editing, the user interface specification must be an integral part of the graphical language. Both aspects are highly interlinked and are treated within a common framework using a small set of basic concepts. The central part of the chapter deals with these concepts.

The task of writing down a specification for a graphical language is very closely linked to specifying a user interface for some application program. Still there are additional problems that arise from the fact that an application interface resembles an editor with two potential simultaneously working users: the human user and the application program. We will discuss the resulting problems at the end of the chapter.

## 8.2 Graphical languages

In this section we sketch out the concept of graphical languages. Analogous to established specification methods for textual languages (e.g. context-free attribute grammars) we introduce techniques capable of treating the much higher variability inherent in graphical representations and the richer interaction possibilities for the user. The concept contains a homogeneous approach for such different problems like linking of subpictures, semantical attributation and multiple representations of the same object.

When using GEGS, the editor designer must be familiar with the concept; the editor user may work with the generated editor without being aware of details of the language specification.

### 8.2.1 Basic ideas

Graphical languages, as introduced here, are based on the concept of directed graphs. The external editing object is the graphical representation, the "picture"; the internal editing object is a directed graph.

The pictures contain single well-distinguishable elements (objects) which may be related graphically and/or semantically. The objects correspond to graph nodes, the relations to edges. This characterizes the graphics that are treatable with GEGS as discrete pictures. Examples are Petri-nets, SADT diagrams, JSD (Jackson System Development) diagrams, flowcharts, PERT charts, data flow nets, electrical wiring plans, or circuit layouts. On the other hand, analog pictures like geographic maps, or scanned photographs are excluded.[1]

We will not be able to treat all subtleties of GEGS's graphical language concept in the appropriate detail. To give an impression of its features, we will make use of Petri-nets as a sample language. We assume a minimal familiarity with the graphical representation of Petri-nets. Hence, this section should make clear how a graphical Petri-net editor would be specified as input to the GEGS system. This chapter contains only excerpts, but the complete formal specification has been done by the author.

We are interested in structure only and not in any dynamic aspects. The graphics contain places and transitions, represented by circles and rectangles. These two classes of elements can be linked by directed arcs. No arc may link two transitions or two places. All elements can be inscripted with short comments. An example picture is shown in Figure 8.1. This figure gives an impression of the projected editor's (simplified) interface. The symbols on the left are icons representing the different types of elements that can be added to a picture. The top line is a drop-down menu bar; the menus appear as soon as the mouse cursor enters the menu head. The figure shows that the editor specification has to cope with picture elements of the "pure" language of Petri-nets, as well as with picture elements that are needed for the

[1] The distinction is by no means well-defined or exact, but instead is based on intuition. Any kind of picture that "obviously" is based on a directed graph is suited for GEGS.

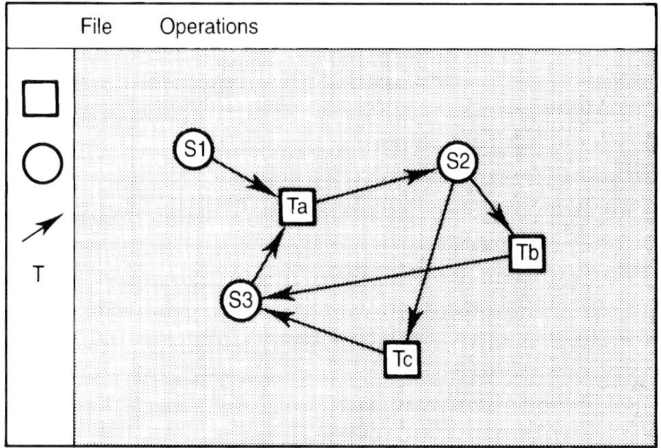


**Figure 8.1** A typical screen of the generated Petri-net editor.

editor's interface only. The separation of these two classes of picture elements is neither sharp, nor is it advantageous to be kept up. The mechanism of selecting transitions or places within the drawing window with the mouse is one of a number of examples that lie between those two classes. It is a feature of the editor's interface, as well as closely linked to the graphical representation of the Petri-nets.

The graphical language specification contains features for defining static pictures, by giving rules about how to connect subpictures to each other. It also contains parts for describing dynamic aspects, like modifications of the graphical representations, and operations modifying the underlying directed graph. All aspects are treated within one common framework, using a few basic concepts.

A graphical language in the sense defined here consists of four main parts: structure declarations, graphical picture information, conditions and core editor macros.

### 8.2.2 Structure declaration

The internal representation of the graphics is maintained as a directed graph. The nodes correspond to the pictures' basic elements, the edges to relations between those elements. The first specification step is the introduction of language-dependent node types. These types correspond to classes of picture elements of the graphical representation.

In our Petri-net example we have, e.g. the node types `Transition`, `Place` and `Arc`, as there are basic picture elements representing objects of these types. In general, every graphical entity visible on the screen must correspond to a node. This applies to the user interface elements as well; so we have the node types `TransitionIcon`, `PlaceIcon`, `ArcIon`, `TextIcon` and node types representing the menu bar elements like `File` or `Operations`.

The node types are not simply listed without any relation between them. Instead they are introduced as symbols of a context-free EBNF grammar. An EBNF (Extended Backus Naur Form) grammar is a means to describe context-free languages with more comfort, by using more complex right sides in the production rules than in ordinary grammars. The grammar author can use iteration (*), non-empty iteration (+), alternatives (|), and can specify optional parts ("{ . . . }"). As seen from language theory the described class of languages is the same as with ordinary BNF (Backus Naur Form) grammars. Using such a grammar the elementary hierarchy relations between the node types are specified. This enables the editor designer to structure the node types in a top-down way and to specify hierarchies inherent to the language.[2]

The grammar for our example language starts with

```
PetriNetEditor     →    PetriNet ScreenLayout;
PetriNet           →    Places Transitions Arcs;
Places             →    Place*;
Transitions        →    Transition*;
Arcs               →    Arc*;
Arc                →    PTArc | TPArc;
ScreenLayout       →    Menus Palette;
Menus              →    File Operations;
File               →    New Open Save;
Operations         →    Copy Cut Paste;
Palette            →    TransitionIcon PlaceIcon;
                        ArcIcon TextIcon;
```

[2] Our Petri-net example contains no "true" hierarchy relations for the sake of simplicity in this chapter. It could easily be introduced by, e.g. allowing a transition to be redefined into a complete new Petri-net.

The `Arc`-node type is refined into two classes: `PTArc` and `TPArc`. With "`PTArcs`" we denote arcs leading from a place to a transition, "`TPArcs`" lead from a transition to a place. These are the only two possible connections between transitions and places. The derivation trees of this grammar will serve as spanning trees for the internal directed graphs. These trees can be seen as special cases of directed graphs with one special type of edges, i.e. the derivation tree edges.

Semantic relations between objects, however are not tree-like in most cases. Additional named edges linking the tree nodes may be inserted to express these non-context-free relations. They are written similarly to grammar production rules. In our example language the linkage of places and transitions by `PTArcs` is expressed by edges of the types `from` and `to`:

```
PTArc          -from->          Place;
PTArc          -to->            Transition;
```

By analogy, we have

```
TPArc          -from->          Transition;
TPArc          -to->            Place;
```

for the `TPArcs`. So far, the internal editing objects of our Petri-net language will contain three different types of edges: the derivation tree edges, the `from`-edges and the `to`-edges.

GEGS uses the concept of attributation of nodes as well, as it proved to be necessary. Attributation means storage of semantical values within the nodes, e.g. strings of integer numbers. However, we stick to the concept of directed graphs to implement attributation.

Apart from the user-defined[3] node types, there are predefined node types. They represent data types that are available for attributation and include **string, integer** and **boolean**. For every possible value of these data types a node is assumed to exist and a node's attribute is assigned that value conceptually by adding an edge from the node to that value node. These edges are called *dot edges* as they are written in the common dot notation.

In our example we have, e.g. a **string** attribute for every `Place`

[3] When speaking of user, we mean the editor generator user, i.e. the editor designer, and not the user of the generated editor!

node, containing a short description or a name for the node. This is specified by writing

```
Place          .Description          string;
```

in the grammar and `.Description` is treated like an edge from a `Place` node to a **string** node. Another example is

```
File           .selected             boolean;
```

where the logical value will be used to store the information concerning whether the editor user has selected the menu `File`.

Furthermore, attribute data types are needed to store graphical objects that are linked to a node type. Therefore the predefined types **point**, **line** and **rectangle** exist; their role will be explained in the next paragraph. Additionally there are some predefined attributes for some node types and even small predefined structures present in all editors generated by GEGS. Some of them will be introduced at appropriate parts of the chapter.

### 8.2.3 Local picture specification

Up to now, graphics have not been entered into the specification at all. Just abstract node and edge types have been introduced. The next step is to assign graphical outlook specifications to the node types representing graphical elements. This is done with the aid of a drawing tool that offers the usual drawing facilities.

In our example, the node types `Transition`, `Place` and `Arc` are assigned a graphical representation; as well as the icon node types, like e.g. `TransitionIcon`, and the node types representing the menus, like `File`. As the editor's screen contains constant graphical elements, these are drawn as pictorial representations of the node type `ScreenLayout`.

The user may draw several pictures for every node type, representing different views on the same object. They are distinguished by assignment of different **integer** numbers. The number of the actually shown view of any node is always given as predefined integer attribute `.View`. So, implicitly we have the structure declarations

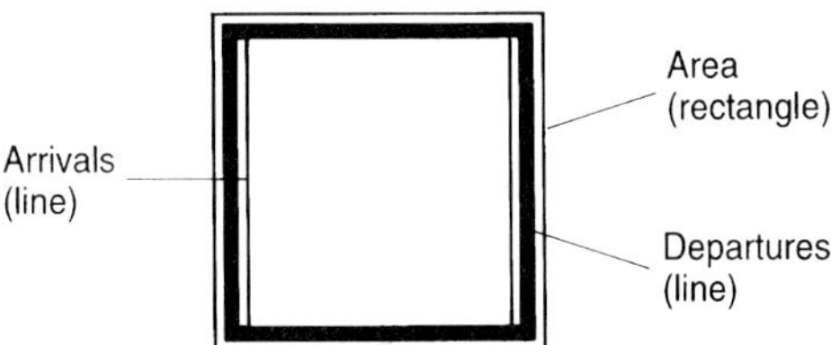


**Figure 8.2** View of node type `Transition`.[4]

```
Place          .View        integer;
Transition     .View        integer;
. . .
```

for all node types with graphical representations.

The single drawings may contain named picture elements of the basic types **point**, **line** and **rectangle**. By naming them they are implicitly added as graphical attributation to the corresponding node type. Figure 8.2, for instance, shows a view on the node type `Transition`, with the named picture elements `Arrivals`, `Departures` and `Area`.

Naming these three picture elements implicitly declares the three attributes

```
Transition     .Arrivals      line;
Transition     .Departures    line;
Transition     .Area          rectangle;
```

for the node type `Transition`. We call these graphical objects *reference objects*. Figure 8.3 shows some more graphical representations containing reference objects. They will be used to link the single pictures together, according to their semantic relations. This will become clear in the next subsection.

Without giving any details, we state that the drawings contain information on how the single pictures may be modified in size or form. Using this information, the final editor will be able to allow the user some freedom in revising his pictures with shift and stretch operations.

[4] The reference objects `Arrivals`, `Departures` and `Area` are drawn slightly out of their proper place to make them visible. The **rectangle** `Area`, for instance, coincides with the rectangle drawn with the thick black line.

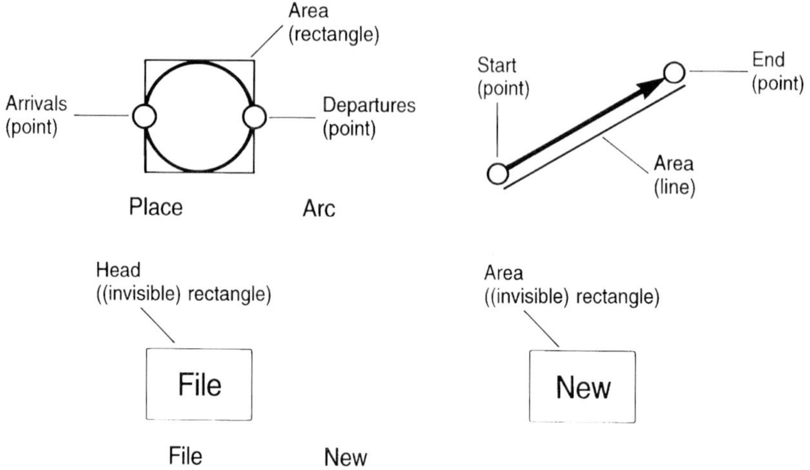


**Figure 8.3** Some graphical representations.

### 8.2.4 Conditions

The previous two subsections explained how value attributes and graphical attributes are linked to the user-defined node types. The only rules that can be specified for edges up to now are the grammar like rules that are concerned with types and not with individual nodes. This is not sufficient and we introduce a second concept, the conditions. Basically, conditions allow the specification of rules for the way a node is connected to its neighboring nodes. There are two types of conditions, called *definitions* and *assertions*.

Definitions actively establish edge relations to hold, by evaluating given relations and adding new or modifying edges; assertions just passively test whether rules hold and signal errors. Both types of conditions are written down as so-called *path conditions*. We will treat the concept of paths first and return to the conditions afterwards.

*Path specifications*

A *path* specifies a wandering through the graph along the edges of the user-defined edge types and the derivation tree edges. This includes steps to attribute value nodes along dot egdes and steps within the value node universe by value calculations. A path `P` can be "applied" to any node `N` of the graph; it results in a set `s(P,N)` of nodes ultimately reached when "following" the path, starting from `N`. A path is written in square brackets as a sequence with alternating elements: it starts with a path restriction that is followed by a path step that again is followed by a path restriction and so forth.

A *path restriction* is a list of node types: the intermediate node set is restricted to nodes of the mentioned types. This may further on be refined by testing attribute values of intermediate nodes. A path in our example that only makes sense for `Arc` nodes, for instance, starts with the restriction to this node type:

```
[Arc . . . ]
```

applying this path to a non `Arc`-type node yields the empty set.

A *path step* in its simplest version means a transition from a node set to the direct neighboring nodes with respect to one edge type. This is specified by writing down the edge type like in the structure declaration. The path

```
[Arc-> TPArc -from-> Transition .Departures line]
```

for instance, starting at some `Arc` yields a one-elementary set, containing the "departures" line of the transition, where the arc is supposed to start. Restrictions may be omitted, if not needed; they were included in the example above for clarity only, as they did not truly restrict the intermediate node sets. In its shortest version this path could be specified as

```
[Arc-> TPArc -from-> .Departures]
```

All edges can also be used in the backward direction, as in the following example:

```
[Transition .Description string Description. Place
<-to- TPArc <-Arc]
```

This path specifies a wandering starting from a `Transition` node, reaching the set of all `Place` nodes with the same description attribute value attached, and finishing with all arcs that end up at these places.

As a path may also start with a constant value,

```
[''ABC'' Description. Transition, Place]
```

yields all transitions and places labeled with the description string `''ABC''`. We will see soon that there are important predefined constants to be used in paths for the specification of user interaction features.

More complicated path steps include searches for a node type along an edge type, as in

```
[ScreenLayout down File .Head]
```

where **down** `File` denotes a search operation along the derivation tree edges towards the leaves until the first `File` node is met. Likewise we have the **up** operator for the opposite direction. The operators **forward**() and **backward**() are used for searches along the specified edge types. Furthermore these search operations can be qualified with * or +, to denote cumulative searches.

The last type of path steps concerns steps within the attribute value universe. They are specified by writing down a calculation that makes sense for the intermediately reached node set.

Assume, for instance, we had an **integer** attribute `.Number` for every `Transition` type node which contains a numbering of all transitions present. The path

```
[Transition .Number ●+1 Number. Transition]
```

would then describe a path that leads from any transition with number $n$ to all transitions with number $n + 1$. The ● is used as pseudo-argument in the expression describing the calculation and stands for the intermediate node set reached. The usual set of operations can be applied to the attribute values, together with some predefined functions.

*Assertions*

An *assertion* is specified by writing down *two* path specifications `P1` and `P2` and some comparison operator. It can either be a set comparison operator ○, like e.g. **subsetof** or == (set equality), which applies to whatever node types are results of the two paths. In this case the assertion `P1 ○ P2` is said to hold, iff for all nodes `N`, we have

```
s(N,P1) ○ s(N,P2).
```

In our example language, we could for instance write

```
[Transition] == [Transition .Number Number. Transi-
tion]
```

to ascertain that no two transitions have the same number.

The comparator can also be a type dependent comparison operator △. In this case the final node types and the comparator type must match. The assertion `P1 △ P2` then is said to hold, iff for all nodes `N` and all elements `E1` of `s(N,P1)`, there is at least one element `E2` in `s(N,P2)` with `E1 △ E2`. This seemingly complicated definition copes with the problem that a relation between values of some data does not automatically induce a likewise relation between sets of values of that type. In practice, either `P1` or `P2` mostly define one-elementary sets, which renders the definition very natural.

An example in our language could be

```
[Transition <-from- -to-> Place <-from- -to-> Transi-
tion .Number]
>
[Transition .Number],
```

which would test whether the numbers of all direct successor transitions of a transition `T` are larger than the number of `T`.[5]

There are other cases, which deal with graphical comparison operators. An example in our language is

```
[Arc.Start] in [Arc -> TPArc -from-> Transition
.Departures]
```

**In** is a comparison operator between graphical primitives, in this example used for comparing a **point** and a **line**. For a **point** `p` and a **line** l `p` **in** l is said to hold iff the **point** `p` lies on **line** l. The asssertion above states that the starting point of any "TP-type" arc (referenced by **.Start**) lies on the `.Departures` line of the transition, where the arc is supposed to "leave" from. Similar rules for the `.Arrivals` points

[5] A rule which could only be fulfilled if no cyclic dependencies existed.

and the `PTArcs` establish the graphical linkage of arcs to the objects they connect.

There is one other graphical comparison operator, called **samespot**, which states that two graphical objects of the same type occupy the same spot on the plane.

With these two operators, not only are pictures linked together, but user interactions depending on, e.g. mouse movements, can be described. The following assertion for instance holds, iff the mouse cursor was shifted into one of the two menu head areas

[**mousepos**] **in** `[File, Operations .Head]`

`.Head` is a rectangle type reference object, as shown for the file menu in Figure 8.3. **Mousepos** is a constant of type **point** which is automatically updated whenever the user moves the mouse.[6] To make proper use of assertions like these, they can be switched on and off, dependent on other conditions. We will treat this aspect later (see "Modes" below).

*Definitions*

Whereas assertions just passively test relations, *definitions* establish them actively. Syntactically a definition is a restricted version of an assertion: The left-hand side may not be an arbitrary path, but may either contain one simple edge step or just dot edges; instead of the comparison operators, only the definition sign `:=` is allowed. For boolean attribute definitions, comparisons on the right-hand side are allowed as well.

Whenever some component of the right-hand side changes somewhere in the graph, it gets recomputed and assigned to the left-hand side. A definition works like an axiom which is stated once and automatically forced to hold from there on. It is used for automatically defining edges from other information present in the graph.

In our example we might have for instance

```
[Transition -next-> Transition]
                                    :=
[Transition .Number ●+1 Number. Transition]
```

[6] Syntactically it is a constant within the definition of path specifications; semantically it is far from being a constant!

for an edge declared by

```
Transition          -next->          Transition
```

to link the transition with number $n$ to the transition numbered $n + 1$. Or we specify

```
[File, Operations .Head rectangle .inverted]
                              :=
[File, Operations .selected]
```

which defines the predefined attribute `.inverted` for the menu heads' rectangles as **true** as soon as the menu gets selected. If `.inverted` is set, the rectangle and its contents are shown inverted on the screen. The following definition shows another feature available within conditions, the **if-then-else-ifend** construct:

```
[File, Operations .View]
                                 :=
if [File, Operations .selected] then [2] else [1] ifend;
```

This definition sets the predefined `.View` attribute for the menus, dependent on their `.selected` attribute: as soon as `.selected` becomes true, the view 2 is shown, otherwise view 1. If the two views 1 and 2 are defined as sketched in Figure 8.4, then this change of view makes the chosen menu "drop down".

The reference objects `NewArea`, `OpenArea` and `SaveArea` are used to place the `Area` rectangles of the single items within the menu. (See Figure 7.3 for the `New` item, for instance.)

*Modes*

The concept of assertions and conditions is further refined by some additional features. First, assertions may have three different qualities, which state their mode of testing. Second, the mode of testing may be specified to vary automatically, dependent on other assertions.

- An assertion may be hard, soft or optional. A non-optional assertion must always hold. If it is hard the editor user is actively hindered to execute operations violating the assertion, or the system enforces

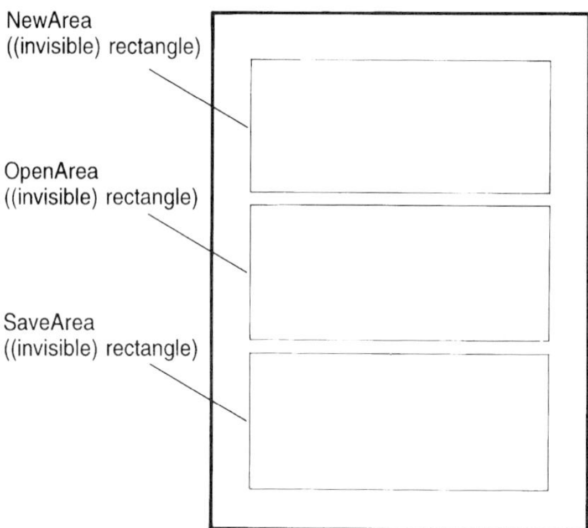


**Figure 8.4** The two views of the `File` menu.

assertions to hold which establish reference object relations; if it is soft the user is not hindered but just signaled. If it is optional, it can be switched on and off by the user explicitly, or by some other assertion implicitly.

- Explicit switching makes sense for assertions which are hard to test and for efficiency reasons should not be checked whenever a component of one of the paths involved changes. Implicit switching is defined via a finite automation that can take different states, in this context called modes. For every mode a set of conditional assertions is listed which are switched on when the automaton enters that mode; all other conditional assertions are switched off. This mechanism allows it to signal interests for certain user action events and to react to them appropriately.

In our example language we could have, for instance the mode definition

```
initialmode InitialMode:
                    on InMenu to MenuMode;
                    on InPalette to PaletteMode;
endmode.
```

The identifiers `Inmenu` and `InPalette` refer to the names of conditional assertions that are defined by

```
optional InMenu: [mousepos] in [File, Operations .Head];
optional InPalette:
                    [mouseup] in
                    [Palette → TransitionIcon
                    PlaceIcon,ArcIcon .Area].
```

**Mouseup**, like **mousepos**, is a constant **point** type object that is updated automatically, as soon as the user releases the mouse key. `InPalette` holds iff the last mouse click happened inside the rectangles denoted by `.Area` linked to one of the icon elements.

In the initial mode `InitialMode` the only events that are of interest are `InMenu` and `InPalette`; no other optional assertions are then tested. After, for instance, `InMenu` has become true, the mode changes to `MenuMode`, as specified. `MenuMode` might be defined like this:

```
mode MenuMode:
                    on InNew to InitialMode;
                    on InOpen to InitialMode;
                    on InSave to InitialMode;
                    . . .
                    on InPalette to PaletteMode;
```

`InNew`, `InOpen`, `InSave` etc. again are optional assertions which hold, iff the user clicks into the item's rectangle denoted by `.Area`. `InNew` for example is defined by

```
optional InNew: [mouseup] in [New .Area]
```

After having detected that the user has chosen, e.g. the menu item `New`, the editor should of course perform some appropriate action. This is the only point where procedural information is added to the specification, which is purely declarative so far.

The action to be taken on an event is entered in the mode definition following the keyword **do**. In our example we might have

```
mode MenuMode:
                on InNew do ExecNew() to InitialMode;
. . .
```

ExecNew() is a procedure written in the core editor's command language. It may modify the internal graph and has to be provided by the editor designer. He or she is the only person to decide what actually has to happen when a "New" command is executed. Only sequences of editor commands or macros and so-called **unmark**-commands (to be explained soon) may be specified in the **do**-parts of a mode definition.

User actions to be treated in this way include all mouse activities and keystrokes. For instance, there is a built-in function called **mouseenter** which takes a rectangle as argument and returns a **boolean** value **true**, iff the mouse cursor enters the rectangle. **Keystroke** is a **string** constant that is internally updated whenever the user touches a key. Using these features, events like

**optional** EnterNewItem: [New .Area **mouseenter**(●)] == [**true**]

and

**optional** EnterLetterA: [**keystroke**] = [''A'', ''a'']

can be defined.

### 8.2.5 Core editor macros

We do not want to go into details concerning the core editor command language (Kaschka and Wimbert, 1987). The language is oriented towards MENTOR's (Donzeau-Gogue *et al.*, 1980) command language. It allows insertion, copying, deleting, shifting and exchanging of nodes and edges. These commands are used to perform any editing operation on the internal editing object.

Basically we allow arbitrary paths starting from constants as parameters to the editor macro calls. This results in specifying node sets as parameters. Although this seems quite powerful, there remains a problem.

It is very common that operations need interactively selected elements as parameters. For instance, the editor user selects some object of the drawing window by clicking it and afterwards executes a menu-chosen operation with that special element. To enable the editor designer to specify effects like these the concept of marking and unmarking is introduced. The idea is quite simple. When writing some assertion, one can request the "marking" of some intermediate node set under a given name. This marking only takes place if the complete assertion holds. Markings are implicitly defined like additional **boolean** type attributes and can be used like those. For instance,

```
optional TransitionSelect:
                    [mouseup] in [Transition mark(T)
                    .Area]
```

holds as soon as any rectangle `Area` linked to a transition `T1` is clicked into. Additionally transition `T1`'s **boolean** attribute `.T` is set to **true**. When the user clicks another transition `T2`, and `TransitionSelect` is still switched on, then `T2`'s `.T` value will be set too. Marks are reset only on explicit request in the procedural **do** parts of mode definition by an **unmark** command. For instance,

```
on TransitionSelect
                    do WorkOnTransition([true
                     T. Transition]);
                    unmark(T)
                    to . . .;
```

could be included in some mode definition. The editor macro `WorkOnTransition` gets the `T`-marked transition as parameter and the marking is reset afterwards.

The editor designer can specify alterations of the graph by both writing definitions and including core editor macros. Although extensive use should be made of the declarative method, it will not in general suffice to express all modifications needed. The insertion of new nodes in particular is impossible with definitions only.

## 8.3 Editor functionality

The input to the editor generator specifies numerous details of the resulting editor's properties. This section gives a short outline of the functions presented to the editor user as implicitly defined by the specification.

1. There are some predefined editor macros that may immediately be linked to the individual specification by the editor designer. They are independent of the supported graphical language and are available for every editor generated by GEGS. Examples are file operations like `New`, `Open`, `Save` or `Save As`, commands for window management like `Close`, `Resize`, `Zoom` or `Move`, and navigation commands to traverse the underlying directed graph. Additionally all editors have mechanisms for applying the graphical transformations "shift" and "stretch" to the pictures shown on the screen.
2. There are language-dependent operations, which only transform the graphical representation and do not touch the underlying directed graph. These operations need not be specified explicitly, as they can be derived from the graphical language specification. Using the transformation limits defined while drawing the single node type representations and the conditions imposed on the reference objects, the editor infers the consequences of graphical tranformations the user applies. It ensures that all graphical assertions and transformation rules hold. Thus, a user operation leading to a violation will not be executed.
3. There is the collection of functions that are provided explicitly by the editor designer for the dedicated language. They are written as core editor macros. The editor designer might be considered to be at a loss as to what functions to provide. However, as the structure graph definition yields a sort of guide for these language-dependent semantic functions, this is not the case. The editor user manipulates the editing object in terms of the language's elements. Hence, the editor designer can scan the list of all language elements `X` and decide whether it makes sense to add an operation `Delete X`, `Insert X`, or `Modify X`. This approach in general does not necessarily cover all operations, but it offers a convenient framework to think about what functions to provide.

With respect to the reaction to errors, GEGS basically follows the philosophy that the generated editors are as fault-tolerant as possible. In particular incomplete states are normal during the creation phase, so the editors have to be able to cope with incomplete editing objects. The system signals states that cannot be corrected by adding information as soon as possible. From experience, however, we enforce that the context-free rules must not be violated at any time in the process. We observed that their role is very central within the editing object. Violations of soft path conditions are accepted.

## 8.4 Implementation

Shortly after development of the concepts of GEGS, a running prototype was created. It contained the core editor for directed graphs called EGO, written in C. This program read a language specification containing structure declarations, assertions and definitions. Editing could be done using the core editor command language, assertions were tested and definitions were performed. EGO represented the central functional kernel of the future to be GEGS system.

EGO, however, worked purely textually and it did not yet incorporate the uniform integration of attributes, and the modes concept as presented here. Work was going on to add a graphical representation component to EGO within the NeWS system on SUN workstations and integrate atributation. However, before this could be finished, the author left the University of Dortmund, where this work had been done, for a professorship at the University of Paderborn. During and after the transition time, work on implementing the GEGS system was not resumed, leaving the concept in the state as presented here.

## 8.5 Conclusions

Specifying a graphical structure editor, as laid out in this chapter, is different from specifying an interface for an application program. Some additional problems come up, as there is a second, "background" user involved – the application.

1. The human user gives input to the application by editing the screen; the application performs some action and shows the results to the human user in the same way. This double-user situation must be coped with in the specification and in the implemented system. Furthermore, the application must be able to gather information

from the internal editing object; appropriate access methods have to be implemented.

2. The problem arises that the editor must be able to visualize progam-generated structures. This problem has to be solved in the "pure" editor as well, because it supports multiple views on the editing object; but it is much harder to solve without any interaction.
3. When using GEGS for interface generation, it should support modular specification. A library of standard interaction technique modules should be provided, which could be tailored to the designer's needs. When implementing GEGS under existing window systems, using the predefined interaction elements, e.g. the class `LiteMenu` within NeWS, would be easily achievable then.

## Acknowledgment

This work was carried out during the author's stay at the University of Dortmund, Fachbereich Informatik, Lehrstuhl I, in Dortmund, Germany.

## References

Backlund, B. *et al.* (1988). LOGGIE – a language oriented generator of graphical interactive editors. Technical Report, Swedish Institute of Computer Science, Kista, August.

Beer, S., Welland, R. and Sommerville, I. (1987). Software design automation in an IPSE. In *Proceedings of the First European Software Engineering Conference (ESEC'87)*, Strasbourg, September, 97–105.

Chiola, G. (1987). GreatSPN user's manual. University of Turin, Department of Computer Science, Draft 8/4/87, Version 1.3, August.

Conklin, J. (1987). Hypertext: an introduction and survey. *IEEE Computer*, September, 17–41.

Donzeau-Gogue, V., Huet, G., Kahn, G. and Lang, B. (1980). Programming environments based on structured editors: the MENTOR experience. INRIA, TR 26.

Göttler, H. (1988). Graph grammars and diagram editing. In *Proceedings of the Third International Workshop on Graph Grammars and their Applications to Computer Science*. Lecture Notes in Computer Science, Springer.

Haberman, A. N. (1978). The Gandalf research project. *Computer Science Research Review 1978/79*, Carnegie-Mellon University.

Hekmatpour, S. and Woodman, M. (1987). Formal specification of graphical notations and graphical software tools. In *Proceedings of the First European Software Engineering Conference*, Strasbourg, September.

Hershey, W. (1985). Idea processors. *BYTE*, June, 337–350.

Kaschka, H. and Wimbert, R. (1987). Considerations about the implementation of structure editors for graphical objects (In German). Master's Thesis, University of Dortmund, Department of Computer Science, August.

Lissandre, M. and de Vaulx, B. (1987). SPECIF-X: a tool for CASE. In *Proceedings of the First European Software Engineering Conference* (ESEC'87), Strasbourg, September, 297–306.

Newbery, F. J. (1988). EDGE: an extendible directed graph editor. Internal Report No. 8/88, University of Karlsruhe, Institut für Programmstrukturen und Datenorganisation.

Rubin, L. F. (1983). Syntax-directed pretty-printing – a first step towards syntax-directed editing. *IEEE Transactions on Software Engineering*, **9**(2).

Teitelbaum, T. and Reps, T. (1981). The Cornell program synthesizer: a syntax-directed programming environment. *Communications of the ACM*, **24**(9): 563–573.

Tichy, W. F. and Newbery, F. J. (1987). Knowledge-based editors for directed graphs. In *Proceedings of the First European Software Engineering Conference*, Strasbourg, September.

# CHAPTER 9

# A multimodal syntax-directed graph editor

*Edwin Bos*

## Abstract

In this chapter I show that a combination of the action mode of interaction with the language mode of interaction can provide a suitable interface for structure-based editors. I present a prototype system that provides a comfortable environment to construct and edit directed graphs. By manipulating graphical representations or by entering commands in formal or natural language, a directed graph consisting of nodes and arcs can be constructed and modified. The grammar of a particular graphical language can be described by rules that apply to node and arc types. The syntactic rules can be incorporated into the system as constraints which prevent the user from making syntactic errors. The prototype system is currently applied to three domains: managing the knowledge base of a natural language dialog system, formulating queries in an experimental graphical query language, and managing files in a file system.

## 9.1 Introduction

Although the first steps in the field of structure-based editing date from the early 1980s (e.g. Donzeau-Gouge *et al.*, 1980; Walker, 1981), today many users still (mis)use general-purpose applications such as text editors and drawing packages for specific tasks such as editing computer programs and drawing flowcharts. Minör (1992) investigates why structure-based editors have failed to attract a wide audience, despite their obviously good qualities. Exploring the possibilities of structure-based editing in the domain of computer program construction, Minör finds that most objections raised against structure-based

Structure-based Editors and Environments
ISBN 0-12-681890-8

editors concern the interface. Bad interface design, he argues, has obstructed the wide introduction of structure-based editors. Minör proposes the *action mode* as the basic mode of interaction in structure-based editing. (Other names for this mode are direct manipulation (Shneiderman, 1982), model-world metaphor (Hutchins *et al.*, 1986; Hutchins, 1989), and graphical paradigm (e.g. Szwillus, 1989)). In this mode of interaction, the user mimics an action by manipulating external objects, which represent internal objects. At any time, the picture displayed on the screen reflects the internal state of the system: what you see is what you have made. The manipulations are performed in a model world, where a task domain metaphor is used to ease interaction, by exploiting the user's knowledge of the concepts and actions of the particular domain. The counterpart of the action mode is the *language mode*, in which actions are described to an intermediary, instead of mimicked.

In this chapter, I argue for the use of *both* action and language mode in the interface of structure-based editors, resulting in a multimodal interface. I present a prototype system[1] for editing graphs with a multimodal interface. The system integrates two subsystems: $Gr^2$, applying the action mode, and DoNaLD (Classen and Huls, 1991), applying the language mode of interaction. The name of the multimodal system integrating these two subsystems is EDWARD (see Claassen (1992), Bos (1992) and Claassen *et al.*, (1993) for several aspects of EDWARD). In this chapter I primarily focus on the action mode, hence I will usually refer to the system as $Gr^2$.

$Gr^2$ is a *graph editor*: graph editors enable users to construct and edit directed graphs consisting of nodes and arcs. Graph editors can be used as a tool in designing and managing computer network topologies, state transition diagrams, Petri-nets, knowledge bases and in many other domains involving structures. $Gr^2$ is a *graphical* graph editor: Graphs are represented graphically on the screen. Nodes are represented by labels and/or icons and arcs are represented by straight arrows or lines. $Gr^2$ is a *structure-based* graphical graph editor. Of course, conventional drawing packages could be used to draw a graph, but then only a picture is obtained. $Gr^2$, however, is capable of dealing with graph objects internally. It can do two things which drawing

[1] The system runs on DECstations and Symbolics Lisp Machines. It is written in Common Lisp with, respectively, Allegro Common Windows and Zetalisp graphics and mouse functions.

packages cannot. First, $Gr^2$ is able to exploit knowledge of the rules of graph theory and of the rules of a particular task domain. This exploitation of the construction of graph representation is faster and safer than is possible by using software packages for drawing. Secondly, the use of $Gr^2$ results not only in an external structure representation but also in an internal structure that can be used by an application program, for instance, a natural-language dialog system or a code generator.

This chapter is structured as follows. I will first give a brief overview of the basic operations of $Gr^2$ (Section 9.2). Then I will elaborate on those components that make $Gr^2$ a structure-based editor (Section 9.3). Subsequently, in Section 9.4, I will briefly describe the results of user experiments conducted with the system's forerunners. In Section 9.5, I will briefly compare the system with related systems. In Section 9.6, I discuss the usability of structure-based editors in general. Finally, in Section 9.7, I will enumerate the many research paths that may be taken from $Gr^2$'s current state.

## 9.2 Short description

Basic operations on objects are *create*, *copy*, *drag* and *delete*. In a typical session, in order to construct a graph, the user first selects nodes from supply nodes from supply stacks that are positioned along the left- and right-hand sides of the working area (see Figure 9.1), and drags them to their desired positions. Next, they are linked to other nodes by means of arcs. To create an arc $(x,y)$, one of the two nodes $x$ and $y$ must be selected, the "create arc" mouse button must be pressed and held down while the mouse cursor is moved into the selection area of the other node, and the button must be released. During arc creation, a dashed arrow or line from the first selected node to the mouse cursor is displayed, indicating that the arc does not yet exist. Of course, while nodes are dragged over the working area, all their incoming and outgoing arcs follow suit. Apart from being created, dragged, copied and deleted, nodes and arcs can also be given another label, their type can be changed, or information on their properties can be retrieved. Arcs can be redirected.

All operations are either directly available under mouse buttons, or listed in object-specific context-dependent pop-up menus. Operations like *undo*, *redo*, *delete all*, *search node*, *print*, *load*, *save* and *set*

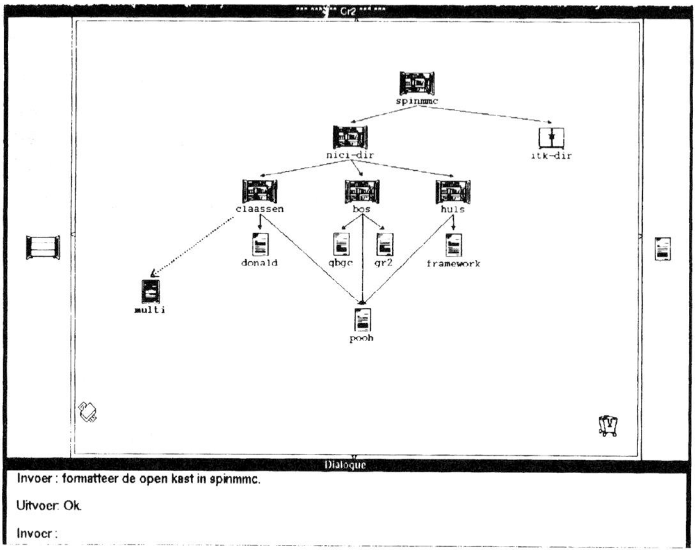


**Figure 9.1** A snapshot taken during the construction of a small graph. Supply stacks of nodes are positioned along the sides of the working area. Nodes are represented by labeled icons. Arcs are represented by arrows. In the Dialog window on the lower part of the screen, the user has entered "format the open closet in spinmmc".

*interface parameters* are provided in the main menu. Operations can be performed on single objects as well as on a set of objects of the same class. To select several objects simultaneously, either the multiple selection key can be used or the objects can be enclosed in a rectangle drawn by means of the mouse. In Figure 9.1, a snapshot is depicted during the construction of a graph.

To navigate through a large graph, the user can move his viewpoint by clicking in the scroll area (possibly while dragging an object), by selecting a *scroll* option, or by moving an icon representing the viewpoint in an overview window. This overview window is opened upon selection of the *navigate* option in the main menu. A welcome feature is the *search* option for delegation of search actions in the model world, for example, to find the node named *pooh*.

All operations can also be described in formal or natural language (e.g. "copy this node" ). This is particularly useful if multiple objects are to be manipulated (e.g. "delete all arcs" ), especially if some of them are not present in the current viewpoint.

## 9.3 $Gr^2$ and structure-based editing

There are two types of knowledge that make $Gr^2$ a structure-based editor. There is knowledge about graphs and there is knowledge about rules that apply in a particular task domain. $Gr^2$'s exploitation of knowledge about graphs is described in Section 9.3.1. I include a brief section on the three domains to which $Gr^2$ currently is applied (Section 9.3.2) in order to illustrate the exploitation of domain dependent knowledge (Section 9.3.3).

### 9.3.1 Exploitation of knowledge about graphs

$Gr^2$ has knowledge about graphs. One (trivial) example of a rule from graph theory applied by $Gr^2$ has already been given in Section 9.2: arcs obligatorily have a start and end node and therefore incoming and outgoing arcs follow nodes if these are being dragged. Analogously, if a node is deleted, all its incoming and outgoing arcs are deleted as well. Other basic rules applied are: the end node of an arc may not be identical to the start node, nor may there be any other arc linking this start node to this end node.[2] If the user manipulates the external objects in such a way that the resulting internal structure would violate these rules, $Gr^2$ does not allow successful completion of this manipulation. Upon entering the selection area of the particular node during arc creation, this node will not be accepted as the end or start node of the arc to be created. $Gr^2$ thus prohibits the user from breaking the rules of graph theory.

A special type of subgraph has been named *successor graph*. A successor graph is defined as a compound object: it consists of several simple objects, both nodes and arcs. A successor graph has a root node; in addition, all the root node's successors, its successors' successors, etc. are part of the successor graph. All arcs that link these nodes belong to the successor graph as well. Analogously, a predecessor graph grouping could be defined. It depends on the domain which of the two comes in most handy.

The use of successor graphs facilitates the construction process. They don't need to be defined by the user but are automatically

[2] Such so-called loops and multiple joins would transform the graph into a multigraph, which currently cannot be handled by $Gr^2$. However, if in a specific domain loops or joins are needed, $Gr^2$ can easily be extended to handle them.

available to him, as soon as he adds successors to a node. Apart from basic operations like dragging, copying and deleting, three additional operations can be applied to successor graphs which do not alter the internal structure of the graph. Instead they just alter the graphical representation on the screen. These operations are closing, opening and formatting (of which the last is described in the section on domain dependent aspects of $Gr^2$ (Section 9.3.3)).

A successor graph can be "closed": all the successors and arcs of the successor graph are hidden in a closed successor graph icon and only the root node of the successor graph is displayed (see Figure 9.2). Usually the domain provides a feasible metaphor for the closed successor graph icon (e.g. closed closets for directories), but if it does not, a triangle is used or, if in the successor graph a circuit can be identified, a circle. By closing a successor graph, the user can abstract from details that momentarily are of no importance and that would just clutter up the screen. Of course, the user can open a closed successor graph whenever he wishes to inspect the hidden details.[3]

### 9.3.2 A brief description of applications of $Gr^2$

Before describing $Gr^2$'s exploitation of task domain knowledge, I will first briefly describe the three domains $Gr^2$ currently is applied to. $Gr^2$ is used for managing the knowledge base of a natural language dialog system, for managing files in a file system, and for formulating queries in an experimental graphical query language. Here I will mainly focus on the third domain.

In the first domain, $Gr^2$ is a tool for the designers of the natural language dialog system DoNaLD (Claassen and Huls, 1991). In order to inspect and adapt the contents of the knowledge base at some stage

[3] Note that a successor graph containing nodes that have predecessors not belonging to that successor graph (e.g. in Figure 9.2, the node <creature> in the successor graph of <concrete-entity> has a predecessor <cognizer> that is not in the successor graph of <concrete entity>) currently cannot be closed. This is to prevent representation problems of arcs that link nodes in a closed successor graph with "the outside world". A possible solution to this representation problem may lie in using lines that end in the center of the closed successor graph icon (e.g. in the closed closet). Conklin and Begeman (1988) suggest, apart from the solution mentioned above, a heavily domain-dependent solution that "upgrades" arc types and changes the problematic arcs into arcs to the visible root of the successor graph.

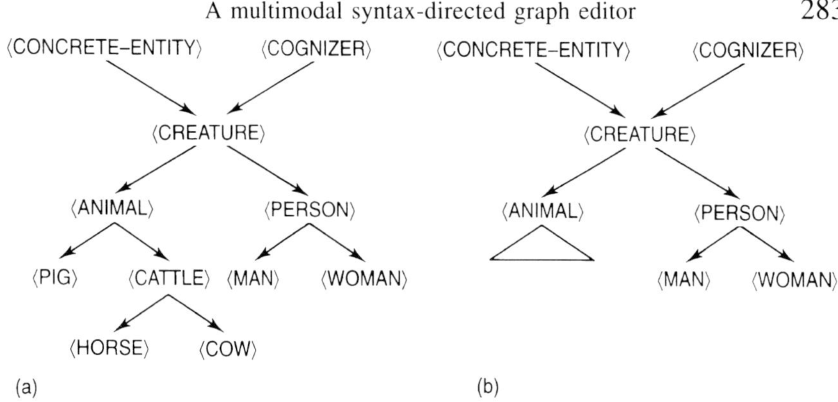


**Figure 9.2** A successor graph with root node labeled <ANIMAL> (a) is closed (b). All details are hidden in a triangle.

of the dialog, the developers can use $Gr^2$. They can add and edit generic and specific information graphically to test the dialog manager and the generation modules of the dialog system. Nodes and arcs are created to link classes of entities and relations with their subclasses and superclasses (e.g. <man> with <person>), and to link specific instances with each other via case role nodes (e.g. *town#4* with *live-in#14* via "place" ).

The second domain, management of files, was proposed by Desain (1988) as an interesting domain for his tree editor named TreeDoctor. Nodes represent files and directories; arcs show the hierarchy. To move files, arcs are redirected.

The third domain is query languages. $Gr^2$ is used in QBGC (Query By Graph Construction), an experimental graphical query language in which a user can express queries (Bos, 1990a). QBGC elaborates on the ideas implemented by McDonald (1975) in CUPID. By constructing a graph by means of $Gr^2$, the user can query a database. QBGC is designed for users who rarely query a database and who do not have a full conceptual model of the contents of the database. By graphically displaying the relations stored in the relational database, the naive user can be aided in formulating queries. He or she can formulate queries by constructing directed graphs that contain nodes of different types linked to each other by arcs. In Figure 9.3 an example of a QBGC graph is given. This graph corresponds to the natural language utterance "What are the names and addresses of all adult female patients with blood type A?"

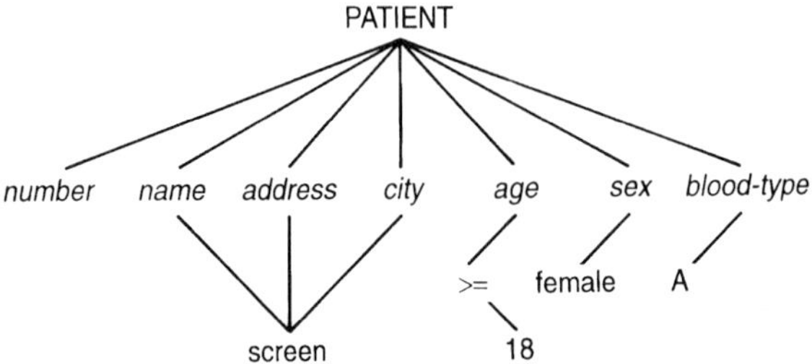


**Figure 9.3** A QBGC graph representing the query "What are the names and addresses of all adult female patients with blood type A?".

As said before, there are several types of nodes in this domain. There are nodes of type relation (for instance the node with label "PATIENT"), attribute (e.g. "address" and "blood-type"), constant (e.g. "A" and "female"), print ("screen"), comparison operator ("$\geqslant$") and set operator. These node types are the word categories of the graphical language. Particular graphs, then, are sentences in the language. The syntax of the language can be described by rules that apply to the node and arc types.

Examples of grammar rules that apply in QBGC are:

1. The graph must have at least one node of type relation an one node of type print.
2. Relation nodes must have a successor of type attribute.
3. Attribute nodes have successors which must be of type print (just one), constant or comparison operator.
4. Print nodes have no successor and have predecessors of type attribute.
5. The graph may not contain any circuits.
6. The graph must be connected.

If the user has finished his query, the graph is translated into an SQL query which can be handled by a regular relational data base management system.

QBGC does not seem to fully satisfy the goal set for it (directly usable by naive users, without a learning period). The graphical paradigm indeed seems to aid the user in making a model of which concepts can be queried, yet suffers from inherent problems when dealing with complex queries (e.g. queries with multiple logical operators). Whereas other independently developed look-alikes (see, e.g. Rohr 1988; Gyssens *et al.*, 1990; Burgess, 1991) rigidly stick to the graphical paradigm and thus suffer from either fairly restricted expressive power or usability, QBGC will in the future be enriched with linguistic components. I think the only way to provide a query language that is not only with maximal usability for naive users, but also with full expressive power is in combining the virtues of graphics and language.

### 9.3.3 Exploitation of domain knowledge

One of the factors that contributes to the surplus value of $Gr^2$ over drawing packages is the ability to exploit knowledge about the particular task domain. Two types of domain dependent knowledge are incorporated into $Gr^2$: knowledge about the external objects, in particular the representation of graphs, and knowledge about the internal objects, in particular the grammar of the graphical language.

*Exploitation of domain knowledge about external objects*

Knowledge about the representation of graphs concerns the domain dependent representation of nodes and arcs. For instance, arrows are not used for arc representation in QBGC because their heads tend to clutter up the screen easily. Therefore, the convention of lines leaving from the bottom of the start node and arriving at the top of the end node is adopted. This convention is adopted from linguistics, where end nodes in parse trees are usually positioned below the start nodes as well. In other domains, the convention of having fixed arriving and departure points for arrows or lines might be adjusted, and, for instance, double headed arrows might be used to represent pairs of arcs $(x,y)$ and $(y,x)$. Likewise for nodes, specific labels or icons might be used to represent them.

Knowledge about the representation of graphs incorporated in $Gr^2$ also concerns the layout of the graph. A format operation is provided for the automatic improvement of the layout of a graph or of a selected successor graph. It improves the comprehensibility of the graph. An additional advantage of using formatted graphs is standardization: other people working in the same domain can pick up information easier. The precise form of the layout is, of course, heavily dependent on conventions used by practitioners in the domain. In some domains, successors are presented below their predecessors, in others to the right of them, in yet others above them. Watanabe (1989) proposed a heuristic algorithm for general layout calculation. When applied to QBGC, however, such an algorithm will not generate the easiest comprehensible layout. A tree-like representation that is used in linguistics is more suitable for QBGC. Therefore, I adjusted the algorithm for tree-layout used in Desain's (1988) TreeDoctor which was based on node label width (see Bos (1990b) for details). The results of the format operation applied to a quickly constructed graph (Figure 9.4(a)) are depicted in Figure 9.4(b).

To make clear what the format operation actually does, animation is necessary. If the reflection of the changed positions would be done by erasing all the elements of the graph at their old locations and, next, drawing them at their new ones, the user would easily get confused. He will have problems determining which new positions correspond to the old ones, or he might even think that the structure itself has been

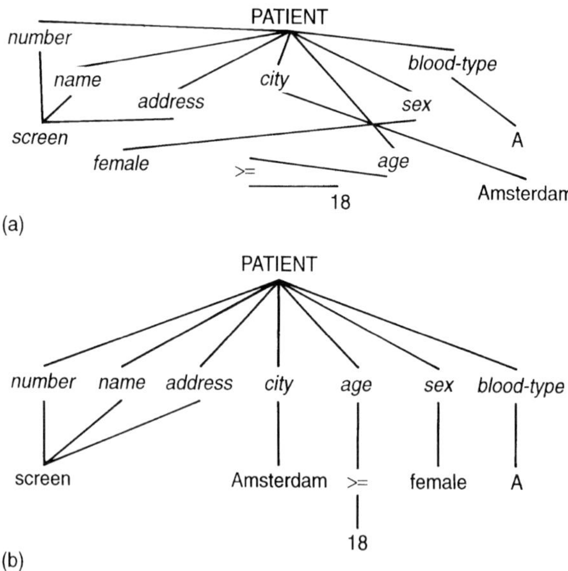


**Figure 9.4** Automatic improvement of the layout of a QBGC graph. Before (a) and after (b) the format operation. Animation is used to have the first representation gradually transform into the second.

altered, that is, that nodes or arcs have been deleted or added. Animation is likely to prevent this, provided that the pace used is neither too slow nor too fast. $Gr^2$ currently uses a top-down, breadth first animation in which nodes are moved to their new position one by one, resulting in a spider-like movement of the graph. Ideally, however, the basic principles of animation named *slow-in-and-slow-out* and *blur* (Ungar, 1992) would be implemented. That is, by drawing the firsts and lasts states of the motion accurately and in normal pace, but using vague *clouds* (known from comic-strips) in intermediate states, it, firstly, will be much clearer to the user which objects move where, and, secondly, will require far less time to perform the animation.

One flaw of the current algorithm is that the formatted graphical representations of the selected successor graph might partially overlap the representation of another object. This substantially eases computation but might partially undo the powerful positive effect of the format operations on comprehensibility.

### *Exploitation of domain knowledge about internal objects*

The second type of domain dependent knowledge concerns the internal objects. The grammar of a particular graphical language can be incorporated into $Gr^2$ to enable the system to prevent the user from constructing syntactically incorrect graphs.

Firstly, there are rules that deal with the links allowed between nodes of certain types. For instance, nodes of the same type are not allowed to be linked in some domains. In other domains such links are the only links allowed, and in yet other domains, nodes of type X are not allowed to be linked to nodes of type Y by arcs of type Z. For example, suppose $Gr^2$ is applied to the QBGC domain and the user selects node "screen" of type print, starts creating an arc from it and enters the selection of node "PATIENT" of type relation. $Gr^2$ will then prohibit node "PATIENT" from being linked to node "screen", because this operation would violate the rule which says that nodes of type relation can only have successors of type attribute (example rule 2 of Section 9.3.2). The node "PATIENT" will not be highlighted and if the user releases the mouse button while in its selection area, no arc is created (see Figure 9.5(a)). Thus, by not allowing successful completion of the manipulation, the system prevents the user from breaking the rule. If the user has opted for the explanation

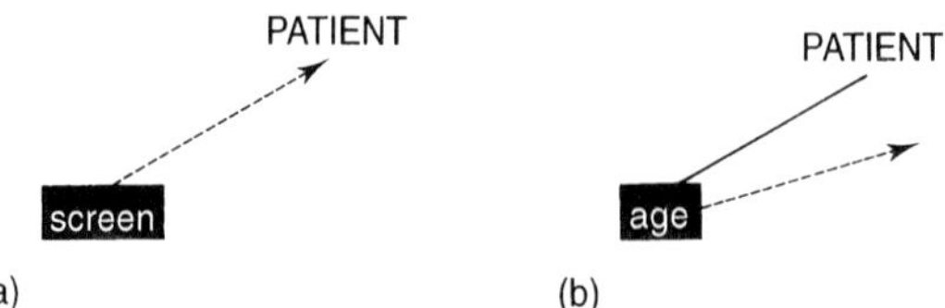


**Figure 9.5** Preventing the construction of syntactically incorrect graphs. (a) The target node during arc creation (i.e. the node labeled "PATIENT" ) is of the wrong type, and thus node "PATIENT" is not highlighted, even though the mouse cursor is in its selection area. The arc ( "PATIENT", "screen" ) is not created. (b) The node from which arc creation is started ( "age" ) has reached its maximum number of predecessors. Therefore, it can only be given a new successor and, consequently, the dashed line representing the arc to be created is drawn from the bottom of "age".

option, $Gr^2$ reports the rule being violated, in terms understandable to the user, for example, "Top-level things cannot be connected to 'screen'".[4] This way the user immediately knows what is wrong. He does not have to worry, for instance, whether he positioned the mouse all right. Furthermore, he will learn the rules faster than when $Gr^2$ would allow him to create ill-formed graphs.

Secondly, other domain dependent rules determine the maximum number of predecessors or successors of a node. If, for instance, attribute node "age" has reached its maximum number of predecessors (viz. 1), the creation of arcs from "age" will be influenced. The representation of the arc to be created will no longer flip around the node if the mouse cursor is circling around it, as it used to do because of the criterion of relative node position (cf. Section 9.4). For now, the arc to be created can only have node "age" as start node and therefore a dashed line is drawn from the bottom of "age" to the mouse cursor, irrespective of the mouse cursor's position (see Figure 9.5(b)). Likewise, if there are no other nodes that can be linked to a particular node, or if that particular node has reached its full complement of predecessors and successors, the *create arc* operation cannot even be started. Again, the reasons for preventing these operations can be reported. For example, "Attribute nodes can have maximally one predecessor".

Thirdly, apart from prohibition rules, there also may be obligation rules in the grammar of the graphical language. For instance, the final

[4] The use of color might come in handy here. It allows for straightforward references to node types (e.g. "Blue nodes can only be connected to red ones" ).

structure has to be a connected graph in QBGC. Violations against obligatory grammar rules cannot be prevented; they can only be detected upon the user's indication of having finished the construction process. Thus, what can be prevented is "handing over" a graph in which obligatory rules are violated. In QBGC such violations are detected as soon as the user indicates he has finished formulating his query by selecting the "find answer" option in the main menu. $Gr^2$ then informs the user which rule has been violated, possibly using a multimodal expression like "The graph is not connected. Connect or remove this print node", with a simultaneous simulated pointing gesture to the particular node.

Fourthly, there are rules that specify the way in which arcs are ordered, thus possibly affecting the semantics of the graph. For instance, in QBGC, the order of incoming arcs of a comparison operator determines which of the two selection criteria *weight* < *length* and *length* < *weight* is meant. The order is clockwise, or to put it more formally, the order is determined by the angle that arc representations make with the upper vertical half line through a node.

## 9.4 User evaluation

Although $Gr^2$ has not yet been subjected to user experiments, I will give in this section some conclusions that can be drawn from experiments done with $Gr^2$'s forerunners and with the multimodal system EDWARD, in which $Gr^2$ is embedded. Desain (1988) designed and implemented a tree editor named TreeDoctor. Hensgens (1989) conducted a series of experiments with TreeDoctor and with four slightly differing prototypes of another tree editor (Bos and Hensgens, 1989), which were derived from TreeDoctor and ran on an IBM PC/AT. Hensgens analyzed and evaluated several individual components of each system with respect to their usability. He made a GOMS analysis (Card *et al.*, 1983) supplemented by an analysis using Norman's theory of action (Norman, 1986). The results of the analysis and evaluation have been used in the design of $Gr^2$.

Hensgens finds users of $Gr^2$'s forerunners appreciating both the functionality and the user interface. Users enjoy working with these structure-based editors. During the short learning period (about half an hour) some users could not resist exploring the system without having completely read the user manual. A significant number of the possibilities offered by the systems were discovered during this independent exploration.

In Hensgens' experiments, the most important design issue for any tree editor or graph editor was studied, viz. arc creation. In a graph editor, linking nodes is the most important action and thus the creation of arcs must be easy and fast. Therefore, the use of a special mouse button is preferred over having the user setting a mode, as is done in drawing packages. The intriguing question is: what is the most convenient way for the user to indicate the direction of the arc to be created? There are at least three alternatives:

1. *The "mouse cursor position within node" criterion.* The direction of the arc to be created is determined at the instant the user pushes the "create arc" button. If the user pushes the button while the mouse cursor is at the upper side of the selected node $x$, then $x$ will be the end node of the arc to be created, else it will be the start node.
2. *The "relative node positions" criterion.* The direction of the arc to be created is determined by the relative positions of the particular nodes. If the user releases the "create arc" button while in the selection area of node $y$, then the relative position of node $y$ with respect to node $x$ determines the direction of the arc linking $y$ with the selected node $x$. If $y$ is above $x$, the arc $(y,x)$ is created, else $(x,y)$.
3. *The "argument order" criterion.* The direction of the arc to be created is determined by the order of selection of the particular nodes. The first selected node will always be the start node[5] of the arc that is going to be created.

Hensgens found, in accordance with our expectations, that the second option (the "relative node positions" criterion) provided the highest usability in terms of number of errors, amount of required attention and user satisfaction. A more general conclusion that can be drawn from Hensgens' analysis and experiments with respect to the interface of $Gr^2$ is that arc creation must support the way arcs are drawn in pencil drawings of graphs on paper in the particular domain. For example, in trees and tree-like graphs such as the QBGC graphs, arcs are drawn downwards; in many process flow diagrams, arcs are drawn downwards at the left-hand side of the graphical representation

[5] By default; a global variable controls whether the first selected is the start or end node.

of a graph and upwards at the right-hand side. Additional supprt for this domain dependent approach can be found in studies on pencil drawings: Sommers (1984) concludes that arrows are not always drawn in the same way because semantic aspects affect the strategy of their production. If, however, in a particular domain no arc drawing convention is available, the criterion of argument order can best be applied. This seems especially the case if arrows instead of lines are used to represent arcs: people start drawing at the start node and finish drawing the arrow head at the end node.

The experiments also provided some indications with respect to the usefulness of having the system prevent violation of rules. The system prevented many violation errors, and in the interview following the experiments all users expressed their appreciation of this supervisor behavior of the system. Based on the impression that initially more violation attempts were made than at the end of a session, it is concluded that this supervisor behavior probably facilitates learning the rules of the domain. A comparative study should verify this.

Further results of the experiments conducted by Hensgens involved indications for the usefulness of the concept successor graph. The concept aims at facilitating working with larger structures. In a tree construction task, initially the concept wasn't used at all. But as the number and size of available successor graphs grew, and the users got more experienced, successor graphs were indeed selected and manipulated successfully. One experiment, involving a large tree of 125 nodes representing classes and sorts of animals, showed how crucial system speed is for usability. In three of the four prototypes, dragging a very large successor graph was annoying and error-prone, since the system could not keep up with the user's mouse movements. The fourth prototype displayed only the contours of the graphical representation of the successor graph being dragged. Despite the fact that it provided less information (the successor arcs and nodes were not displayed), it showed a higher usability.

A pilot study run with EDWARD (Huls, *et al.*, 1993) provided some information about multimodality. Ten novices, unfamiliar with EDWARD, were given several tasks they had to perform without any learning period (i.e. no manual, no training session, not even a demonstration how things work). It turned out that however easy-to-use the concept of arc redirect was once it was known to them, most subjects simply did not even try to select an arc. It did not come to their minds. In order to perform tasks involving arc redirect successfully, they switched from the action mode to the language mode, entering commands (e.g. "move the copy to new-dir" ). This shows the necessity of the presence of the language mode in interfaces for novice users. In a future experiment with more experienced users, we hope to show the usefulness of the language mode for these kinds of users as well.

## 9.5 Related systems

Many graph editors have been designed. Most of them have been designed especially for the management of a particular knowledge base, for example, KNAPS (Lee, 1989), SB-Graph (Kalmes, 1990), GROW (Kindermann and Quantz, 1988), KREME (Abrett and Burnstein, 1987), SMGC (Moulin and Kabbaj, 1990), gIBIS (Conklin and Begeman, 1988) and SemNet (Fairchild, 1985). The main difference between these systems and $Gr^2$ is their dependency on a domain. $Gr^2$ is a domain-independent system while the other ones mentioned are not. Domain independence allows reusability of the design and implementation of the system when applied to another domain. If $Gr^2$ is applied to a new domain X, only the domain dependent modules have to be revised. The main task is enabling $Gr^2$ to apply the rules of X. The remaining tasks deal with the graphical representation of the objects in X.

Other structure-based editors aiming at domain independence are described by Szwillus (1989), Göttler (1986, 1992) and El-Kassas (1991). An important distinction between these systems and $Gr^2$ is the way they deal with the syntax of a particular graphical language. $Gr^2$ is a rule-embedded system: it does not have a parser generator at its disposal in which the grammar can be fed. The parser has to be written by the designer/implementer of $Gr^2$'s application to the particular domain. The research prototypes that Göttler, El-Kassas and Szwillus describe can be seen as metatools: tools that support the development of tools that support the construction and editing process of graphs. They are able to read a graph grammar and generate a structure-based editor for the corresponding graphical language.

Comparing automatically generated graph-editors with $Gr^2$, I focus on two distinct points:

1. Which is the best (i.e. fastest, least error-prone, etc.) way to create a structure-based graph editor for a graphical language X?
2. Which way yields the best structure-based graph editor?

Because actual tests go beyond the scope of my current research project, I have to confine myself here to a short theoretical analysis. With respect to the first question, I expect $Gr^2$ to lose, because the rules have to be encoded in Lisp instead of in a language especially designed for writing graphical grammar rules in. With respect to the second question, I expect $Gr^2$ to win, because of its flexibility. $Gr^2$ offers its users a world in which they can create graphs in a bottom-up method. All automatically generated graph editors I know of are either top-down (e.g. the derivation engine which El-Kassas describes), or do not allow their users to interact in a direct way (e.g. the PAGG system of Göttler, in which users have to select options like "insert next cupboard" from a given pop-up menu; this creates a new object and automatically places it next to the last created object). Furthermore, feedback on why actions are illegal is given much more attention in $Gr^2$, or even more generally, the entire user interface has been more in focus than in the metatools. In the long run, the two should converge, that is, $Gr^2$ should be equipped with an easy-to-use parser generator, and the metatools should be provided with a fully-fledged interface.

## 9.6 Discussion of structure-based editing

As mentioned in the introduction, structure-based editors have, until now, failed to gain broad acceptance, despite the fact that they are purported to enlarge usability. Minör (1992) lists and sharply analyses several objections raised in the past against structure-based editing of computer programs, particularly those of Waters (1982). In this section I briefly discuss the objections that seem relevant to structure-based editing in general.[6]

1. "*Top down construction constrains the user*". Almost all structure-based editors, including those of Göttler, El-Kassas and Szwillus

[6] In computer program editing, the issue of representation is controversial: text representation vs. syntactic trees. This issue is less relevant in many other domains, including the domains $Gr^2$ is applied to.

described in the previous section, impose a strict top-down construction strategy upon the user. The typical interaction pattern is (1) select a node in the hierarchical structure constructed so far; (2) give an expand command; (3) select an alternative from the menu of appropriate expansions. This kind of interaction indeed constrains the user. However, such constraints are not inherent to the concept of structure-based editing. $Gr^2$, for example, allows for bottom-up construction. Moreover, Minör proposes the use of an additional, unconstrained workspace (the "clipboard" ) for temporary storage of fragments, in combination with a constrained workspace where only legal structures can exist, and traditional expansion-by-menu facilities, to support both bottom-up and top-down construction techniques.

2. "*Enforced consistency limits the user's freedom*". In many domains, experienced users often "express a desire to fiddle around with the textual [read: non-structured] representation and tolerate inconsistent states while the structures is modified from one consistent form to another. They do not want to be scrutinized; they know what they are doing" (Minör, 1992, p. 403). This desire seems to come from, again, bad interface design. It is not an objection against structure-based editing itself. "There is actually no reason for *wanting* an inconsistent state at any time. What we really want is an editor that allows programs [read: structures] to be manipulated conveniently with maintained consistency. Whether this is a realistic goal, or if users actually cannot avoid inconsistent states in their work constructing a consistent program, is, however, still an open question" (p. 403).
3. "*Experts do not need guidance*". Once more, this objection stems from bad interface design. Even if it would be true that experienced users never need guidance (which can be challenged if they enter areas near the border of the scope of their expertise, or if they suffer from lapses in concentration or memory retrieval), guidance need not be as importunate as in traditional structure-based editors. By highlighting nodes, $Gr^2$ might, for example, indicate appropriate successor or predecessor candidates. Hence, the guiding information can easily be discarded by experts, but is available to novices (and, of course, also to experts with a knowledge black-out).

## 9.7 Conclusions and future research

In this chapter, I have shown that a combination of the action mode and language mode of interaction can provide a suitable interface for structure-based editors. Structures, regardless whether they are flowcharts, computer programs, document structures, or anything else, can easily be represented in a graphical model-world, in which their representations can be manipulated. A core action in editing structures is moving structures, and the action mode allows for easy-to-understand ways to move objects. The action mode by definition means mimicking, and mimicking implies that the form of an expression is mapped onto its meaning by "a nonarbitrary relationship" (Hutchins *et al.*, 1986, p. 111). The language mode comes in handy if, for example, elements have to be collected from out-of-the-way places of the model world, far beyond the current veiwpoint, or if the way actions are to be performed is unknown.

In the future, the design loop for improving the user interface and functionality of $Gr^2$ must be continued. Implementations of recommendations obtained in the last evaluation study (Hensgens, 1989) must be evaluated and other new features can be designed and implemented. Some possible extensions and research topics are: dynamic grid for nodes (Desain, 1988); alignment of selected nodes; cut and paste actions to a visible clipboard; recursive zoom-in; context-sensitive help on all issues concerning the interface language, the functionality and the domain; macro-editing for dynamical chunking of operations by demonstration; parameter adjustment by demonstration; an icon-editor for on-line specification of node and arc icons; on-line setting of event-command bindings; use of color, as successfully applied in $Gr^2$'s forerunner (Bos and Hensgens, 1989; Hensgens, 1989); formatting graphs in general (see e.g. Watanabe, 1989); navigation in large graphs (see e.g. Fairchild *et al.*, 1988); isolating embedded rules: development of a parser generator (cf. Section 9.5).

From the many research paths that may be taken from here, I have chosen the one that aims at full integration of the action and language of interaction. Integral compensation of the disadvantages of the action mode, and full exploitation of the advantages of the language mode, in particular of natural language, have already yielded some promising initial results.

## Acknowledgments

I wish to thank Peter Desain, Jan Hensgens, Wim Claassen and Carla Huls for their contributions to the realization of $Gr^2$. I thank Tony Jameson, Gerard Kempen, Alice Dijkstra and Koenraad De Smedt for their comments on earlier versions of this chapter.

## References

Abrett, G. and Burnstein, M. H. (1987). The KREME knowledge editing environment. *International Journal of Man–Machine Studies*, **27**: 103–126.

Bos, E. (1990a). Query by graph construction. SPIN-MMC Research report #3. Nijmegen: NICI. (In Dutch.)

Bos, E. (1990b). $Gr^2$: a graphical graph-editor. SPIN-MMC Research report #8. Nijmegen: NICI.

Bos, E. (1992). Some virtues and limitations of action inferring interfaces. In *Proceedings of the Fifth Annual Symposium on User Interface Software and Technology*, Monterey, CA 15–18 November. ACM Press, New York.

Bos, E. and Hensgens, J. (1989). Directe manipulatie van ontleedbomen. (Direct manipulation of syntactic trees). In Pijls, F. and Sandberg, J. (editors) *De computer als expert en didacticus.* Coutinho, Muiderberg. (In Dutch.)

Burgess, C. G. (1991). A graphical, database-querying interface for casual, naive computer users. *International Journal of Man–Machine Studies*, **34**: 23–47.

Card, S. K., Moran, T. P. and Newell, A. (1983). *The Psychology of Human–Computer Interaction.* Lawrence Erlbaum Associates, Hillsdale, NJ.

Claassen, W. (1992). Generating referring expressions in a multimodal environment. In Dale, R., Hovy, E., Rösner, D. and Stock, O. (editors), *Aspects of Automated Natural Language Generation*, Proceedings of the Sixth International Workshop on Natural Language Generation, Trento, Italy, 5–7 April, Springer, Berlin. 247–262.

Claassen, W. and Huls, C. (1991). DoNaLD: Dutch Natural Language Dialogue system. SPIN-MMC Research report #11. Nijmegen: NICI.

Claassen, W., Bos, E., Huls, C. and De Smedt, K. (1993). Commenting on Action: A Continuous Linguistic Feedback Generator. In *Proceedings of the International Workshop on Intelligent User Interfaces*, Orlando, FL, 4–7 January. ACM Press, New York.

Conklin, J. and Begeman, M. C. (1988). gIBIS: A hypertext tool for exploratory policy discussion. *ACM Transactions on Office Information Systems*, **6**, 303–331.

Desain, P. (1988). TreeDoctor: A software package for graphical manipulation and animation of tree structures. In: Mulder, G. and van der Veer, G. (editors) *Human–computer Interaction: Psychonomic Aspects.* Springer, Berlin.

Donzeau-Gouge, V., Huet, G., Kahn, G. and Lang, B. (1980). Programming environments based on structure editors: the MENTOR experience. In: Barstow, P. R. (editor) *Interactive Programming Environments.* New York: McGraw-Hill, pp. 128–140.

El-Kassas, S. (1991). Visual languages, their definition and applications in system development. In Nuñez, A. (editor) Euromicro '91: Hardware and software design automation. *Seventeenth Symposium on Microprogramming*, Vienna, September.

Fairchild, K. M. (1985). Construction of a semantic net virtual world metaphor. *Technical Report No. HI–163–85.* Micro Electronics and Computer Technology Corporation, Austin TX.

Fairchild, K. M., Poltrock, S. E. and Furnas, G. W. (1988). SemNet: three-dimensional graphic representations of large knowledge bases. In: Guindon, R. (editor) *Cognitive Science and its Applications for Human-computer Interaction.* Lawrence Erlbaum Associaties, Hillsdale, NJ.

Göttler, H. (1986). Graph grammars and diagram editing. In: Ehrig, H., Nagl, M., Rozenberg, G. and Rosenfeld, A. (editors) Graph-grammars and their application to computer science. *Third International Workshop*, Warrenton, VA, December, Springer-Verlag, Berlin.

Göttler, H. (1992). Diagram Editors = Graphs + Attributes + Graph Grammars. *International Journal of Man–Machine Studies*, **37**: 481–502.

Gyssens, M., Paredaens, J. and Van Gucht, D. (1990). A graph-oriented object model for database end-user interfaces, In Garcia-Molina, H. and Jagadish, H. V. (editors) *Proceedings of the 1990 ACM Sigmod International Conference on Management of Data.* SIGMOD Record, **19**:(2). ACM Press, New York. 24–33.

Hensgens, J. (1989). Een theoretisch en empirisch onderzoek naar de gebruikers-interface van een boomeditor. (A theoretical and empirical research into the user interface of a tree-editor). Master Thesis Cognitive Science, University of Nijmegen. (In Dutch.)

Huls, C., Bos, E. and Damen, H. (1993). Fully integrated multimodality: a case study. *Abridged proceedings HCI International '93*, 8–13 August, Orlando, Florida.

Hutchins, E. L. (1989). Metaphors for interface design. In: Taylor, M. M., Neel, F. and Bouwhuis, D. G. (editors) *The Structure of Multimodal Dialogue.* Elsevier Science Publishers B.V., Amsterdam, North-Holland.

Hutchins, E. L., Hollan, J. D. and Norman, D. A. (1986). Direct manipulation interfaces. In: Norman, D. A. and Draper, S. W. (editors) *User Centered*

*System Design: New Perspectives on Human Computer Design*. Lawrence Erlbaum Associates, Hillsdale, NJ.

Kalmes, J. (1990). SB-Graph. Eine graphische Benutzerschnittstelle für die Wissensrepräsentationswerkbank SB-ONE. (A graphical user interface for the knowledge representation workbench SB-ONE). SFB 314 (XTRA), Memo Nr. 44. Universität des Saarlandes. (In German.)

Kindermann, C. and Quantz, J. (1988). GROW: Graphik-orientierte Wissensrepräsentation für KL-ONE. (GROW: graphic-oriented knowledge representation for KL-ONE). KIT-Report Nr. 63, TU Berlin. (In German.)

Lee, N. S. (1989). Graphical knowledge programming with KNAPS. *International Journal of Man-Machine Studies*, **31**: 611–641.

McDonald, N. H. (1975). CUPID: a graphic oriented facility for support of non-programmer interactions with a database. Memo ERL-M563, Ph.D. Thesis, University of California, Berkeley, CA.

Minör, S. (1992). Interacting with structure-oriented editors. *International Journal of Man–Machine Studies*, **37**: 399–418.

Moulin, B. and Kabbaj, A. (1990). SMGC: A tool for conceptual graphs processing. *CC AI, The Journal for the Integrated Study of Artificial Intelligence, Cognitive Science and Applied Epistemology*, **7**: 23–47.

Norman, D. A. (1986). Cognitive engineering. In: Norman, D. A. and Draper, S. W. (editors) *User Centered System Design: New Perspectives on Human-computer Interaction*. Lawrence Erlbaum Associates, Hillsdale, NJ.

Rohr, G. (1988). Advanced Information Management Prototype. In: Blaser, A. (editor) *Bi-annual Report 1986/87*, IBM Germany Research Scientific Center, 87–116.

Shneiderman, B. (1982). The future of interactive systems and the emergence of direct manipulation. *Behavior and Information Technology*, **1**: 237–256.

Sommers, P. van. (1984). *Drawing and Cognition.* Cambridge University Press, Cambridge.

Szwillus, G. (1989). Editing graphical structures. In: Smith, M. J. and Salvendy, G. (editors) Work with computers: organizational, management, stress and health aspects. In *Proceedings of the HCI International '89*, Boston, MA., Elsevier Science Publishers B.V., Amsterdam, North Holland.

Ungar, D. (1992). Applying the principles of motion pictures in user interface animations. Panel on Animation, UIST'92, 15–18 November, Monterey, CA.

Walker, J. H. (1981). The document editor: a support environment for preparing technical documents. *Proceedings of the ACM SIGPLAN/SIGOA Symposium on Text Manipulation, SIGPLAN Notices*, **16**: 44–50.

Watanabe, H. (1989). Heuristic graph displayer for G-BASE. *International Journal of Man–Machine Studies*, **30**: 287–302.

Waters, R. C. (1982). Program editors should not abandon text oriented commands. *SIGPLAN Notices*, **17**: 39–46.

# Part IV Editors of alternative structures



Parts I and II dealt with structure editors for programming and their evaluation; Part III dealt with a natural generalization towards graphical structure editors. Part IV presents approaches applying the concepts of structure editing to non-programming problem fields: editing of formal software specifications, document structure recognition and construction of graphical user interfaces.

In Chapter 10, MacLean and Brazier's Formaliser tool focuses on supporting formal specification languages such as Z with an appropriate structure editing tool. The project provides a universal structure editor system using an underlying attribute grammar. The grammar rules are central to the construction process, they not only define selection menus, but are explicitly shown to, and can be utilized for selection and creation of nodes by the user. The attribute system is utilized for defining type and scope rules in formal languages. This system favors the structure editor approach against a batch-style because of the importance of small edit–check cycles necessary for successfully applying formal techniques in software-engineering tasks.

In Chapter 11, document structure recognition is the domain of Marovac's structure editor system. The approach is to recognize the organizational structure of a printed, published text through analysis of layout and typographic information. This enables the reuse of published material in electronic libraries of hypertext documents. The working system determines partly automatically and partly interactively the structure of a scanned document or parts of it. It visualizes the structure elements with color decorations on the screen, displays the document structure grammar and the document's derivation with respect to the grammar. In this system the duality of working explicitly with the underlying grammar and with the text structured according to the grammar is very obvious.

In Chapter 12, Tarumi *et al.* describe Canae, a framework for implementing interactive, graphical applications. It is a system for the construction of graphical user interfaces to all kinds of applications; systems like these are commonly referred to as application builders. The system's practical success proves that a large number of user interface problems can be pinned down and solved by applying structure-editing concepts. The basic idea of Canae is to present a collection of "small" editors for basic types of information structures together with powerful features for combining these components. There are editors for text, graphics, bitmaps, networks (graphs), tables and hierarchies. All editors are configurable towards a specific application and can be combined optically and semantically to a high degree. This includes multiple views on one internal structure, as well as highly interdependent nesting of structures and their representations. The author shows that a large proportion of the implementation effort can be spared in a huge variety of applications, when making use of Canae.

These final chapters not only deal with entirely different application fields but also show a great variety in how to make use of structure editing concepts. Chapter 10 stays within the concept of classical structure editing: a structure is defined with an attributed grammar and a universal structure editing device, loaded with the specification, gives the end user access to a structured document. The innovative aspect of this is the support of formal specification techniques as opposed to conventional programming. Chapter 11 has an emphasis on flexible work with the underlying grammar structure of a printed document; the grammar is not fixed, but in contrast can be considered to be the real object under consideration. Chapter 12 makes the point that many incarnations of sophisticated graphical user interfaces can basically be constructed by combining basic editors and structure editors for canonical structures such as tables, graphs and hierarchies. All three chapters show that structure-editing concepts show up in different types of systems with different facets, making it worthwhile – as shown throughout this book – to expose the principles and concepts explicitly.

# CHAPTER 10

# Formaliser – tool support for formal notations

*Roy MacLean and David Brazier*

## Abstract

Logica UK's *Formaliser* tool allows syntax-aware structure editing of documents in formal specification languages such as the Z notation. A direct-interaction WYSIWYG interface is combined with powerful checking of type and scope constraints. This is achieved by using an extended-attribute grammar for the language concerned. The tool is generic, allowing grammars to be created or modified using the tool itself, since grammars are themselves documents expressed using a grammar-definition metalanguage.

## 10.1 Context

Formal notations are being increasingly used for software specification and design, particularly in the development of systems that are safety-critical or security-critical. Their use can provide a greater degree of rigor in the development process than is possible with free-text or diagrammatic representations. There are a number of ways in which use of formal notations can improve the development process:

- Documents in a notation that has precise semantics are a more effective means of definition and communication than documents in a natural language, however "structured" the latter may be.
- The process of constructing a formal document is itself a useful activity, revealing gaps and confusions in the writer's intuitive understanding of a problem.
- There is scope for investigating properties of a system at the level of specification or design.

Structure-based Editors and Environments
ISBN 0-12-681890-8

- Onward development of a formal specification or design (towards executable code) can be more rigorous than when natural language representations are used.

Formal notations for specification and design are similar in some ways to conventional programming languages: they have *syntaxes* definable by context-free grammars, and may also have *type systems* and *scope rules*. In the case of a programming language, correctness with respect to syntax, type and scope is usually checked by a compilation system. Construction of programs by free-text editing, with the possibility of error that this allows, is often acceptable because of this subsequent checking. In the case of a formal notation used for specification or design, however, documents are not input directly to a translation process that performs checking, so it is important that basic correctness is ensured by the construction process. It might seem that a batch-mode tool, analogous to the front-end of a compiler, would be sufficient for this purpose. However, we believe that constructing a specification is in large part an exploratory activity, in which editing and checking need to be closely integrated, and a high level of interaction maintained. A structure editor with built-in checking and query facilities is therefore an appropriate kind of tool support.

Logica's *Formaliser* tool was developed to provide support for the construction of documents in a variety of formal notations. The aims of the tool are:

- to provide structure editing of documents, ensuring syntactic correctness at all times;
- to support a simple, direct style of interaction, and WYSIWYG[1] presentation;
- to integrate editing and checking facilities as closely as possible;
- to allow contextual properties, such as type and scope, to be defined as part of a notation's grammar, rather than being "hard-wired" into the tool;
- to make the tool as generic as possible, supporting multiple notations, and allowing grammars to be added and modified.

A further aim was to investigate the use of object-oriented techniques for the design and implementation of this kind of tool, primarily

[1] What You See Is What You Get.

through use of the Smalltalk programming language. Although these object-oriented techniques are not immediately apparent to a user, they have helped make the tool both highly interactive and generic.

The tool is now being used in both training and development contexts. In particular, we believe that such a tool can be an important aid to the teaching of formal notations, and "technology transfer" generally. At present, there is little knowledge of how people construct specifications, particularly in the very early stages of development when the activity involves creating and expressing an abstraction of some real-world domain. We hope that *Formaliser* will provide a vehicle for investigation of this issue, and can itself be improved in the light of the findings.

At Logica, we are particularly interested in the formal notation Z (Spivey, 1989), and so the primary instantiation of *Formaliser* supports Z. The screen examples in this chapter are of Z documents, but the details of the notation are not important here. It is enough to know that Z allows various kinds of formal paragraph, interspersed with free text paragraphs; formal paragraphs contain declarations and predicates (constraints); these may in turn contain expressions involving identifiers and set-theoretic constructs (sets, tuples, mappings, etc.). The overall structure of a document is a sequence of paragraphs.

The next section looks at the user view of *Formaliser*, and the kinds of interaction involved; various issues concerning structure editing are discussed. Then the grammars that underlie *Formaliser* are described, in particular the attribute systems that they embody, and how they extend the capabilities of the tool. The final section discusses directions in which the tool might evolve: improving document and library structuring, improving navigation around documents, and treating attribute values as structures.

## 10.2 The user view

### 10.2.1 Presentation and navigation

The top level of this system is represented by a library browser which displays the names of all documents currently in the *Formaliser* library. Operations on whole documents (remove, rename, copy, etc.) are available at this level. The library can contain documents in

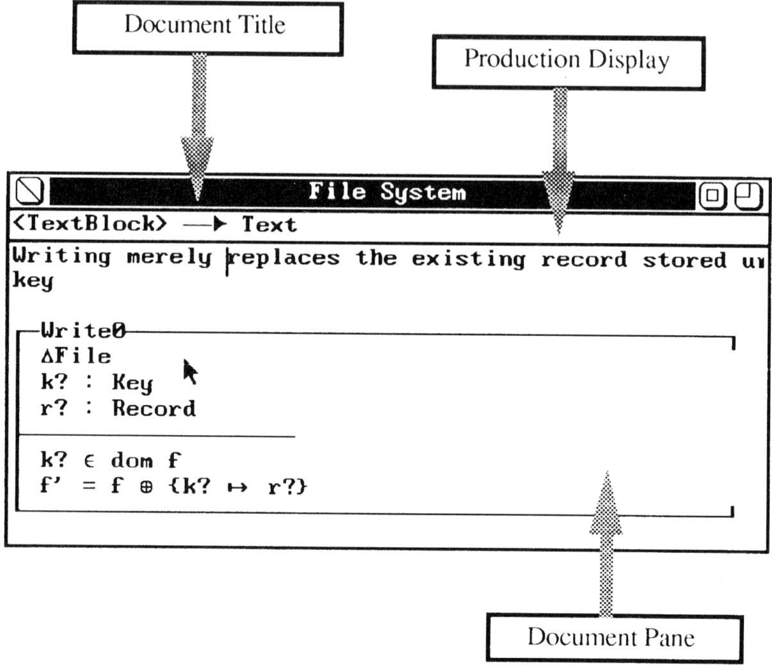


**Figure 10.1** A *Formaliser* document open for editing.

different notations – the appropriate grammar is used whenever a document is opened for editing.

When a document is opened for editing, a scrollable, resizable window containing the document is presented. The user sees a document essentially as it would appear on a printed page; Figure 10.1 shows an example. Within the document pane, the box labeled "Write0" is a formal paragraph; above that is some free text (a TextBlock paragraph). Immediately below the document title is the production display; this shows the production rule for the currently selected node in the parse-tree of the document. Since the text cursor is currently in the free text paragraph, the production display indicates a TextBlock node, which consists simply of text (i.e. a string).[2]

What the user sees of a document is actually a projection of the

[2] In production rules, the arrow separates left and right sides of the production; a vertical bar separates choices.

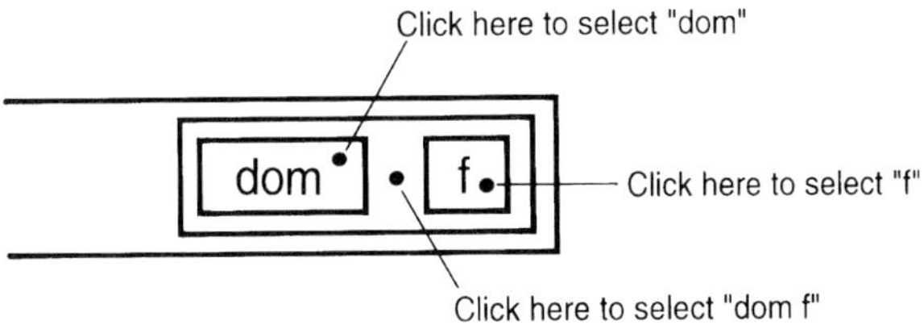


**Figure 10.2** Node selection by mouse clicking.

abstract syntax parse-tree in terms of one particular concrete syntax. Each node in a parse-tree is associated with some area of the screen. Conversely, every point within the document pane is mapped on to some node in the parse-tree; clicking the mouse cursor on a point selects the corresponding node. For example, clicking anywhere on the word "dom" will select the corresponding VarRef (variable reference) node; the area of the document pane corresponding to the parse-tree below this node (the whole of the word "dom") will be highlighted. Alternatively, clicking *between* the words "dom" and "f" will select the node corresponding to "dom f" (an ExpressionSequence).[3]

Because of the hierarchical relationship of nodes in a document, a point on the screen corresponding to a node will fall within the rectangles corresponding to *all* ancestors of the node. Only one node can actually be selected, and it seems sensible to choose the lowest level node within whose rectangle the point lies. However, this means that finding areas of the screen that correspond to an intermediate-level node but do not correspond to any of its sub-nodes can be tricky (e.g. clicking on the gap between "dom" and "f" – see Figure 10.2).

There are two ways round this problem. Firstly, repeated clicking on a node "walks up" the parse-tree. For example, repeated clicking on the word "dom" selects the following nodes in turn, up to the paragraph level:

`dom` ➪ `dom f` ➪ `k?` ∊ `dom f` ➪ . . .

[3] "dom" is a generic function that returns the domain of a relation – in this case of the function "f".

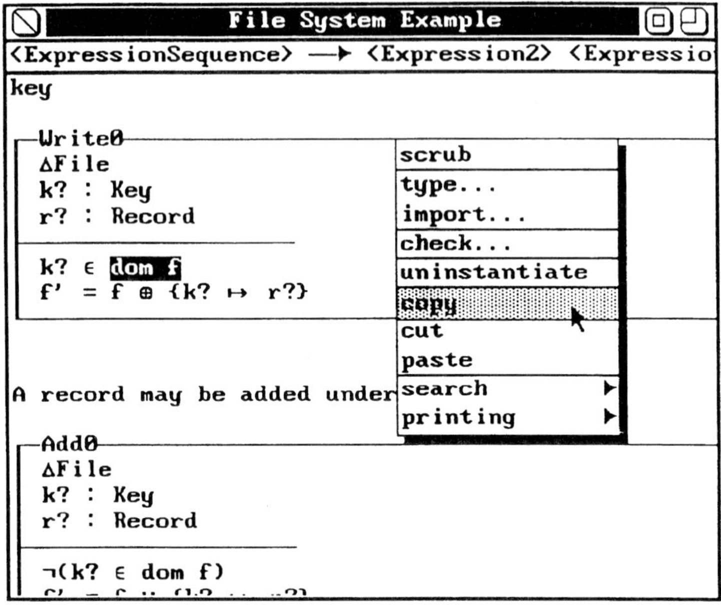


**Figure 10.3** Operations on a selected node.

Secondly, depressing a mouse button and dragging the cursor selects the lowest level node within whose rectangle both ends of the drag lie. This "wiping" technique is useful for selecting higher-level nodes such as whole paragraphs or lists of items.

These are the basic mechanisms of node selection: point-and-click, wiping and repeated clicking; in addition, a scroll-bar can be used to bring a target node into the document pane. Node selection using these mechanisms is more direct and more convenient than selecting explicit navigation operations from a menu (e.g. Up/Down/Left/Right).

The screen example in Figure 10.3 shows the "dom f" node selected. The production display displays the rule:

<ExpressionSequence> → <Expression2> <Expression3>

This indicates that the selected node is an ExpressionSequence, and that its parts are an Expression2 and an Expression3. Expression2 is the node corresponding to "dom"; Expression3 is the node corresponding to "f".[4] Rather than selecting these sub-nodes by clicking on their representation in the document pane, they can be selected by clicking on the appropriate part of the production display. Thus, clicking on <Expression2> in the production display will select the "dom" node.

This is a useful way of selecting sub-nodes, particularly if they are difficult (or impossible) to point to. For example, a SchemaBox – such as that labeled "Write0" in Figure 10.3 – can optionally have a list of formal parameters following its name. This list is usually empty, in which case it has no concrete representation – no area of the screen corresponds to it. If we want to insert an element into the empty list we must first select the node. However, it is clear that we cannot do this by clicking within the document pane. Clicking on the name of the syntactic component within the production display provides an alternative mechanism.

In the next section we will see that the production display can be used effectively when constructing a document, for alternating selections and instantiations.

Having selected a node, a variety of operations can be performed on it, depending on the kind of node and its status. These operations are available in a pop-up node menu obtained by clicking a mouse button (see Figure 10.3).

### 10.2.2 Construction and editing

The document in the screen example above is fully constructed – that is, the parse-tree goes all the way down to terminal nodes that hold literal values (symbols and names). In order to arrive at this state, we need to instantiate non-terminal nodes. Initially, a document consists of just one uninstantiated <Paragraph> node (Figure 10.4).

Since the selected node is uninstantiated, the production display holds a *choice rule* (as opposed to the sequence rule in Figure 10.3); this is read as "a Paragraph can be a TextBlock or an Inclusion or . . ." (the production display scrolls horizontally).

The <Paragraph> node can be instantiated in two ways. Firstly, the operation **instantiate** can be selected from the node menu (Figure 10.3 shows **uninstantiate** because the selected node is already instantiated).

[4] An Expression2 is a simple expression; an Expression3 can be either an ExpressionSequence or an Expression2. This defines left associativity of multiple function applications.

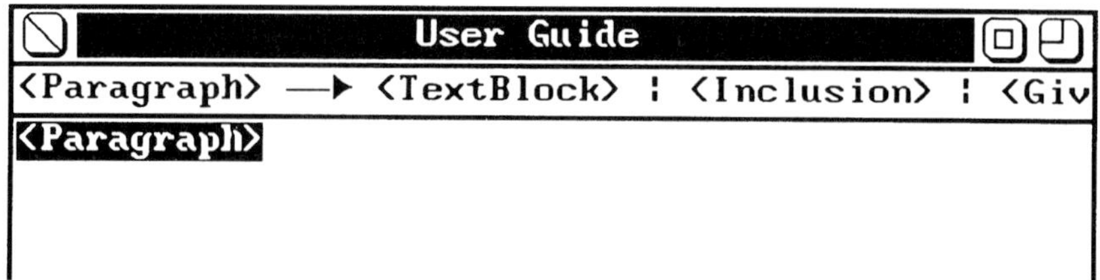


**Figure 10.4** A new document.

A menu containing "TextBlock", "Inclusion", etc. will then be presented, and the user can choose one of the possible instantiations. This mechanism is common in other structure editors. Note that the choice menu is derived from the grammar for the document's notation – it is not hard-wired into the tool.

The second way to instantiate the <Paragraph> node is to click on one of the options in the production display, e.g. "<Inclusion>". This mechanism can be used effectively in conjunction with node selection via the production display (see above): a user can repeatedly select and instantiate sub-nodes without invoking the **instantiate** operation explicitly. This technique is particularly convenient when the instantiation decisions are standard choices, and so do not require inspection of the document text.

Having chosen an option, the node is replaced by a construct consisting of one or more sub-nodes. These sub-nodes are either terminal nodes that require a value from the user, or "choice-rule" nodes that require a decision as to how they are to be instantiated. For example, suppose we had instantiated a paragraph to be an Inclusion, we would then have the construct:

```
Include bindings from ''<DocumentName>''
```

<DocumentName> is a terminal node for which a value is required; the rest of the construct is concrete syntax.

Going back to Figure 10.4, our document consists of a list of paragraphs, and it is clear that we may want more than one. In the node menu (for list nodes) there are operations to **insert** (before), **append** (after) and also **delete**. Typically, we would construct the higher levels of a document by a mixture of list operations and instantiations. A document can be constructed depth-first or breadth-first (or a hybrid strategy), depending on how list operations and instantiations are interleaved.

In addition, **copy**, **cut** and **paste** can be used to perform structural manipulations. For example, if we have a paragraph that has been instantiated through several levels, and we know that other paragraphs will have the same form, then we can **copy** the instantiated paragraph and **paste** it on to other uninstantiated paragraphs as necessary. A paste buffer holds the cut/copied node as a parse-tree, not as a string, so that re-parsing is not required when the contents are pasted. However, the contents of the paste buffer must be a valid instantiation of the target node; if this is not the case, the operation is aborted with an error message.[5]

When instantiation reaches terminal nodes, actual strings are required from the user. With a terminal node selected, a value can be supplied simply by typing the value and hitting the RETURN key. The same mechanism is used to change the value of an already instantiated node. Typed strings are parsed and rejected if incorrect (with the option to edit and resubmit). In the case of terminal symbols, parsing consists of lexical analysis – for example, imposing restrictions on the non-alphabetic characters that can occur in a name. Non-ASCII symbols can be chosen from a separate Symbol Palette, or keyboard equivalents typed (using ALT, CONTROL and Function keys).

The same "direct typing" technique can be applied to non-terminal nodes: a typed string is parsed according to the syntactic class of the node and accepted only if correct. This technique is most useful for single-line statements or expressions (Figure 10.5 shows an example).

In this example, a <BasicDecl> node has been selected, and the user is part-way through typing a declaration such as

```
x : Boolean
```

which will be parsed as a SimpleDeclaration. Note the "I-bar" text cursor at the current typing position.

[5] Ordinary cut, copy and paste – extracting and inserting character strings – are available within free-text paragraphs.

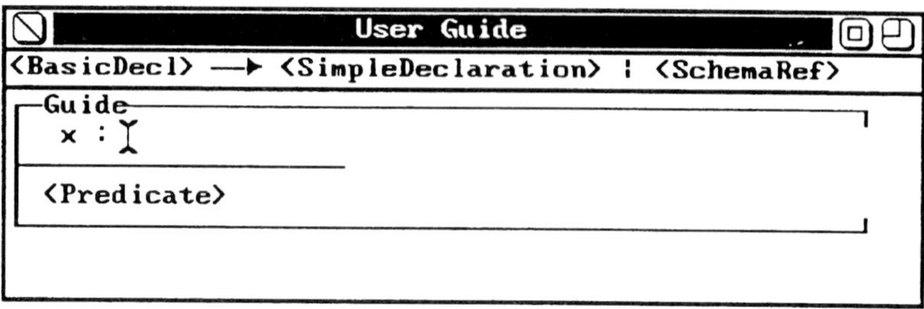


**Figure 10.5** Typed input.

### 10.2.3 Structure editing issues

We can now summarize the mechanisms by which a document is constructed and manipulated:

- node selection by point-and-click, wiping and repeated clicking on portions of document text;
- node selection by point-and-click on component names within the production display;
- node instantiation by explicit **instantiate** operation and selection from a menu of syntactic choices;
- node instantiation by point-and-click on choice-rule options in the production display;
- node instantiation (and re-instantiation) by direct typing at a selected node.

There is redundancy in this set of mechanisms, in that there is usually more than one way to perform node selection and instantiation. However, this redundancy allows alternative strategies for document construction. In particular, we have found that users use explicit instantiation (using choice menu or production display) at the higher syntactic levels, and directly typed input at the lower levels. In general, we have found that users prefer not to use menus, where there is an alternative mechanism. "Pop-up" menus are visually disruptive, since they (temporarily) obscure the document text.

One problem with structure editing is the degree to which a user has to be aware of the grammar of the notation being used. Grammars for non-trivial notations can be quite complex: the Z grammar, for example, has about 125 syntactic classes (including terminals). Although the classes have "meaningful" names, it is hard to characterize them precisely – the names may seem obscure to the novice user. Clearly, this is a problem where the structure editor is being used to teach a formal notation.

As we have seen, instantiation of a node involves selection from a set of options which are the names of syntactic classes in the grammar. There are two problems with this. Firstly, the user needs to know what constructs are denoted by the names and when they should be used. Secondly, common and rare options are presented together, complicating the selection of "standard" options.

It may also be that selection of an instantiation option at one node effectively constrains the instantiation of another related node. For example, given a node of class A:

A → B B
B → C | D | E | F

instantiating the first B node to an E of an F may make it likely that the second B node will also be an E or an F. When it comes to instantiating the second B node, it would be helpful to prioritize the E and F options in some way (ordering the menu, offering a default). Such context-dependency could perhaps be derived from previous usage, but it might be better expressed in some ways as part of the relevant grammar.

Multiple syntactic classes may be defined to restrict the instantiation options in particular contexts (i.e. to implement context dependencies in the grammar). For example:

A → B2 B2
B1 → C | D | E | F    B2 → E | F

However, this has now restricted the sub-nodes of A to be *only* Es or Fs – a stronger restriction than in the previous example. Also, from a user's point of view, it can be hard to remember the significance of the variant classes such as B1 and B2.

There is another way in which grammars do not reflect common usage. Optional sub-nodes which are only rarely needed appear in every occurrence of their parent node, and must be explicitly deleted or made empty. For example, a SchemaBox – such as that labeled "Write0" in Figure 10.3 – can optionally have a list of formal parameters following the name. This list is usually made empty, in which case it has no visible representation. However, there is currently no means of indicating that this construct is normally empty and setting a default instantiation for it.

Having syntactic classes to hold optional constructs can also lead to apparently unnecessary levels of structure. For example, a variable reference in Z consists of an identifier and optional parameter values, and an identifier consists of a word and an optional decoration (e.g. a prime), both of which are terminal strings. An unparameterized, undecorated variable name (the normal case) thus appears to have three levels of structure, all with the same visible representation.

The problems of explicit instantiation mentioned above are mitigated by allowing directly typed input. Typically, document construction consists of explicit instantiation at the higher syntactic levels, leading on to typed input for lower level nodes (at or below the "statement" level). At the higher levels, instantiation tends to be more obviously progressive. For example, in Z, instantiating a <Paragraph> node to be a SchemaBox generates four sub-nodes in a graphical box extending over five lines; in contrast, instantiating an <Expression2> node to be an Expression3 produces no apparent increase in concreteness. Also, the significance of the different classes of Expression is unclear (it is to do with nested subexpressions). Shifting from explicit instantiation to direct typing thus provides a more ergonomic mechanism for instantiating lower syntactic levels, and helps to hide some of the syntactic complexity at these lower levels. In general, the more expert the user, the more complex the constructs that can be entered (successfully!) by direct typing.

It has been found that some users construct partially instantiated "templates" for commonly occurring structures. These can be repeatedly copied/pasted and then instantiated fully. A possible enhancement to the tool would be to provide a "palette' for such user-defined template structures, rather than having them within the document itself.

Modifying an already constructed document also presents problems. The basic mechanism is deletion or uninstantiation followed by re-entry. Cut, copy and paste can be used to preserve components of a modified node, although this is often somewhat tedious. Consider for example the grammar:

$$A \rightarrow B \mid C \qquad B \rightarrow D\ D \qquad C \rightarrow D\ D \qquad D \rightarrow \ \ldots$$

Now suppose we have the tree

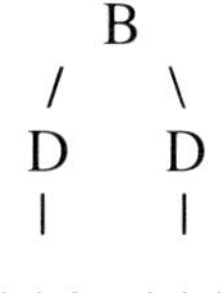


(within a list of A nodes) and we decide that it should really be a C node, then we would like to do a simple substitution of C for B retaining the two D nodes. This seems reasonable, since B and C are syntactically compatible in terms of their components. However, in the present *Formaliser* tool, it is necessary to (1) create a new A node by appending to the B node; (2) instantiate the new A node to a C node; (3) instantiate the new D nodes by copying and pasting the components of the B node; (4) delete the B node. As is apparent from this example, this is a complex operation for an apparently minor change.

One solution would be to "unparse" the tree rooted at B into a string, and allow text editing of the string. The resultant string would then be parsed on submission, just as if it had been typed directly. Alternatively, we can conceive of a facility that would allow substitution of syntactically compatible nodes, where compatibility is interpreted either as exact equivalence or as a looser matching. For example:

| | | | | | | |
|---|---|---|---|---|---|---|
| A | → | C D | can replace | B | → | C D |
| A | → | C D E | can replace | B | → | C D |
| | | | | (an uninstantiated E node is created) | | |
| A | → | C D | can replace | B | → | C D E |
| | | | | (the E node is lost) | | |
| A | → | C C | can replace | B | → | C C |
| | | | | (the C node ordering is retained) | | |

A variant of this problem occurs when we want to change the association of subexpressions within a larger expression. For example:

In addition, we can envisage generalized structural transformations, such as exchanging two nodes in a list of nodes.

### 10.2.4 Querying and checking

In addition to the editing mechanisms described above, *Formaliser* allows interactive querying and checking on a per node basis. These facilities provide access to the underlying attribute system which is defined as an extension to the basic context-free grammar. This aspect of *Formaliser* is discussed in detail in the next section, but the user view is described briefly here.

In the case of documents in the Z notation, the attributes of primary interest are those to do with type and scope. These attributes can be derived from the syntactic structure of a document – for example, the type of an expression is synthesized from its component sub-expressions, starting with basic types of elementary components. The rules for deriving attribute values are defined as part of the grammar for a notation – they are not "hard-wired" into the tool.

Checking attributes such as type and scope allows a level of correctness beyond mere syntactic correctness to be attained. The facilities are interactive because querying and checking perform an important exploratory role in constructing formal documents. In many cases, attribute errors can be traced back to context-dependent syntactic errors: use of a construct that is acceptable in terms of the grammar, but is inappropriate in the particular context.

Checking a Z document is typically incremental: checking paragraphs as they are added during construction. If a paragraph check fails, then exploratory querying and checking is done on sub-nodes of the paragraph in order to isolate the source of the error. This is considered to be preferable to batch-mode checking that generates "compiler style" error messages.

## 10.3 Grammars

The support provided by *Formaliser* is based on an *attribute grammar* for each supported notation. A grammar consists of a set of *rules*, each of which has the following components:

- an *abstract syntax production* – sequence, choice, list, terminal, etc.; terminal productions are associated with lexical checking blocks;[6]
- a set of *attribute equations* defining attributes, guard conditions and checks (with associated error message templates);
- a set of *unparse expressions* for one or more unparse schemes (concrete syntaxes).

The way that the abstract syntax productions are used in document construction and editing has been described above. This section will look at attribute equations and unparse expressions, in order to show that they greatly increase the power of a grammar-driven tool.

### 10.3.1 Attribute systems

In an attribute grammar, each class of node is associated with a set of named attributes. The values for these attributes are defined in terms of other attributes of the same node, attributes of related nodes (parent, children, etc.) and constant values. Attributes defined with reference to child nodes are called *synthesized attributes*; those defined with reference to the parent node are called *inherited attributes*. Knuth (1968) shows how an attribute grammar can be used to evaluate multi-digit binary numbers: a "value" attribute is synthesized, and a "scale" attribute is inherited. Reps (1983) applies attribute grammars to language-based editors.

In a *Formaliser* grammar, attribute values are defined by attribute equations associated with each production rule. Evaluation takes place through unification of attribute expressions. As an illustration, we will consider the rules for a SimpleDeclaration node (from the Z grammar). The production rule is:

SimpleDeclaration $\rightarrow$ DeclList Expression

where DeclList is a list of names being declared, and Expression is the type of the declared items. For example:

```
light1, light 2   :   {Red, Green, Blue}
---------------       ----------------
DeclList              Expression
```

[6] A piece of Smalltalk code – for example, checking alphanumericity.

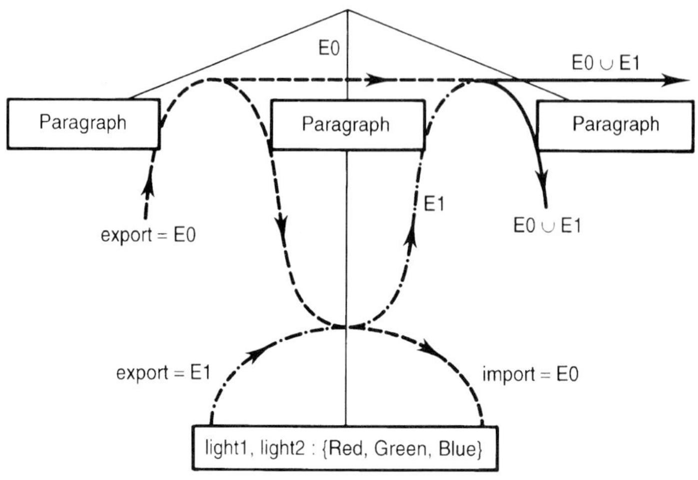


**Figure 10.6** Synthesis and inheritance of scope attributes.

where Red, Green, Blue are values of type Color.

A SimpleDeclaration has an *export* attribute which is an environment containing the pairs `(light1, color)` and `(light2, color)`. This environment will pass up through the parse-tree (as a synthesized attribute) to the Paragraph level, where it will be amalgamated with other environments containing other name–type pairs, and passed on to the succeeding paragraph (as an inherited attribute).

A SimpleDeclaration also has an *import* attribute, which is an environment containing all the name–type pairs that are in scope at this point. The name–type pairs `(Red, Color)`, `(Green, Color)` and `(Blue, Color)` must be in this imported environment.

Environment E0 (Figure 10.6) needs to contain entries for Red, Green and Blue; E1 will contain entries for light1 and light2; the amalgamation of these environments (E0 ∪ E1) will be passed to the following paragraph (and indirectly to the rest of the document).

The attribute equations for SimpleDeclaration are as follows (with added comments in italics):

Import is defined by Parent

*The imported environment is inherited from the Paragraph level, via the Parent node.*

ExpType = Child2.Type

*ExpType is the type of the second child node Child2 – the Expression. In our example, ExpType = Setof Color.*

Setof DeclType = ExpType

*DeclType is the type of ExpType "minus" the set constructor. In our example, DeclType = Color.*

Export = Inject (Child1.NameSet,BindEmpty,BindAddDecl)

*Export is an environment in which each name in Child1.NameSet (i.e. the names in DeclList) is associated with the type DeclType. Inject adds an entry for each name into an initially empty structure using the function BindAddDecl. In our example, an environment is created containing the name–type pairs* `(light1, Color)` *and* `(light2, Color)`.

**check** Setof SomeType = ExpType
**else**

"The type expression should be a set"
"However, the type expression is of type:@n@1" with ExpType

*ExpType must be a set of some other type (i.e. DeclType). This needs to be checked – if the two sides of the equation cannot be unified, an error message is presented. An erroneous value for ExpType is put in the error message template at @1; @n is a Newline character.*

Note that these attribute equations are essentially declarative, even those that include functional expressions, like that for Export.

Using groups of attribute equations, like those above, complex attribute systems can be defined as part of a grammar. In the case of the Z notation, the attribute system concerns primarily type and scope attributes, but it should be emphasized that the underlying facilities are completely generic.

In customizing the tool to handle a particular grammar, it is possible to identify those attributes that are required to be externally visible – that is, can be queried by a user of the tool. In the case of the Z grammar, the "external" attributes include Import, Export and Type. If the currently selected node in a document has any external attributes, then entries corresponding to them will appear in the node menu. In Figure 10.3, for example, the "dom f" node has external attributes Type and Import; selecting one of these entries will cause the attribute to be evaluated and the result presented to the user.

Querying, say, a Type attribute will cause other attributes to be evaluated, both at the selected node and at related nodes (e.g. its children). Since the amount of computation involved in such a chain of evaluations may be large, attribute values are cached: every node instance maintains an environment which maps attribute identifiers to attribute values. Subsequent queries simply retrieve attribute values from the local environment. Only if structural changes are made to a document are attribute values "scrubbed" from affected nodes. At the moment, this is done naively – attribute values at all nodes "below and after" a changed node are scrubbed.

The **check** operation (see the node menu in Figure 10.3) is completely generic in that it simply forces evaluation of all attributes of the current node. In doing so, it will encounter checks such as that for SimpleDeclaration (see the example above), which will produce error messages on failure. Note that checks are defined declaratively within a grammar; they are not hard-wired into the tool.

### 10.3.2 Unparse schemes

An unparse scheme defines a concrete syntax for a grammar. A document is displayed by projecting its abstract syntax parse-tree through the unparse expressions for a particular scheme. Each production in a grammar is associated with an unparse expression which consists of (1) literal characters, and (2) the projections of its child nodes. To illustrate this, consider the following very small grammar:

```
A → B B C      '@1;@n@2;@n@3.'
B → D          '( @1 )'         – a D in parentheses
C → D          '{ @1 }'         – a D in braces
String: D      'X'              – a literal
```

The unparse expressions appear in single quotes after the production rules. @1, @2 and @3 are the first, second and third children of the current node; @n is a newline; everything else (including whitespace) is a literal character. The projection of an A node in this grammar is then:

```
(X);
(X);
{X}.
```

The language for unparse scheme expressions includes codes for handling list nodes, for controlling indentation, spacing and linebreaking, for repeating characters and for marking positions. Literal characters may be graphics characters and symbols, as well as alphanumerics and punctuation. This allows considerable flexibility in the presentation of documents.

A grammar may be defined to have more than one unparse scheme, and a user can switch between them at will while editing a document. This separation of abstract from concrete syntax can be used to provide alternative representations of documents. For example, alphabetic keywords can be offered in place of symbols; a block-structured "begin . . . end" representation can be used instead of bracketing, and so on. Changing schemes also provides a simple translation facility. This is used, for example, to convert a Z document to a form suitable for input to the LATEX text-formatting program (Lamport, 1986).

Unparse expressions can also be defined for "expanded" and "compressed" schemes. Changing between these schemes on a node-by-node basis allows a user to expand and compress complex nodes at will. For example, suppose we have a node class A, which consists of a simple name and some complex contents:

```
A  →   Name Contents
       Unparse scheme Compressed is  '@1'
       Unparse scheme Expanded is    '@2'
Contents  →   . . .
```

Selecting the Compressed scheme for an A node would result in only the Name being displayed; selecting the Expanded scheme would result in the full Contents structure being displayed. Schemes could be defined for different levels of compression. This mechanism gives a user considerable control over the level of detail at which a document is presented.

### 10.3.3 The grammar grammar

A grammar is defined by a set of rules consisting of syntactic productions, attribute equations and unparse expressions, as described above. The notation used to write down a grammar can itself be defined by a grammar – a "grammar grammar". Just to give a flavor of the grammar grammar, here are the top-level productions:

| | | |
|---|---|---|
| Grammar | → | Introducion GrammarBody Conclusion |
| GrammarBody | → | {GrammarRule}+ |
| GrammarRule | → | SequenceRule | ChoiceRule | ListRule | List1Rule | StringRule | TerminalRule | TextRule |

The grammar grammar is expressed using the grammar definition notation, just like any other grammar. It therefore defines its own grammar: the rules given above are a SequenceRule, a List1Rule[7] and a ChoiceRule respectively.

*Formaliser* supports the grammar grammar, allowing grammars to be constructed, viewed and modified as *Formaliser* documents. In the current version of *Formaliser*, grammar documents appear in the library alongside ordinary documents, and are used to create new documents. For example, a Z specification would be created by invoking **new** on the Z grammar document; a grammar document would be created by invoking **new** on the grammar grammar document.

When the tool uses a grammar to constrain the editing of a document it needs the grammar in a readily accessible form – the parse-tree representation of a grammar document would not be suitable. It is therefore necessary to "compile" a grammar into an internal form – basically a complex table, implemented as nested Smalltalk dictionaries. This compilation is done using a second unparse scheme for the grammar grammar. The expressions for this unparse scheme translate grammar rules into blocks of Smalltalk code that when evaluated build a structure representing the defined grammar. This translation is performed by a **compile** operation on grammar documents – the user does not have to change schemes explicitly. A further phase of compilation translates the attribute equations into a more efficient form; this can be done all at once or incrementally on demand.

It is thus relatively easy to construct and modify *Formaliser* grammars using the tool itself. Various kinds of modification might be made to a grammar: unparse expressions could be altered or new unparse schemes added; attribute systems could be extended; syntax could be extended (by adding ChoiceRule options, for example). Existing docu-

[7] A non-empty list.

ments can be promoted to use a new version of a grammar, providing that they remain structurally compatible with it. Defining a completely new grammar is a substantial task, and for any non-trivial notation will require multiple iterations of an edit–compile–test cycle. This can be done entirely within *Formaliser* – testing a compiled grammar would involve editing test documents created from the grammar.

## 10.4 Future directions

### 10.4.1 Document and library structure

At present, a *Formaliser* document is associated with one particular grammar – the grammar constrains the form of the entire document. In the Z grammar, for example, this means that the structure of a complete Z specification document, including free-text components, needs to be described. For simplicity, the structure of a document is kept simple (a sequence of paragraphs); titles, headers and section structuring are not specified. In practice, a *Formaliser* Z specification document would form a section of a larger document assembled outside the tool.

This suggests that a separation should be made between a grammar defining overall document structure and a grammar defining a particular notation for use at or below the paragraph level. This would allow a notation to be used in different kinds of document, each with potentially complex structures. The interface between the two levels of grammar would have to cope with the propagation of attribute values between paragraphs (as in a Z document).

Generalizing this idea, a grammar could allow nodes in a document to be parse-trees conforming to different grammars. This would allow component sections of a document to use a variety of notations, without the need to define a single grammar encompassing all the required notations. In the sphere of specification, it would be useful to be able to interleave different notations – for example, state-based and event-based descriptions. In the programming sphere, it would be useful to have specification fragments as annotations to code segments.

In addition to handling complexity at the document level, we also need tool support at the library level. At present, a *Formaliser* library is simply a set of documents (grammar documents and ordinary specification documents). In the Z grammar, there is a kind of paragraph that "includes" name-type environments from other documents – an Inclusion paragraph (see Section 10.2.2 for an example). The Export attribute of the Inclusion paragraph is defined to be the name/type environment exported from the named document (which has to be a Z document, as well). The Export of a document is simply an attribute of the root node of its parse-tree.

The effect of Inclusion paragraphs is to link Z documents into directed acyclic graphs, with name–type environments accumulating through a graph. For example, suppose document B includes document A, and C includes B: inside C, all names declared in A and in B will be in scope and can be referenced. This allows large specifications to be partitioned into separate, but linked, documents, allowing the editable units to be kept manageably small. However, embedding Inclusion statements within documents means that the dependencies between documents are not visible at the library level. The tool needs facilities for presenting inclusion graphs to a user, so that the overall structure of a large, partitioned specification can be seen.

### 10.4.2 Navigation

Within a *Formaliser* document, navigation is achieved by scrolling through a document, and using the local point-and-click mechanisms described earlier in this chapter. There is a simple string-search facility, but this does not work across documents, and cannot distinguish references from definitions, since these are linguistic concepts.

In Z specifications, a user often wants to see the definitions corresponding to particular references (this is equally true of many programming languages). However, references and definitions may be far removed from each other – they may even be in different documents. Considerable effort can be expended in trying to view related parts of a specification. A user would therefore benefit from search-and-display facilities that allow related pieces of document to be viewed in juxtaposition.

At the point where a name is referenced, the name must be in scope, and it must therefore be in the current node's Import environment. It should therefore be possible to extend name–type environments with a "where defined" attribute for each item. This attribute would be queryable for all nodes whose class is a reference. Querying the attribute could display the name of the document or construct containing the definition, but it would be more useful if an edit window on the appropriate piece of document could be presented. This would allow a kind of "hyper-text" navigation, stepping through chains of definition and reference. This would be particularly useful for notations, such as Z, which allow complex dependencies within and between documents.

### 10.4.3 Attribute domains

The domains over which attribute values may range are currently implemented as Smalltalk classes. These include sets, tuples and so on. Several primitive operations on these values are hand-coded into the system. While being fairly efficient as an implementation, this does result in a loss of generality of the system: a new language might have attribute values that do not map neatly onto any of the available domains. A more sophisticated approach under consideration is to use *Formaliser* grammars to describe the domains of attribute values. Thus, in general, definition of a new language for *Formaliser* would require building at least two grammars: one for the language itself and one or more for its attribute domains. It is probable that many of these attribute domains are common to a range of languages: a small library should cover most cases.

An important application of this more general approach would be to treat Type attribute values as structures. For example, suppose we have a number of type definitions:

```
Point        ==  Integer × Integer
Polygon      ==  Set Point
Transform    ==  Polygon × Polygon
```

An item of type Transform is thus a pair of sets of integers. At present, *Formaliser* will display the type as:

```
(Set (Integer × Integer)) × (Set (Integer × Integer))
```

This can be regarded as the "base type". While this base type is important for checking purposes, we have lost the information present in the naming of intermediate types: the base type gives no clue as to the intended usage of the type, or how it was constructed out of previously defined types.

Now, if we had a *Formaliser* grammar for types:[8]

| | | | |
|---|---|---|---|
| Type | → | NamedType \| UnnamedType | |
| NamedType | → | TypeName UnnamedType | '@1@2' |
| UnnamedType | → | BasicType \| Tuple \| Set | |
| Tuple | → | {Type}+ | '(@+@n@0×@n@d@n)@-@e' |
| Set | → | Type | '{@+@n@1@n}@-' |

Terminal: TypeName,BasicType

the type could then be a tree-structure, unparsed as:

```
Transform (
      Polygon {
            Point (
                  Integer ×
                  Integer
                  )
            } ×
      Polygon {
            Point (
                  Integer ×
                  Integer
                  )
            }
      )
```

where "{ . . . }" indicates "setof" and "×" indicates association in a tuple. Interestingly, the base type can be defined to be an attribute of this type structure, synthesized upwards from the BasicType nodes. This attribute would be queryable at all nodes. For example, we could select the TypeName node "Transform", and query the BaseType attribute, getting the value:

```
(Set (Integer × Integer)) × (Set (Integer × Integer))
```

## 10.5 Conclusions

Structure editing using only menu-driven navigation and selection has been found to be unergonomic; users prefer a more direct "point-and-

[8] This grammar is rather simpler than would actually be necessary.

click" style of interaction. Text entry by direct typing, with automatic parsing of input strings, provides an alternative to explicit instantiation at lower syntactic levels, which is suitable for more experienced users.

A problem with structure editors for formal notations (including programming languages) is that a user needs visibility of and familiarity with the details of a large, complex grammar. The structure of a grammar may not always coincide with a user's intuitive structuring of a notation. Provision of template structures and context-dependent defaults may go some way to alleviating this problem. There is also a need for structural transformation and substitution operations, in addition to the basic editing and instantiation facilities.

There is a growing need for tools to support the construction of specification documents that combine material in formal notations with material in natural language. An important property of formal notations is that they have attributes that can be checked and queried, and tools should support this as well as supporting basic document construction.

Attribute grammars allow equations for deriving attributes to be associated with abstract syntax production rules. This allows attribute values to be derived and queried, and checks performed, during construction of a document. For the purposes of specification, interactive, exploratory querying and checking is preferable to "batch mode" analysis of a completed document. Large specifications need to be partitioned into conveniently sized documents, linked together so as to allow propagation of attribute values (e.g. name–type environments) between them.

The separation of abstract and concrete syntax is necessary for effective tool support of formal notations. Allowing multiple concrete syntaxes makes it simple to provide alternative styles of presentation, variable levels of detail and other kinds of translation.

Further information on *Formaliser* can be found at:
http: //public.logica.com/~formaliser/.

## Acknowledgements

We would like to thank Mike Flynn and Tim Hoverd for their part in bringing *Formaliser* into the world, and developing it to its current level.

## References

Knuth D. E. (1968). Semantics of context-free language. *Mathematical Systems Theory*, **2**: 127–145.

Lamport, L. (1986). *LATEX: a Document Preparation System.* Addison Wesley.

Reps, T. W. (1983). *Generating Language-Based Environments.* ACM Doctoral Dissertation Awards series, MIT Press.

Spivey, J. M. (1989). *The Z Notation – A Reference Manual.* Prentice Hall.

# CHAPTER 11

# Environment for document structure recognition

*Nenad Marovac*


## Abstract

Document recognition is a process in which a document in its physical (output) form is transformed into an author-oriented structured format in which the document can be manipulated, i.e. edited, restructured and reformatted to change the output appearance.

In most cases full document recognition is not possible. However, it is possible to develop a semi-automatic document structure recognition method that uses two passes. In the first pass, which is automatic, an intermediate structured document format is produced. This intermediate format is acceptable in most applications. If not then the user can use an interactive editor to edit the structure of the document.

The objective of this chapter is to describe an interactive environment for document recognition, its graphical user interface and a document structure editor incorporated into the environment.The graphical user interface has three purposes. It is used to present overall recognized document structure, the target document grammar and the local recognized structure of a document, i.e. detailed representation of the structure for a section of the recognized document structure around an arbitrary user-selected point within the document. The interface also provides the means to control the document recognition process and to edit the recognized structure of a document.


## 11.1 Introduction

Document recognition (Marovac, 1990) is a process of transforming a document from its *physical presentation (output)* form into a *structured author-oriented* format. The physical presentation form can be either bitmaps of document pages, or encoding of these pages in a graphics or

Structure-based Editors and Environments
ISBN 0-12-681890-8

a page description language. The author-oriented format is a format allowing for document manipulation, i.e. addition to the document, editing of the document, modification of the layout of the document and reformatting of the document to alter its physical appearance.

Document recognition shows promise for a number of diverse applications. Some of these may be summarized as follows.

- Often we have a printed document and would like to modify it or extract a portion for another use without going through the effort of re-typing the document into a computer.
- To automatically insert a published document into an electronic library we need to determine the topics the document addresses in order to classify the document. We also need to identify the abstract in order to extract it and place it into a catalog. If the document is to be interactively browsed and read then we have to recognize the document components and store them as directly accessable segments. Document recognition can assist us there. The document can be scanned and stored into a computer and the document recognition process applied to it. Once the structure is recognized and reconstructed, the key elements required for the automatic classification of the material can be identified, abstracts extracted and inserted into reference catalogs, and the document can be reformatted to a form suitable for interactive browsing, etc. In fact, the author is supervising a research project on similar lines, in which document recognition is used for automatic conversion of printed (flat) documents into hypertext documents.
- Document recognition can also be used as a means for open system architecture for an electronic publishing environment in a large organization (Marovac, 1989).

The definition of different document structures and the detailed discussion of the process of document recognition is presented in Marovac (1990, 1992). A brief summary is given below in order to make this chapter self-sufficient. A document can be represented in a number of structural forms which we refer to as its structures. We identify four levels of such structures:

1. *Physical structure* – output format of a document either in the form of printed pages, or an encoding which would generate these pages when sent to a printer.

2. *Typographic structure* – the encoding of the document structure and content in terms of typographical elements, e.g. text strings in particular fonts in particular places on a page.
3. *Layout structure* – a description of the document as a sequence of components, e.g. lines of text, blocks of text, pages, etc., without concern about geometrical and typographical properties of these components (ISO, 1987).
4. *Logical structure* – the representation of the document in terms of objects the author of the document is concerned about, eg. title, author name, abstract, sequence of sections (ISO, 1987).

Document recognition is the task of processing either a physical or a typographical structure of a document in order to generate (or rather to identify) the logical structure for that document.

At present, most of the effort in document recognition is in the area of low level recognition, i.e. separation of text and graphics, character and text recognition (Wilcox and Spitz, 1989; Spitz, 1990; Wordscan, 1991). Some research projects were handling analysis of document structures for specific applications (Ingold, 1989). However, there are a very few projects in which a general high level document recognition was attempted (Marovac, 1989).

Fully automatic document recognition is not possible (Marovac, 1990, 1992). However, it is possible to produce a semi-automatic document recognition strategy. The recognition strategy consists of two stages. The first stage is automatic and processes the typographical structure of a document. This stage is a table driven task. The table is called the *recognition driving table* and it stores *document structure recognition rules* which the system uses to make decisions in order to recognize logical constructs within the document. The automatic task analyzes the document and makes identifications of logical document objects, e.g. title, sections, paragraphs, etc., on the basis of the decision rules in the recognition driving table. As a result, an *intermediate document structure* for that document is produced. This intermediate document structure actually contains encoding for two structures derived as the result of this automatic task, i.e. the logical and layout structures of the document. The intermediate document structure is pictorially presented to the author/user on the screen, either for adoption or for interactive editing. At the same time the user can also modify the content of the paper. This is, however, of secondary importance here.

Document recognition decision based on a possible document grammar (known or assumed) alone will not result in correct recognition of a document structure (Marovac, 1990, 1992). For the document recognition process to be successful the document recognition rules must be based on information about:

- document grammar;
- typographical structure;
- the relationship between the logical structure and the layout and typographical structures for the document, i.e. the transformation process from the logical structure of a document to its layout and typographical structures during the composition and typesetting task for the document in order to be printed.

The process would be deterministic and therefore successful if complete information was available concerning:

- the original grammar for the document logical structure;
- the document composition and the transformation rules from the document logical structures to its layout and typographical structures.

However, this information is not available and therefore the document recognition rules must be general and heuristic. This is the reason for the need for the second stage i.e. the interactive inspection and editing of the document logical structure.

The pictorial representation of the document is obtained by compiling the intermediate structure of the document via the automatic recognition process. In the process, two structures are obtained: typgraphical and physical. To distinguish these from the original, or *initial* physical or typographical structures, we will refer to them as *derived* structures. It is obvious that one of the necessary (but not also sufficient) conditions for the recognition process to be successful is that derived and corresponding initial structures match.

On top of the displayed derived physical representation of the document, pictorial features which depict the obtained logical structure for the document are overlayed. These features are a part of the graphical user interface for editing logical and layout structures of documents. They are discussed in the next section.

It was already mentioned that when the automatic task is completed, the recognition method enters the second stage, i.e. the interactive editing of the structure for the document. At the completion of this stage, the layout and logical structures encoded within the intermediate document structure become the actual (accepted) logical and layout structure for the document.

One should comment at this point that for most documents the first stage i.e., the automatic stage, would be more or less sufficient, and the second stage, i.e., the interactive structure editing stage, is reduced to a preview of the document. This holds true particularly for documents which are produced using systems with fixed and known rules of transformation from their logical structures into their layout, typographical and physical structures, for example the LATEX document preparation system (Lamport, 1985). In fact, efforts are under way to produce a completely automatic document recognizer for documents produced using the LATEX system.

This chapter describes an environment used to interactively edit the logical structure of a document. The internal functionality for the system is described in a paper detailing the document recognition method (Marovac, 1992). Here, we will concentrate on the environment and its graphical user interface for such a system and its use. Such an interface has three purposes; to depict the actual local structure of a document, to indicate the target grammar for the recognized document class, and to show the overall recognized document structure as it is being derived. It also is used to modify the logical structure of a document.

## 11.2 Graphical user interface for document recognition

Figure 11.1 shows the screen environment for document structure recognition and editing. One can recognize four distinctive areas in the environment. They are:

1. the *local structure area* (LSA) in the center of the figure;
2. the *document grammar area* (DGA) on the left;
3. the *document structure area* (DSA) on the right;
4. the *menu area* (MA) at the top.

### 11.2.1 The local structure area

The central and the largest area shows the logical structure for a user selected portion of a document together with the content for that

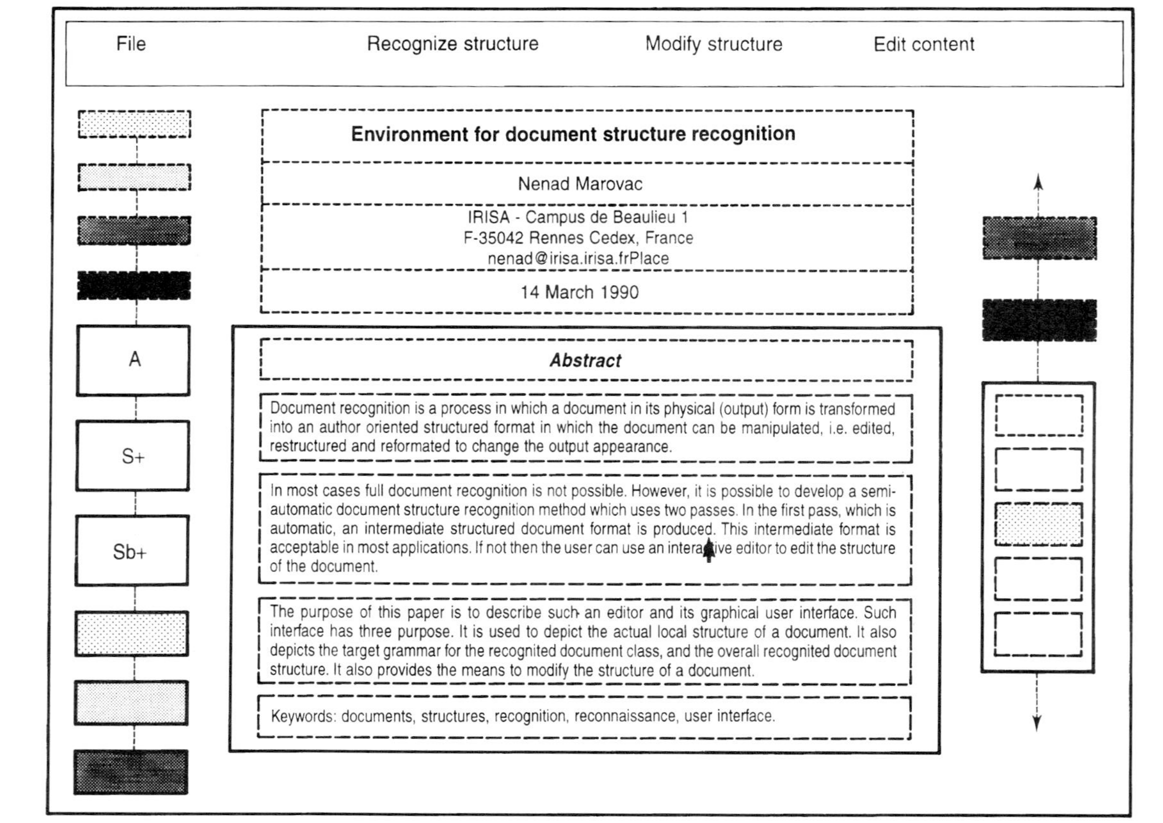


**Figure 11.1** Pictorial environment for document structure recognition and editing.

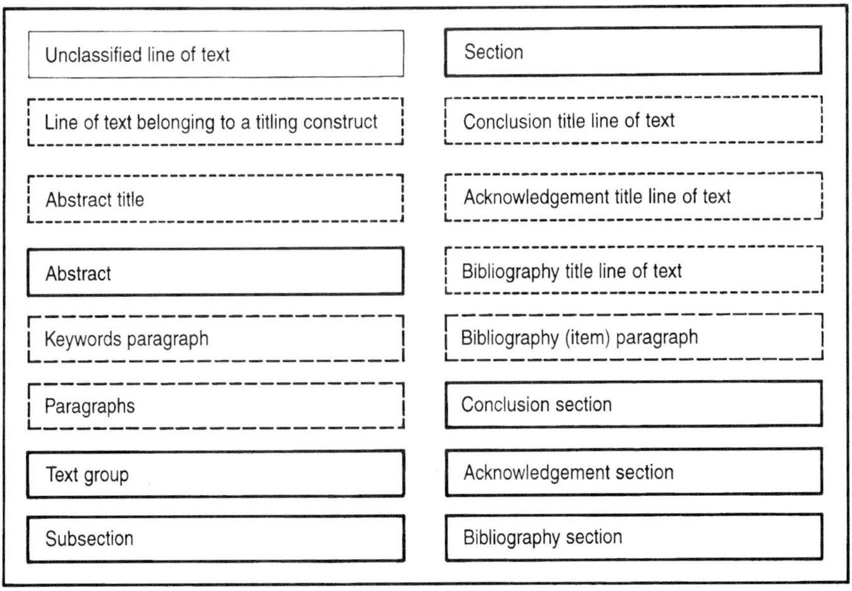

**Figure 11.2** Visual features used to depict the structure of a document.

portion. The structure is depicted using visual features, e.g. frames with different kinds of border and different frame color shadings.[1] The visual features which the system uses are shown in Figure 11.2. From Figure 11.2 we can see that all objects forming the document title page are shown by using dash–dot line frames; the differentiation is made by shading the frame using a different color. All paragraphs in the document are depicted using dash line frames, including paragraphs in the abstract, the keywords paragraph, paragraphs in the document body, conclusion, acknowledgements, and the bibliography entries. Again, different colors are used to differentiate between different types of paragraphs. All text bodies are shown by using double-line frames. Again the differentiation is made by shading a frame with a different color. In Figure 11.1 we show only section and subsections as main document body constructs, therefore implying only two levels of hierarchy. However, the method is general. If we have a construct on

[1] Unfortunately, since color was not available, the different titles are left white. Use of shading to simulate colors leaves the text unreadable.

a lower level, again it will be depicted with a double-line frame, but shaded with a different color. In figure 11.1, in the local structure area in the center of the figure, we can see the recognized title page of the document. It shows the four title objects (the document title, the author, the author address, and the date of creation), the abstract within which we can identify the abstract title and three paragraphs, as well as the keywords paragraph.

### 11.2.2 The document grammar area

This area is used to show a target document grammar for a document class. When a document is being recognized, if a grammar for the document logical structure is known in advance, the user can preselect that grammar. During a document recognition task, the system uses that grammar as the reference. It attempts to fit the logical document structure, as it is deriving it, to fit the predefined document grammar.

If a grammar is not prespecified, then the system will display a document grammar which can be identified from the actual document logical structure obtained in the recognition process.

In Figure 11.1 in the document grammar area, we can see the grammar for the associated document class. The four title frames are shaded with different degrees of shading. We then identify five bodies: the abstract marked as A, section sequences marked as S+ (+ indicating one or more occurrences), subsection sequences within each section marked as Sb+, the conclusion, acknowledgements and bibliography. The last three are again identified by different shadings. We can also see in that area the use of different frame sizes to indicate where items belong in the hierarchy.

### 11.2.3 The document structure area

This area is used to show the generated logical structure for the document being recognized. The user can select display of the entire derived logical structure, or only a part of it. If only a part of the entire structures is displayed, this part is associated with the portion of the document shown in the central (local structure) area of the environment. In either case the construct which the user selects in the central area, i.e. the document component pointed to by the cursor, will be identified and stressed within the document structure shown in the document structure area.

In Figure 11.1, in the document structure area we see a local logical structure. We can identify two title objects (the address and the date) and the abstract. Within the abstract body we can see the abstract title, three paragraphs and the keywords paragraph. We also note that the second paragraph in the abstract is indicated as the current paragraph (via shading), i.e. as being selected by the user.

### 11.2.4 The menu area

This area is used as a two-way communication channel between the user and the system during the interactive processs of document structure recognition and editing. It is used by the system to indicate to the user what structure preview/editing options are available at any time. It is used by the user to direct the document recognition process and later to edit the produced document logical structure. The menu area will be discussed in more detail in the next section while the interactive options are presented.

## 11.3 Interactive document structure preview and editing

The user–system interface language for the document structure recognition, preview and editing processes is context oriented. This reduces the number of environment menu objects on the screen at any one time, and it also simplifies the understanding and identification of users' requests by the system. The second point will be further amplified when discussing the Modify Structure and Edit Content options.

In the menu area in Figure 11.1 we notice four main select menu buttons:

1. File;
2. Recognize Structure;
3. Modify Structure;
4. Edit Content.

The menu button marked *File* is used to select a document to be recognized and to store the derived document grammar, logical and layout structures.

The menu button *Recognize Structure* is used to start the automatic document recognition process and set its parameters. There are two main parameters the user can select at this point. They are the document recognition driving table and the flow mode for the recognition task. When the user is using the system for recognition of a document he can select either that the recognition flows automatically until completed, or he can step through it from stage to stage. Each individual stage is concerned with recognition of certain aspects of the document. The first stage, for example, recognizes individual lines of text and their typographical properties. The second stage recognizes the document titles and individual paragraphs. The third stage recognizes the text groups, i.e. titled sequences of paragraphs. Finally, the fourth stage completes the hierarchy within the document structure by identifying the document sections and their subsections, and the lower level text groups if there are any.

The menu button *Modify Structure* allows the user to edit the derived document logical structure. Here the user has a number of choices to make. He can opt to edit the preseleced or derived document class grammar and restart the automatic recognition method. Alternatively, he can modify the derived document logical structure. He can change colors of a text body to change its designation from, for example, a section to a subsection. The new subsection is then automatically incorporated into the preceding section. He can convert a subsection into a section and specify its extent, i.e. which of the following subsections, if there are any, should be incorporated into the new section. All such logical structure change requests can be specified either on the central local structure area or within the document structure area on the right. As soon as a change is completed by the system, it is depicted in both structure areas. When the user identifies a logical structure object, e.g., a paragraph, a subsection, etc., by pointing at it, the system displays menu buttons which correspond to the actions the user can perform on the selected object.

The menu button *Edit Content* allows the user to edit the content of the paper without changing the logical structure of the paper. Since this activity is of passing interest here, we will not dwell on it.

The use of two menu buttons to select either Modify Structre of Edit Content options makes an explicit differentiation between modifying the structure of a document from editing content for the document requests. It, therefore, also separates the corresponding sub-environments. This is done for easier identification of items on the screen by the user and by the system. When the user selects modification of the structure and points to a block of text, he is selecting the corresponding paragraph or text group, and not the text item, i.e. a character, a word or a sentence, as would be the case if the user selected the edit content option. This reduces the necessity for the user and the system to go through the sometimes tedious selection of the right object and the right action.

## 11.4 Conclusion and future work

An environment and a graphical user interface for recognition and editing of document structures is presented. However, one important aspect of the interactive document recognition and structure editing remains uncompleted. This is a method to allow a user to interactively specify document recognition rules, both when creating a completely new document recognition driving table, or when modifying an existing rule within or adding a new one into a document recognition driving table which is being used at a recognition task in process. This needs to be done when the user realizes that the recognition process has not produced a completed and satisfactory result and he is aware that the reason is that a recognition rule is missing, or a rule was not correctly formulated. Currently, the document recognition rules and the associated driving table is generated "off line". However, the objective of the current research is to formulate a language for definition of recognition rules which would be suitable for interactive specification of rules interactively by a user on the screen.

## Acknowledgments

I would like to thank Jacques Andre for inviting me to visit and work at INRIA-IRISA at Rennes, and INRIA and Ministry of Research and Technology of France for their support during my stay. Some ideas described in this chapter originated during my stay at IRISA.

# References


Ingold, R. (1989) *Text structure recognition in optical reading*. In Andres, J. et al. (editors), *Structured Documents. The Cambridge Series on Electronic Publishing, Cambridge University Press.*

ISO (1987). *Information Processing – Text and Office Systems – Office Document Architecture* (ODA) *and Interchange Format*, parts 1–8. ISO DIS 861 1987-07-16.

Lamport, L. (1985). *Latex – a Document Preparation System.* Addison-Wesley.

Marovac, N. (1989). *Open system architecture for electronic publishing systems.* In *Online-89, Electronic Publishing and Print Conference*, London.

Marovac, N. (1990). *Document structures and document recognition.* In *RAE – Reconnaissance automatique de l'acrit*, Le Havre, May 1990, Proceedings BIGRE No. 68.

Marovac, N. (1992). *Principles of document recognition.* Office Information Systems, ACM.

Spitz, L. (1990). *Recognition processing for multilingual documents.* In *Proceedings EP-90, Cambridge University Press.*

Wilcox, L. D. and Spitz, L. (1989). *Automatic recognition and representation of documents.* In *Document Manipulation and Typography, Nice, April 1988*. EP-88, The Cambridge Series on Electronic Publishing, Cambridge University Press.

Wordscan (1991) *Wordscan User Guide*, Calera Recognition system, Santa Clara, CA.

# CHAPTER 12

# Canae's structure-based editor components: rationale, description and field-tested applications

*Hiroyuki Tarumi, Jun Rekimoto, Masaru Sugai, Go Yamazaki, Takeshi Mori and Chuzo Akiguchi*

## Abstract

*Canae* (鼎) is a platform for development of a user-interface (UI) with the X Window System.[1] Canae provides structure-based editors as software components *(editor components)*. In this system, structure-based editors are not domain specific tools like language-based editors, but UI toolkits. Although Canae also provides a hyperlink mechanism, a UI builder, and an interactive Lisp environment, this chapter focuses on the UI toolkit.

Canae's editor components are offered in C or C++ libraries for editing six kinds of media – text, picture, image, table, graph (i.e. network structures) and hierarchy (i.e. tree structures). The rationale behind these media choices comes from experience gained during the development of Computer Aided Software Engineering (CASE) tools.

Canae's editor architecture is based on Smalltalk-80's[2] MVC model, with modification for delayed drawing.

Canae's standardized data architecture, called *Fragment*, is defined for files and for other common operations. Fragment can encapsulate data from any of the available media, and embed it into another media fragment or data model.

Canae first came on the market in June 1990. It has already exceeded

[1] The X Window System is a trademark of MIT.

[2] Smalltalk-80 is a trademark of ParcPlace Systems, Inc.

Structure-based Editors and Environments
ISBN 0-12-681890-8

all expectations: it has been applied to more domains than were initially anticipated. In some of the approximately 50 domains to which it has been applied in the field, Canae's modular software has halved the number of program codes needed.

## 12.1 Introduction

Canae (鼎)[3] (Tarumi *et al.*, 1990) is a platform for development of a user-interface (UI) with the X Window System. Canae's innovative software components provide structure-based editors as a set of UI toolkits, in other words as software components.

In the past, UI toolkits were usually general-purpose, while a structure-based editor depended on specific structures. Experience in developing Computer Aided Software Engineering (CASE) tools stongly suggested the advantages in software components that could handle select data structures. The rationale is that a few data structures are very important in some application areas. This is especially true in CASE applications.

For example, there is an enormous number of CASE tools supporting dataflow charts. Even if general-purpose UI toolkits are available, programmers who develop such CASE tools must furthermore create routines for maintaining the network structure of the chart, routines for tracking the input from a mouse, such as dragging, routines for making nested charts, etc. In the past, these kinds of routines were regarded as application-oriented. It is true for routines for implementing a specific style of dataflow chart. However, any style of dataflow chart is a kind of network structure. If there is a UI toolkit for implementing general network structure editors it will be helpful. For the same reason, other structure-based editors, like table editors or tree editors, are helpful if they are provided as UI toolkits.

The advantage of providing UI toolkits for structure-based editors is not simply in the reduction of development size. Most of the application programmers, such as CASE tools programmers, are not good at the special UI programming techniques, like event handling, window

[3] 鼎 is a Chinese character which means a tripod kettle. This character looks like a window placed on a support. In China, the tripod kettle is also a symbol for the emperor. We can find many huge examples in the old palace in Beijing. (Canae and 鼎 are trademarks of NEC Corporation.)

operation, mouse tracking, etc. These techniques have nothing to do with these programmers' main interest. Hiding such techniques into toolkits, they can be free of troublesome UI programming.

The effect of a structure-based editor toolkit can be compared to the high-level programming languages. The functions of conventional UI toolkits are limited to input/output Boolean values (e.g. toggle buttons), selections from several items (e.g. choice buttons, menus), text strings, and real number values (e.g. sliders). This limitation is similar to that found in assembly languages. In contrast, high-level languages are more encompassing because they can handle complex data structures such as arrays and trees. Canae's toolkit is a high-level UI, in the same sense. It can handle such complex data structures.

For all the above reasons, we have designed Canae to provide a structure-based editor toolkit. Canae has six kinds of editors: *text* (including Japanese), *picture*, *image*, *table*, *graph* (i.e. network structures) and *hierarchy* (i.e. tree structures). We call these six data formats "media" in this chapter. Three of the editors, table, graph and hierarchy, are structure-based editors.

The toolkit is provided as C or C++ libraries based on the X Toolkit (Nye and O'Reilly, 1990). It came on the market in June 1990, and has been applied to more than 50 kinds of application software. It has been applied to more domains than were initially anticipated. These include monitoring tools, simulation tools, conferencing systems and electronic computer manuals.

This chapter describes the editor architecture of Canae's toolkit, and our experiences with some examples of field applications.

## 12.2 Configuration of Canae

Although the toolkit is the primary focus of this chapter, an overall description of other advantages cannot be omitted for a complete understanding of Canae.

Canae consists of the following components and tools:

- *A UI toolkit*: this is made up of six C libraries: `libCanae.a`; `libPe.a`; `libFe.a`; `libGe.a`; `libHe.a`; and `libIe.a`. The relative relationships among these libraries, X libraries and applications

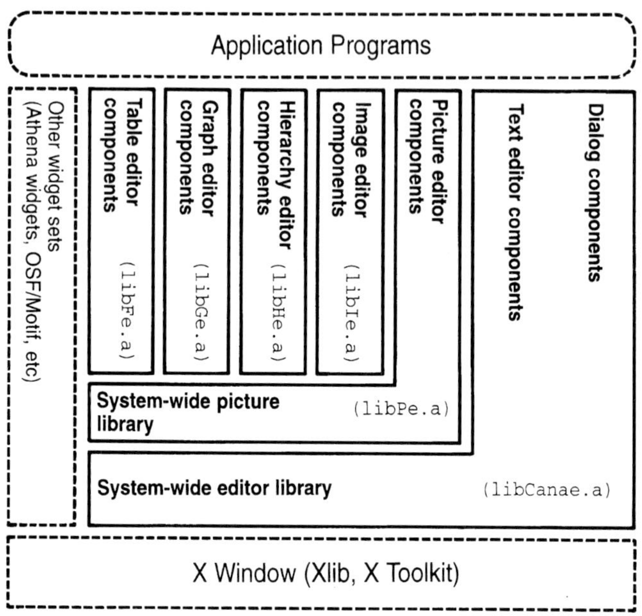


**Figure 12.1** Configuration of Canae.

is shown in Figure 12.1. There is also a C++ version of Canae libraries.

- *A Lisp interpreter*: the language is called "Canae-Lisp", which is a compact Lisp interpreter developed by NEC. From the Lisp environment, the user can interactively call Canae's C functions. An interactive environment similar to the emacs (Stallman, 1981) editor can be developed.
- *YUZU*: A UI design tool, which supports the arrangement of components and the definition of actions, on the graphic display. The designed UI can be tried interactively. YUZU can generate C or Lisp source codes realizing the designed UI.

### 12.2.1 Dialog components

Dialog components are included in `libCanae.a`.

In the X Toolkit environment, a UI component is realized as a form of *widget* data structure. The widget is an object-oriented module that

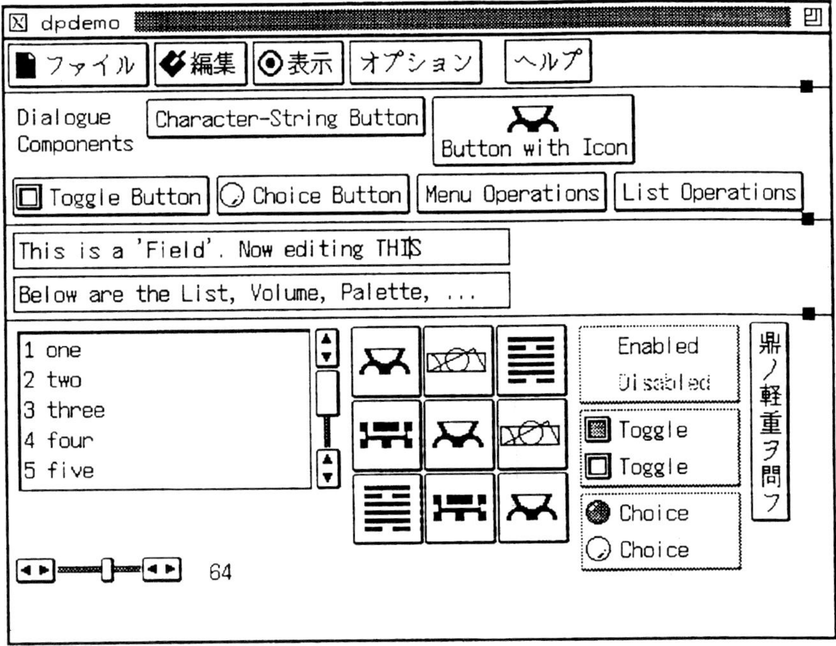


**Figure 12.2** Dialog components.

has predifined methods activated by events. It also has an inheritance mechanism.

Canae's dialog components are of course widgets, which include *Buttons*, *Menus*, *Palettes*, *Lists*, *Fields* (a text input field) and *Volumes* (sliders). These are shown in Figure 12.2.

Dialog components have been designed to develop a total environment with other components of Canae, without the use of other toolkits. While other widget sets, such as OSF/Motif[4] are considered to be alternatives to dialog components, Canae's dialog components have been designed to be part of a total environment.

[4] OSF/Motif is a trademark of Open Software Foundation, Inc.

### 12.2.2 Editor components

*Editor Widget*

Canae uses only one kind of widget for editors. This widget is found in `libCanae.a`, and is called the *Editor Widget*. This Editor Widget is not the editor in itself. An editor is only realized by the combination of Editor Widget and View.

One widget for all editors allows it to be compatible with all media. If a text editor and a picture editor are implemented as different kinds of widgets the text editor cannot become a picture editor. However with Canae's modular design, the text editor can become a picture editor by changing the View data structure.

The Editor Widget performs common functions among media, e.g. event handling, scrolling, scaling, magnifying, etc. The Editor Widget has approximately 50 functions. In the case of the widget for dialog components, the number is less than 10.

*Functions of each editor*

This subsection is a survey of the functions of each Canae editor. Some examples of how these editors are used are also included in this subsection. It must be stressed that these examples are not Canae itself. They are simply examples of the versatility of Canae's toolkit. Canae is not a set of editors; it is a package of components and tools that develop application programs with editor components embedded in them.

*Text editor components.* The text editor components are included in `libCanae.a`. These handle multifont text, including Japanese. Japanese text entry is performed with *Canna*, which is a Japanese text conversion system developed by NEC.[5]

An example of the text editor is shown in Figure 12.3.

*Picture editor components.* The picture editor components are in `libPe.a`. These handle lines, rectangles, ovals, arcs, polygons, Bezier curves and other geometric forms. The data model is based

[5] Canna is denoted to the X Window System Version 11 Release 5 distribution.

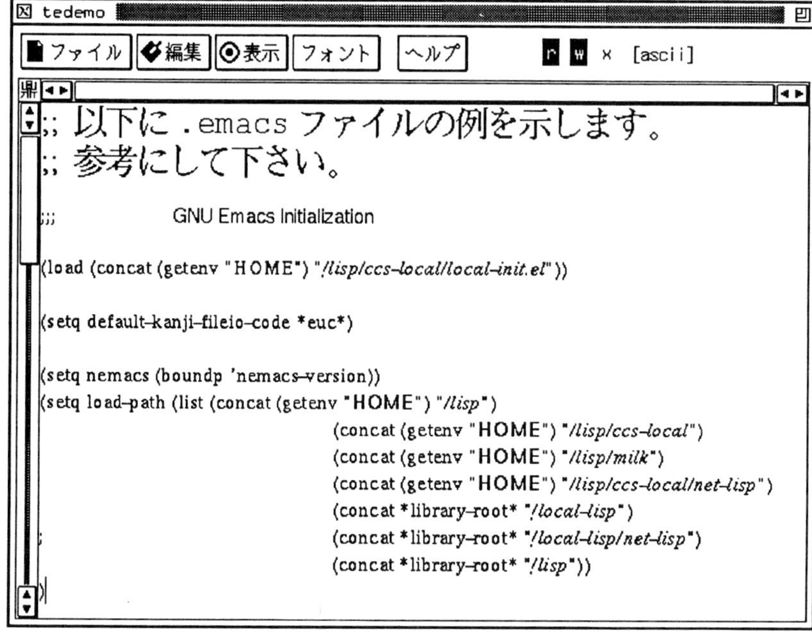


**Figure 12.3** Text editor example.

on that of PostScript (Adobe Systems Inc.,1985) [6]. Some of the functions of these components are filling, grouping, making halftone patterns, rotating and making arbitrary linear transformations. Each picture item can be independently colored or filled with patterns.

The picture editor components are also used as a lower library for image editor components, table editor components, graph editor components and hierarchy editor components. A description of these follows.

An example of the kind of work that can be done using picture editor components can be seen in Figure 12.4.

*Image editor components.* The image editor components, in `libIe.a`, interact with bitmap data. They can read some major image formats such as GIF[7] and xwd. Their functions include scaling, rotating

[6] PostScript is a registered trademark of Adobe Systems Incorporated.

[7] GIF is a trademark of Compuserve, Inc.

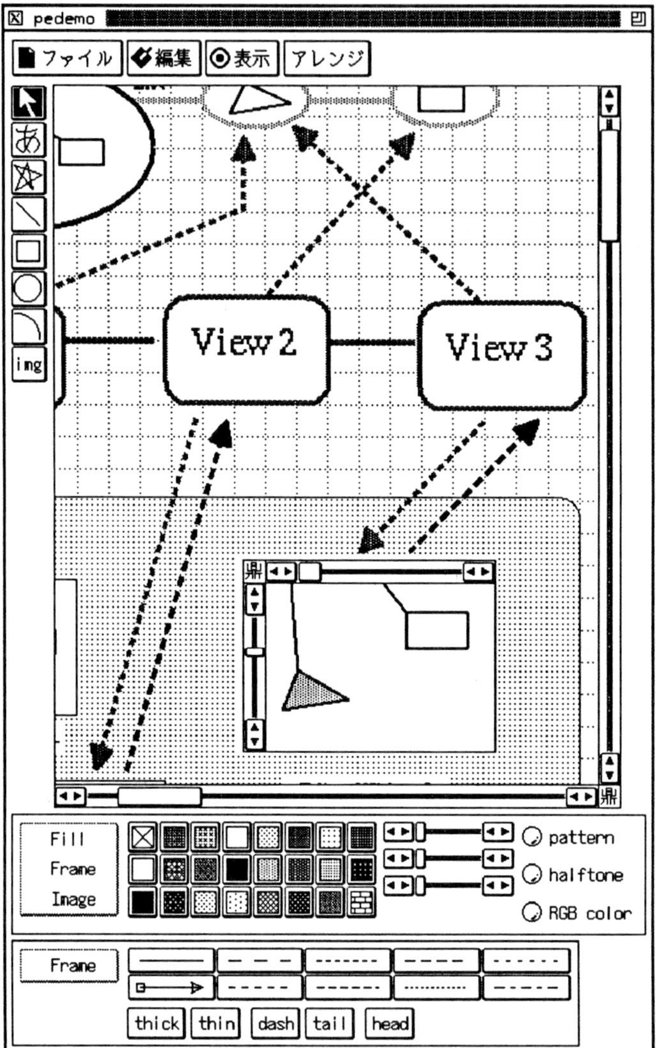


**Figure 12.4** Picture editor example.

and colormap management. Tools associated with these components edit small icons and display scanned still pictures such as portraits.

An example of image editor use is shown in Figure 12.5.

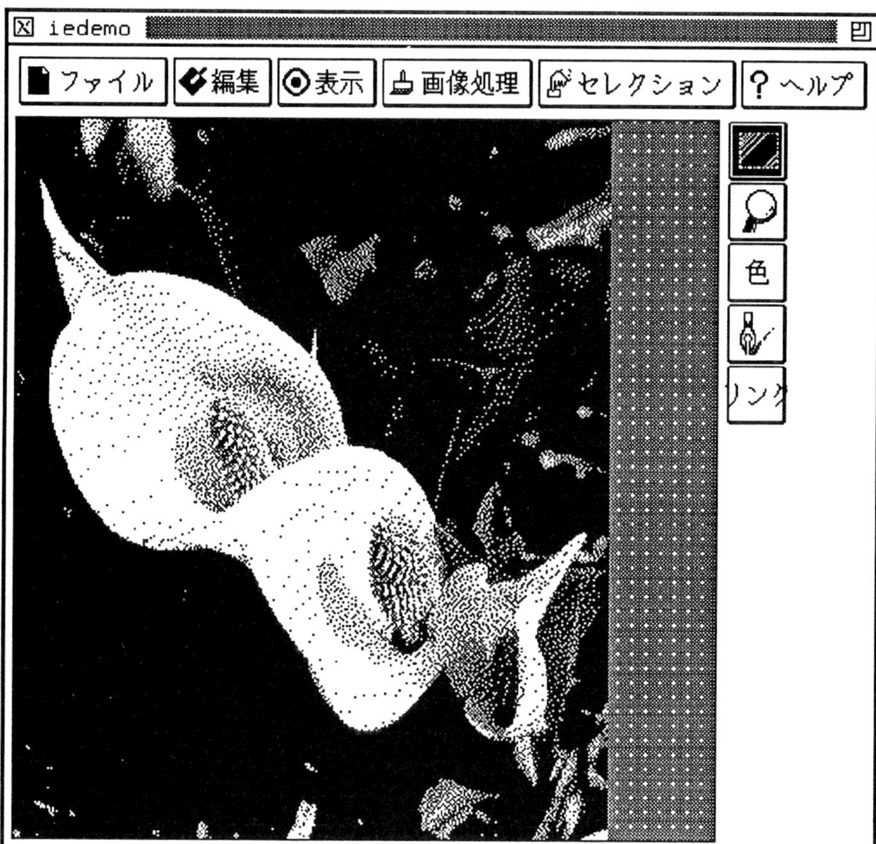


**Figure 12.5** Image editor example.

*Table editor components.* The table editor components of `libFe.a` manage the matrix structure. Each element of the matrix can be embedded as an item in any other medium. Some of the editor components' functions include changing the size of rows and columns, inserting and deleting rows and columns, and changing the position of each element within a cell. However, functions considered application tasks are not done by the table editor components. For example, the table editor components do not compute the total of each column.

Among other uses of table editor components are UI with relational databases, and spreadsheet applications.

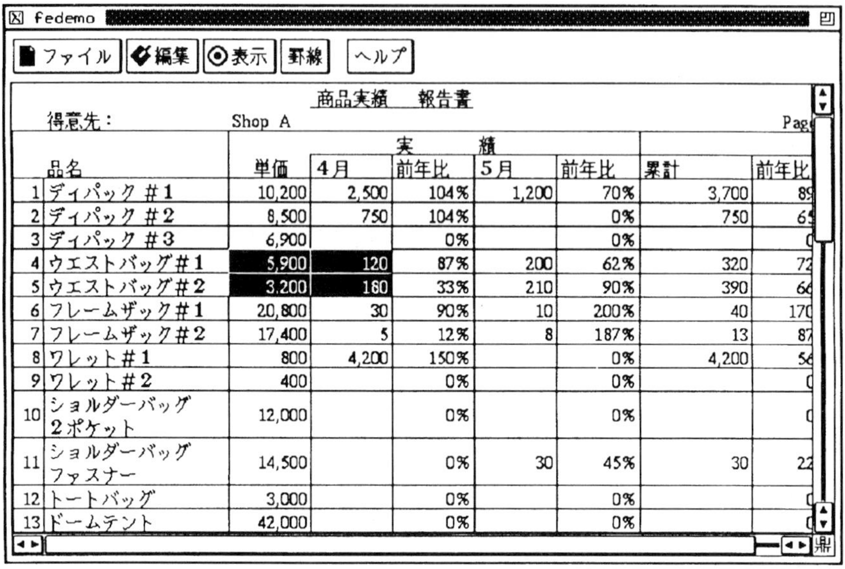

fedemo

ファイル 編集 表示 罫線 ヘルプ

商品実績　報告書

得意先：　Shop A　Pag

| | 品名 | 単価 | 4月 | 前年比 | 5月 | 前年比 | 累計 | 前年比 |
|---|---|---|---|---|---|---|---|---|
| 1 | ディパック #1 | 10,200 | 2,500 | 104% | 1,200 | 70% | 3,700 | 89 |
| 2 | ディパック #2 | 8,500 | 750 | 104% | | 0% | 750 | 65 |
| 3 | ディパック #3 | 6,900 | | 0% | | 0% | | 0 |
| 4 | ウエストバッグ#1 | 5,900 | 120 | 87% | 200 | 62% | 320 | 72 |
| 5 | ウエストバッグ#2 | 3,200 | 180 | 33% | 210 | 90% | 390 | 66 |
| 6 | フレームザック#1 | 20,800 | 30 | 90% | 10 | 200% | 40 | 170 |
| 7 | フレームザック#2 | 17,400 | 5 | 12% | 8 | 187% | 13 | 87 |
| 8 | ワレット#1 | 800 | 4,200 | 150% | | 0% | 4,200 | 56 |
| 9 | ワレット#2 | 400 | | 0% | | 0% | | 0 |
| 10 | ショルダーバッグ 2ポケット | 12,000 | | 0% | | 0% | | 0 |
| 11 | ショルダーバッグ ファスナー | 14,500 | | 0% | 30 | 45% | 30 | 22 |
| 12 | トートバッグ | 3,000 | | 0% | | 0% | | 0 |
| 13 | ドームテント | 42,000 | | 0% | | 0% | | 0 |

**Figure 12.6** Table editor example.

An example of Canae table editing is shown in Figure 12.6.

*Graph editor components.* The graph editor components found in `libGe.a` deal with general network data structures composed of nodes and arcs. Some graph editor component features are listed below.

- Shapes of nodes can be arbitrarily defined for each application.
- Constraints related to the points where nodes and arcs are connected can be defined according to the application. For example, a triangle-shaped node can be defined so that arcs can only be connected at its vertices.
- Labels can be defined for each node.
- Nodes can be grouped. Thus, a group of nodes can become a single new node.
- Styles of arcs can be defined for each application. For example, thick, thin, dotted, broken, and arrow lines are available.
- Arcs can be bent acutely, or curved as Bezier curves.
- Labels and nodes can become data for other media.

The structure-based graph editor components define all node-arc connections and the relative position of all nodes and their labels. All are automatically maintained as a unit unless they have been explicitly changed. Therefore, if a node moves, its label(s) and related arcs follow it.

Some typical applications for graph editor components are in data flow diagrams, flow charts, and state transition diagrams.

An example usage of graph editor components is shown in Figure 12.7.

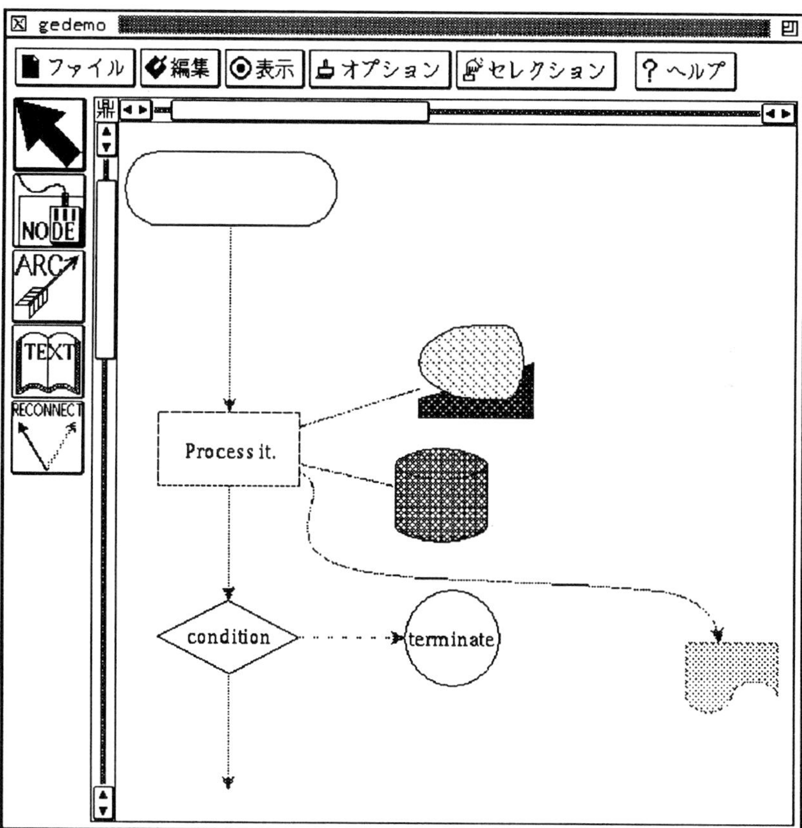


**Figure 12.7** Graph editor example.

*Hierarchy editor components.* The hierarchy editor components are included in `libHe.a`. These components build tree data structures. The nodes of a tree are arranged left to right, and top to bottom. The shape of each node and arc can be defined for specific applications. Nodes can be moved upward (visually to the left, towards the root node of the tree), downward (visually to the right, away from the root node), or arbitrarily to any place on the tree. Each node can contain objects pulled from any other media.

Some typical applications for this library are in outline processing, and in module hierarchy diagramming.

An example of tree drawing is shown in Fig 12.8.

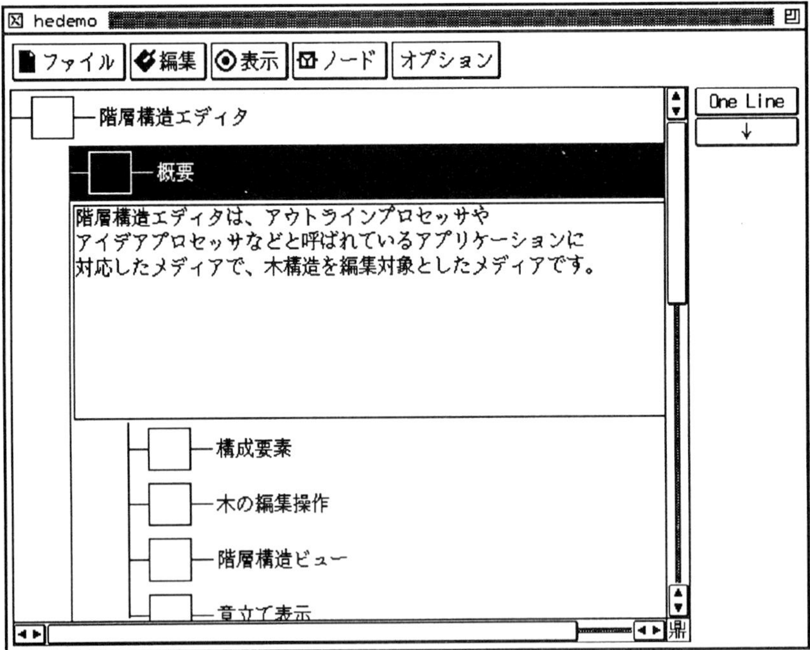


**Figure 12.8** Hierarchy editor example.

### 12.2.3 Event handling

Canae has special mechanisms which simplify and standardize event handling throughout all of the media. These mechanisms are called *KeyMap* and *EventMap*

A KeyMap can be described as a table which shows key input sequences and their corresponding actions. An EventMap can be described as a table which shows mouse-related event types (press, release, etc.) and their corresponding actions.

Some of the key assignments are common among editor components. For example, control-W is defined as the cut operation, on every editor. Consequently, all Canae applications automatically have the same user interface unless the application programmer suppresses it.

Moreover, the editor components for each medium have standard definitions for keyboard use and for mouse use. For example:

- The text editor components define the standard key assignment for Japanese text entry, e.g. control-O for toggling the Alpphanumeric mode and Japanese mode.
- The graph editor components have standard definitions on mouse usage, e.g. press for selecting a node, drag with control key for curving an arc.

These standard definitions are embedded in the editor components in default. The application programmer, however, can suppress them and make alternative definitions.

KeyMaps and EventMaps are hierarchically attached to all widgets (Figure 12.9). For example, if an Editor Widget is used for the text medium, it has a system-wide KeyMap definition (e.g. control-W). This is found at the bottom and has the lowest priority. A medium-dependent standard definition (e.g. control-O) is at the next level up. The user can add KeyMaps on these standards for any given application. The additional definition can then override the standard.

The overriding definition can be described simply. For example, if the action for doubleclicking the mouse button is defined as C function `foo()` it would be described as

```
CxAddEventMap(w,
  CxCreateEventMap(CX_DOUBLECLICK, foo, NULL, NULL),
  NoEventMask, 1, NULL);
```

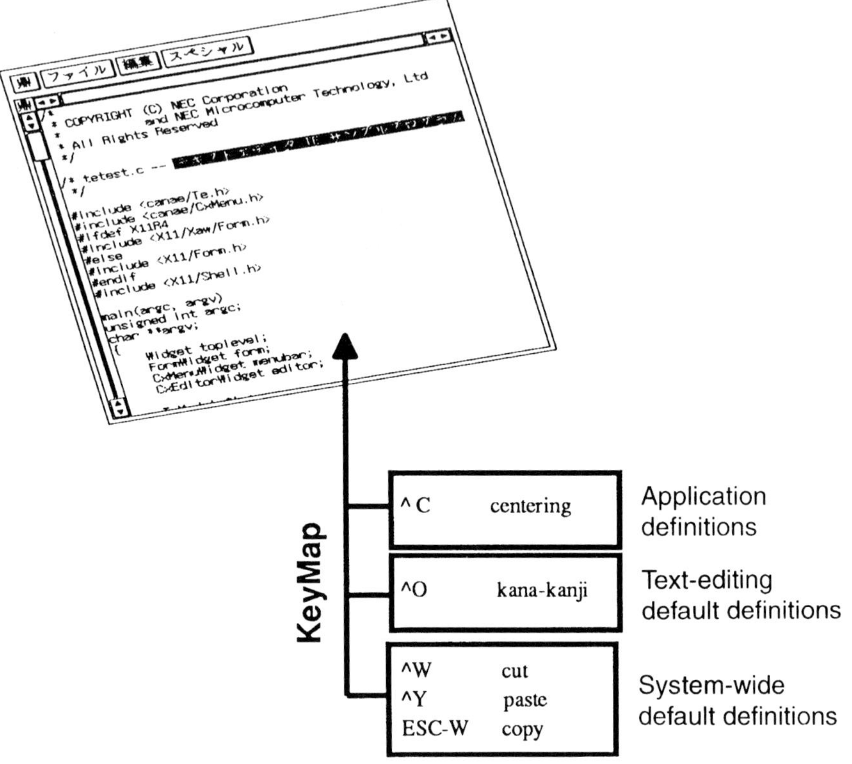


**Figure 12.9** Hierarchy of KeyMap.

Here, `CxCreateEventMap` is the function to create a new EventMap, and `CxAddEventMap` attaches the EventMap to widget `w`.

## 12.3 Editor architecture

### 12.3.1 Modified MVC model

Canae's editor architecture design is based on Smalltalk-80's MVC model (Goldberg and Robson, 1983). It has been modified to fit the X Toolkit environment and to add to its effectiveness. Our model consists of Model, View and Editor Widget.

The Canae design format is shown in Figure 12.10. The oval in the upper left corner is Model. It represents the data being edited. In this case, the Model is a Graph Model. The Model has a View List with three Views: View 1, View 2, and View 3. They are Graph Views, which can handle only Graph Models. Each View is connected to an Editor Widget. A View watches the Model and draws the Model on the connected Editor Widget. An Editor Widget receives events (e.g. exposure events) from the X Server, and sends it to the View. The Model also has a Selection List. Each Selection represents some part of the Model. For example, View can have a Selection to represent the current editing point. That Selection is called the View's *current selection.*

In the following subsections, the modified MVC model is described in detail.

*Model*

A Canae's Model represents the material being edited. A Model is medium-specific–i.e., there are six types of Model corresponding to each medium. A Model can be watched by multiple Views with multiple Editor Widgets.

In each medium, functions to modify Models have been designed. If a Model is changed by functions or by user interaction, the change is reported to the Views.

Some functions are common to six media. The interface of these functions are uniformly defined. The functions are, for example:

- allocate memory for a Model, or a Selection;
- convert a Model to a Fragment;
- undo changes;
- cut/copy/paste a Selection from/to a Model.

*Selection*

A Selection represents a part of the Model. A Model can have more than one Selection. They are maintained as a Selection List.

There are also six types of Selections for each medium. A Model can only have Selections of the same medium. For example, a Text Selection represents a continuous set of characters. A Graph Selection represents a set of nodes, arcs, and labels.

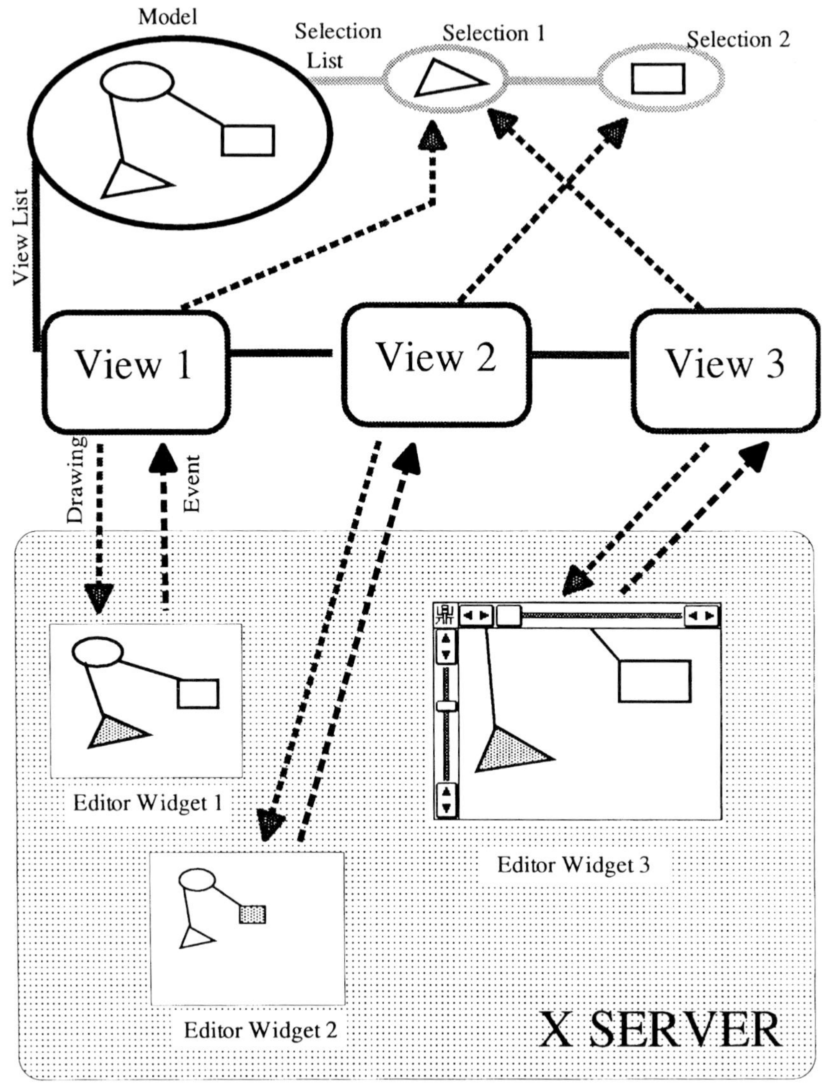


**Figure 12.10** Editor architecture.

Selection is used for View's current selection, and the connection point of hyperlinks.

*Editor widget*

All Canae editors are implemented with only one kind of widget, i.e. the Editor Widget. The Editor Widget is a kind of X Toolkit Widget, i.e. it corresponds to a window (or a subwindow) on the display screen.

The Editor Widget realizes the following common features among all editor libraries.

- *Event handling*: when the widget receives key, mouse, exposure and other events, the Editor Widget sends it to the View.
- *Scrolling*: when the size of drawing of Model is larger than the widget itself, scrollbars are automatically attached to the Editor Widget. Scrolling functions are automatically called when the user manipulates the scrollbars.

  Scrolling is realized independent of medium, because scrolling can be implemented as a bitmap copy on the window. An exposure event is created to redisplay the newly emerging part of the Model. The exposure event is later processed by each medium's redisplay function.
- *Scaling*: the Editor Widget manages the scaling information. The Model is displayed according to the scaling factor. For scaling, the Editor Widget simply changes the scaling factor. Each medium's redisplay function later refers to the scaling factor, and draws the Model according to that factor.
- *GC management*: Graphic Context (GC) (Nye, 1990) is important data for graphical functions, defined by the X library. GC for the editor is managed by the Editor Widget. Such GC functions as the foreground color and style of lines can be independently defined by each editor.

Though the Editor Widget corresponds to Smalltalk's controller, it has more functions than the controller. Smalltalk's controller can only handle events. The Editor Widget has the features described above, including scrolling and scaling. In Smalltalk-80's MVC model, these features are assigned to view. While Smalltalk-80 has many types of controllers according to the type of view, there is simply one kind of Editor Widget in Canae.

*View*

A View structure connects a Model and an Editor Widget. A Model can have two or more Views, which are linked as a View List. There are six types of View, corresponding to each medium. A Model can only have the same kind of Views.

The View watches the Model. When a Model is changed by some editing operations, the changes are accumulated at the View. The changes are not reflected on the Editor Widget until other events have been processed. This mechanism avoids annoying frequent redrawing on the display.

The View maintains coordinate information. When the Model is larger than the drawable size on the Editor Widget, the View draws only a part of the Model. The View changes the data of coordinates when the user drags the scrollbar, when a scrolling function is called, when the size of Editor Widget is changed, or when the scaling factor is changed.

As described above, the View recieves some events from the Editor Widget. KeyMaps and EventMaps are managed by View. When key events and mouse events are received, corresponding actions defined by these maps are invoked.

Some of the functions involved with View are common among all media. Redisplay, initialization and model handling functions are examples. These functions have a compatible interface among media.

### 12.3.2 Fragment

Fragment is the common data format in the Canae system that represents any Model. It is defined as an encapsulated byte-stream format which is independent of machine architectures. Fragment is used to file Models, to transfer media data via networks, and to mix media in a document.

A Fragment has a *shell* and a *body*. The Fragment shell has information including (1) the type of Fragment (i.e. media type), (2) the size of the Fragment itself and (3) the drawing size. The Fragment body represents the rest of the information, the Model data itself. A Fragment can be nested in another Fragment (Figure 12.11).

Figure 12.12 illustrates a typical usage of Fragment.

When a node of a graph is cut to the Clipboard of X Window Server, the cut node is encoded into a Graph Fragment. It can be pasted onto

Fragment shell

Fragment type

Fragment size

Drawing size

Fragment body

= Model data

Nested fragment

**Figure 12.11** Structure of Fragment.

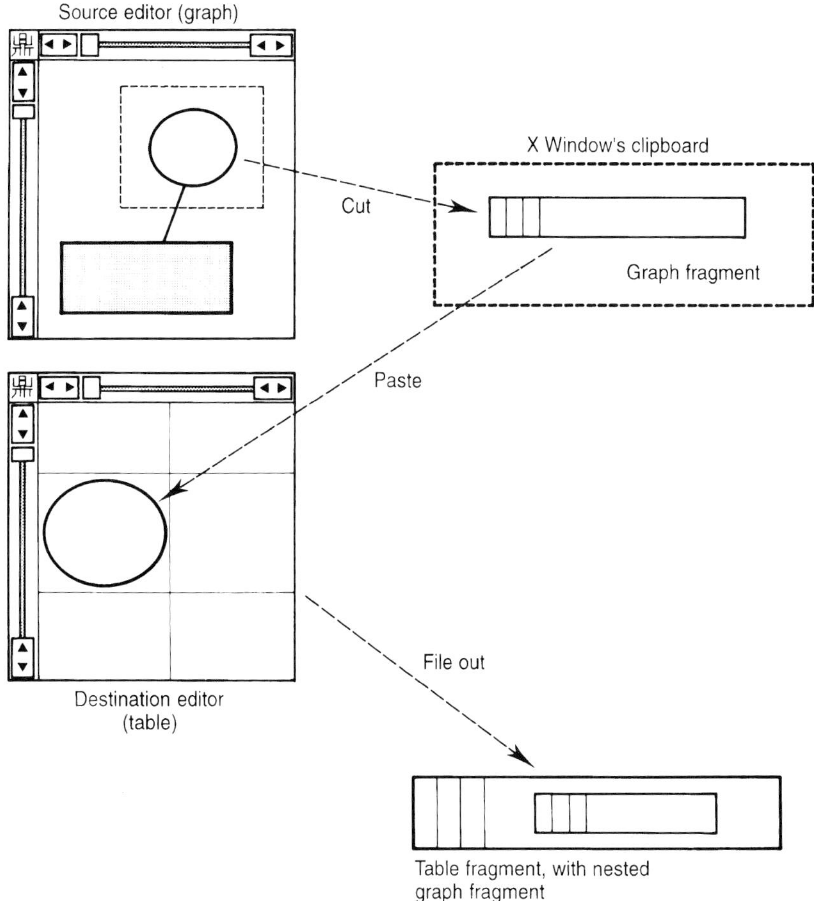


**Figure 12.12** Usage of Fragment.

another Model, a Table Model in this example. On the paste operation, the Table Model gets the Graph Fragment from the Clipboard, and insures that there is enough space in the cell for the size of the drawing Fragment. The Fragment is drawn (the oval in the figure) by the *draw_fragment* function, which is one of the common interface functions among all media. This means that any Graph Models or Graph Views are not necessary at this moment.

If the user wants to edit the nested data, the oval graph node in this example, (s)he can do it by double-clicking the target data. Then the nested Fragment is decoded to a Model, and a View and an Editor Widget is created at the location. Figure 12.13 is another example. This is a hardcopy of the editor in this situation. in this figure, a Hierarchy Model is nested into a Table Model and is now editable with the nested Editor Widget.

When the nested editing is done, the user should click somewhere out of the nested editor. The nested Model is then encoded back to a Fragment again.

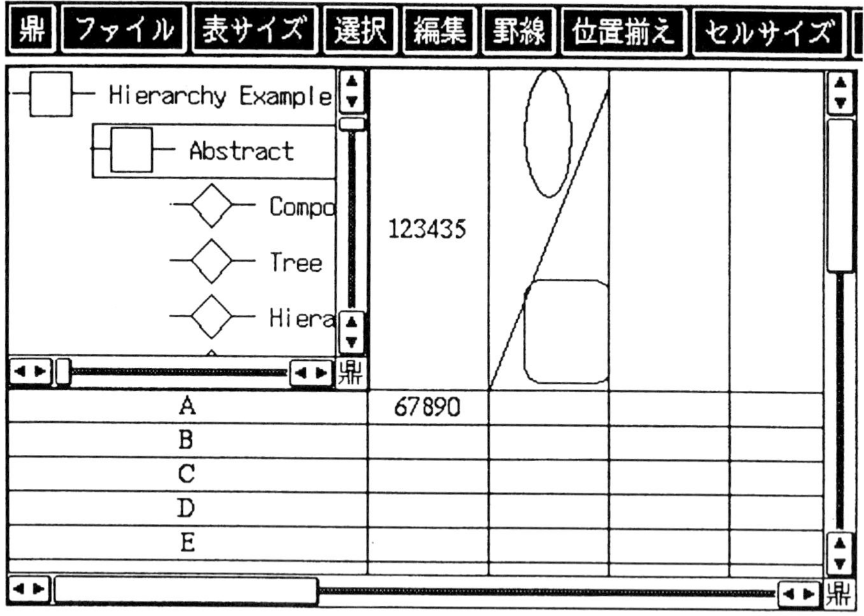


**Figure 12.13** Nested media.

Fragment is also used for filing. Look back to Figure 12.12. When the Table Model is filed out, the Table Model is encoded into the Fragment format, with a nested Graph Fragment in it.

A Model can become a Fragment and vice versa in the Canae system. Although the computational cost of conversion is an important factor to be considered, no problems have been observed.

### 12.3.3 Hyperlink

Canae's hyperlink feature is implemented by simply giving additional data to Selections. If the additional data is a file pathname, the file is linked to the Selection. If the additonal data is a pointer to another Fragment, the hyperlink is connected to an on-memory Fragment.

Generally, the Selection representing hyperlink is not a current selection of any View. View also has a customizable function to highlight this kind of Selection.

To traverse a hyperlink, the user should double-click the linking point. The function to traverse is defined at the EventMap.

Canae's hyperlink has another convenient feature. By giving a Lisp expression as the additional data, the traversing function just executes the Lisp expression.

### 12.3.4 Printing

Canae prepares functions to encode Fragments into PostScript. Consequently, any document can be printed, even if it has nested-media structure.

## 12.4 Practice experience

### 12.4.1 Application examples

The following list shows some of the applications for which UI's have been implemented or designed with Canae, according to information from users:

- a project schedule management system;
- software specification design systems;

- a database management system for software development (*Kyoto*DB (Matsumoto and Ajisaka, 1991);
- a machine translation system;
- knowledge editors;
- CAM systems;
- CAD systems (LSI, circuit);
- an instruction manual with hyperlinks;
- a support system for database design;

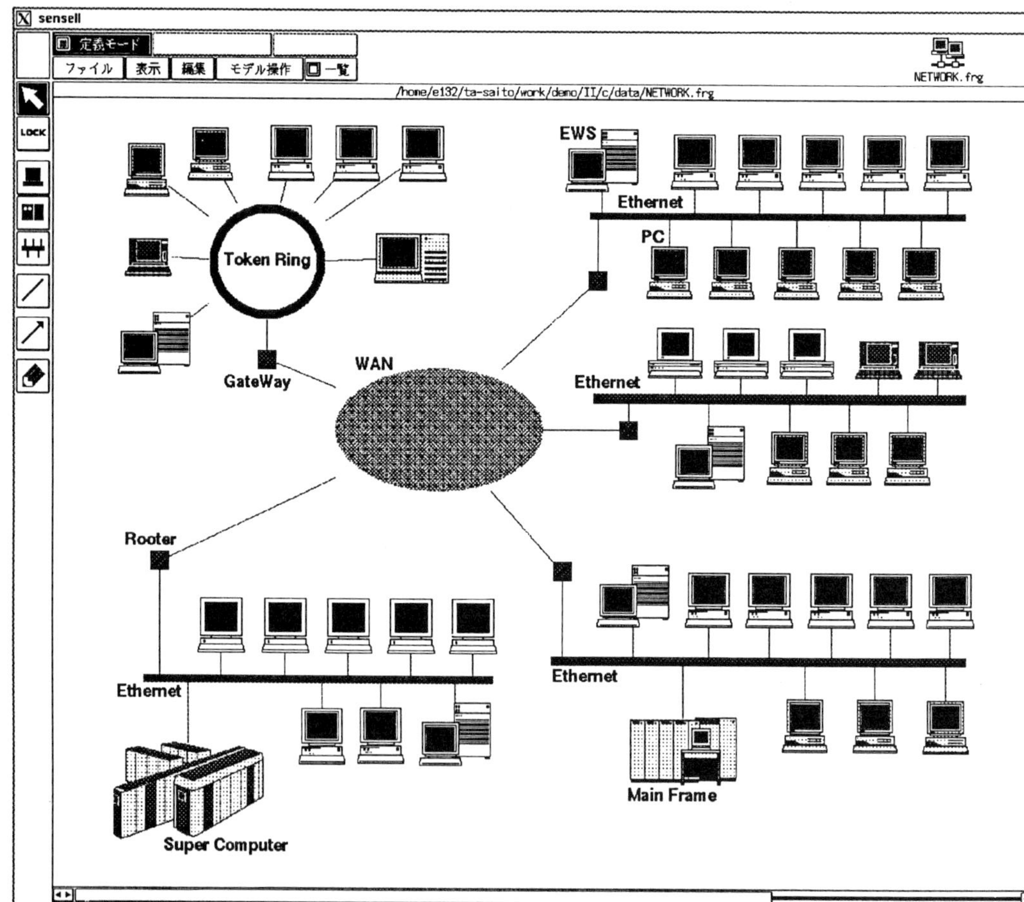


**Figure 12.14** Application sample 1.

- a text correction support system for co-authoring;
- UI for expert systems;
- management and monitoring systems for communication networks;
- a network simulation system;
- a monitoring system for switching systems;
- a multimedia remote conferencing system (MERMAID (Watabe *et al.* 1990);
- program visulization tools;
- a graphic tool for displaying the order of amino acids in protein.

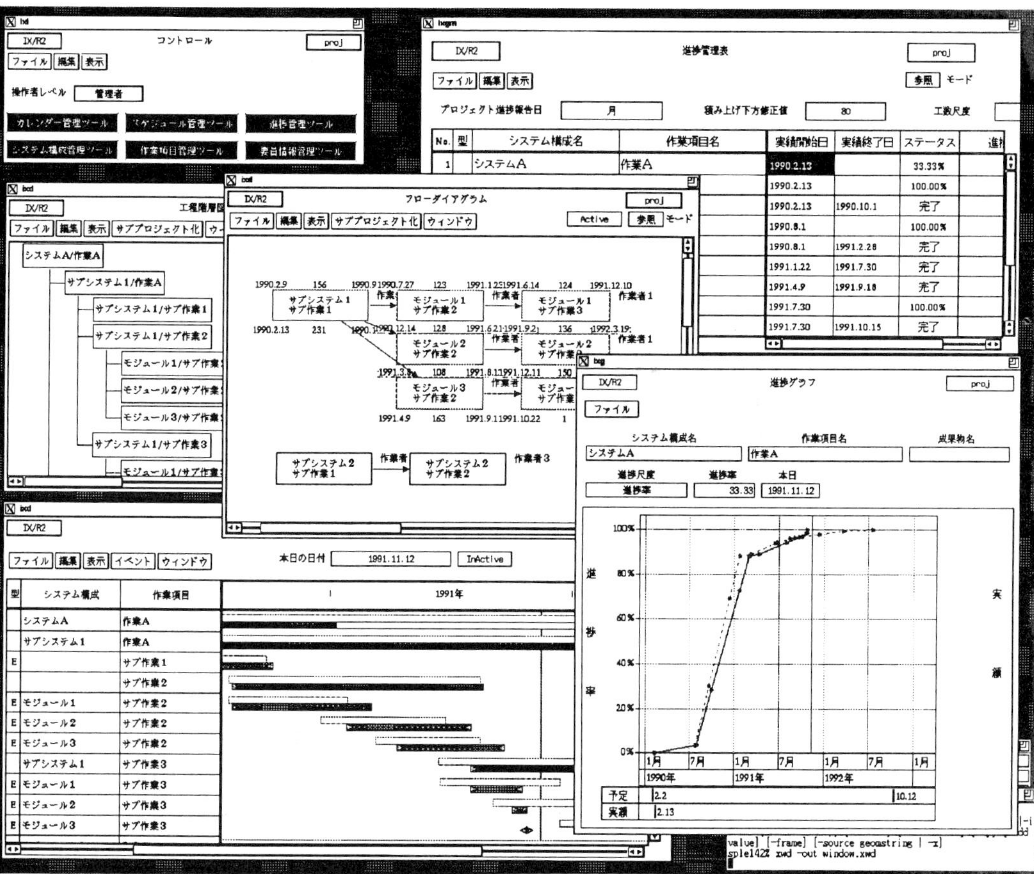


**Figure 12.15** Application sample 2.

Sample windows from some of the above systems are shown in Figure 12.14 and 12.15.

### 12.4.2 Effectiveness as software components

We have made a survey of the users' subjective measurement of cost effectiveness of Canae. Project leaders of Canae users commented that 33–50% of development cost is saved by adopting Canae. This number suggests excellent user satisfaction.

We have also made an objective measurement of cost effectiveness by counting source code lines.[8] We picked up three CASE applications of Canae and counted source code lines of application codes and linked Canae codes.

The ratio $r_1$ below shows the reusability of Canae's toolkit:

$$r_1 = \frac{c}{a + c}$$

where $a$ is the number of application code lines and $c$ is the number of Canae module lines linked to the application.

The investigation showed that $r_1$ was 52–70%. Canae's library code is written with redundancy to make it modular. Taking this redundancy into account, the investigation still shows that the programmer would have to write about double the number of code lines without Canae. It also shows that UI programming acconts for a large part of tool development.

Another investigation was undertaken to clarify how reusable editor components are.

$$r_2 = \frac{e}{c + x}$$

where $c$ is the number of Canae module lines linked to the application; $e$ is the number of source code lines for editor components in each medium $c$'s subset; and $x$ is the number of Athena Widget and X Toolkit code lines linked to the application.

$r_2$ represents the ratio of editor codes in all the UI codes, and was

[8] A detailed description of this investigation is given in Tarumi *et al.* (1990).

33–39%. This means that providing editors as software components, which is our key idea, is successful, at least in the developoment of these three CASE tools.

Of course, the success is not limited to these three tools. In the subjective measurement above, project leaders commented that at least one third of development cost is saved by Canae. This number and the wide application areas shown in Section 12.4.1 made us believe the success of Canae.

### 12.4.3 Example codes

Figure 12.16 is an example of a code with Canae functions, showing the power of this UI Toolkit.

The function `DoMagScale` is a menu item callback function. In other words, when a menu item to which this function is attached is selected by the user, this function is called.

`DoMagScale` increases the scaling factor by 1.25 times. `Cx-EditorWidget` is the data type representing the Editor Widget. The first function `CeSetScaleOpt` declares that the Editor Widget allows any scaling factor. `CeScalePower` retrieves the scaling factor of the `editor`, and `CeSetScalePower` sets the scaling factor. The redundant argument `d2` is specified because the menu's callback functions take three arguments, but it is not used here.

Nothing more is necessary to change the scaling factor, because View has a drawing function referring to the scaling factor of the Editor Widget.

Figure 12.17 shows how to make a graph editor with Canae.

```
void DoMagScale (widget, d1, d2)
Widget widget;
XtPointer d1, d2;
{
    CxEditorWidget editor = (CxEditorWidget)d1;

    CeSetScaleOpt(editor, CX_ANYSCALE);
    CeSetScalePower(editor, CeScalePower(editor) * 5/4);
}
```

**Figure 12.16** Example of a code with Canae functions.

```
CxModel     *model;      /* the Model */
CxView      *view;       /* the View */
Widget      editor;      /* the Editor Widget */
Widget      form;        /* parent widget of editor */
CxSelection *selection; /* a Selection*/

/* after some initialization . . . . */

/* create the editor widget */
editor = CxMakeManagedWidget
        ("GeEditor", /* name of the
         widget */
         cxEditorWidgetClass, form,
         XtNwidth, 600, /* editor's size */
         XtNheight, 600,
         NULL);

/* create graph model, graph view, and graph selection */
model = CxCreateModel(CX_GRAPH);
view = CxCreateView(CX_GRAPH);
selection = CmCreateSelection(model);

/* connect editor, model, and view */
CeSetMV(editor, model, view);
/* define current selection of view */
CvSetCurrentSelection(view, selection);
```

**Figure 12.17** Example of a graph editor with Canae functions.

# 12.5 Discussion

## 12.5.1 Field of applications

Canae is not a general purpose system, because it supports only six data structures. Actually Canae was designed with the assumption that its application area would mainly be in CASE and OA tools. Even though the actual field of application has been much wider than we had originally anticipated, it is still limited.

This limitation is not a shortcoming for Canae, because it is due to Canae's focus that it is so successful at supporting high-level management of structured objects.

### 12.5.2 Canae as UIMS

In a broader sense, Canae is a User Interface Management System (UIMS) because Canae includes UI toolkits.

Conventional UIMSs only cover the tasks of managing user inputs, and displaying the results of computations. Editing operations and management of editable objects are regarded as responsibilities of application codes. In the survey by Hartson and Hix (1988), there is no UIMS which manages editable objects. According to Young, UIMS should not assume data structures of applications (Young *et al.*, 1988).

However, editable objects are stongly related to UI. For example, dragging a graph node changes data both at the UI level and at the application level. A UIMS with editors is, in this sense, quite reasonable.

This method brings about duplicated management of the objects. Canae can handle only Canae's Model data structure as editable objects. Applications have their own data structure. This problem is resolved by providing a hooking mechanism. The programmer can define hook functions, which are called before and after the modification of Canae's Model. In addition, Canae provides many other kinds of hooking functions, to encourage modularized and object-oriented design.

Constraint-based UIMS (Nelson, 1985; Myers *et al.*, 1990) is another effective approach to realize higher level UIMS. With constraints, relationships among objects can be described generally, and the data structure is kept to be shown as relationships among objects. A graph editor like Canae's can be implemented with constraints, by creating constraints between nodes and arcs.

However, to describe constraints for complex data structures is an additional burden for application programmers. Constraint is not considered to be a pragmatic approach for a large body of complex data. Although the structure-based editor approach is not as generally applicable as the constraint approach, our experience with Canae's great versatility demonstrates its pragmatism.

### 12.5.3 Productivity

As we demonstrated earlier, Canae reduces the number of required application codes. This apparently improves the productivity (including maintenance cost) of application software.

However, there are many factors related to the issue of productivity. One of the most important factors is the cost of learning to use the X Window System, the X Toolkit and Canae. It is difficult to get meaningful data on these costs, because each programmer has different experiences with such systems. If Canae is not applied to the applications we have shown, appliction programmers must at least learn the X Window System and the X Toolkit, provided they want to use the X Window System.

We estimate that Canae may possibly increase the total learning cost, but not seriously, because Canae cuts the costs of learning the Xlib level primitive functions. With Canae, application programmers do not have to learn these. In fact, in the survey of subjective measurement of Canae's effect shown in Section 12.4.2, project leaders commented that 33–50% of development cost is saved by Canae, taking the learning costs into account.

Practically speaking, in most cases application programmers learn Canae by reading simple example application programs provided with the Canae libraries. Such examples have made not only Canae, but also the X Window System, easier to learn.

## 12.6 Conclusion

Canae's editor components have been designed, implemented and adapted to many applications. Designing editors as reusable software components appears to be a successful approach to software engineering.

The editor architecture of Canae is flexible, extensible and customizable as software components.

The editor components can be used from a Lisp environment, or a Lisp environment can be embedded into Canae's libraries. A UI designing tool, YUZU, has also been designed and implemented.

## Acknowledgments

The authors would like to thank Mr Tetsuo Saya, Mr Kenji Kawakita and Mr Osamu Shigo for granting this research project and for their helpful suggestions. The authors would also like to thank all the project members who developed Canae applications for giving the authors quantitative data and helpful comments. We thank Mr Tateki Sano, who is a member of the Canae project, for preparing the survey for subjective measurement of Canae's effect. Finally we appreciate the contribution of all the project members of Canae.

## References

Adobe System Inc. (1985). *PostScript Language Reference Manual*. Addison-Wesley.

Goldberg, A. and Robson, D. (1983). *Smalltalk-80, The Language and its Implementation*. Addison-Wesley.

Hartson, H. R. and Hix, D. (1988). Human–computer interface development: concepts and systems for its management. *ACM Computing Surveys*, **21** (1): 5–92.

Matsumoto, Y. and Ajisaka, T. (1991). A computer-aided software requirements engineering environment: Kyoto DB-1. In Ohno, Y. (editor), *Distributed Environments – Software Paradigms and Workstations*, pp.20–38 Springer-Verlag,

Myers, B. A., Guise, D. A., Donnenberg, R. B., Zanden, B. V., Kosbie, D. S., Pervin, E. and Marchal, P. (1990). Garnet: comprehensive support for graphical, highly interactive user interfaces. *IEEE Computer*, **23** (11), 71–85.

Nelson, G. (1985). Juno, a constraint-based graphics system. *ACM Computer Graphics*, **19** (3): 235–243.

Nye, A. (1990). *Xlib Programming Manual for Version 11 of the X Window System*. O'Reilly and Associates, California.

Nye, A. and O'Reilly, T. (1990). X Toolkit Intrinsics Programming Manual OSF/Motif 1.1 Edition for X11, Release 4. O'Reilly and Associate, California.

Stallman, R. M. (1981). Emacs: The extensible, customizable, self-documenting display editor. In *Proceedings ACM SIGPLAN/SIGOA Conference on Text Manipulation*, Portland, OR, pp.147–156.

Tarumi, H. Rekimoto, J., Sugai, M., Yamazaki, G., Sugiyama, T. and Akiguchi, C. (1990). *Canae – a user interface construction environment with editors as software parts. NEC Research and Development*, No. 98: 89–98.

Watabe, K., Sakata, S., Maeno, K., Fukuoka, H. and Ohmori, T. (1990). Distributed multiparty desktop conferencing system; MERMAID. In *Proceedings of Computer Supported Cooperative Work (CSCW)'90, ACM,* pp. 27–38.

Young, M. Taylor, R. N. and Troup, D. B. (1988). Software environment architectures and user interface facilities. *IEEE Transactions on Software Engineering* **14**(6): 697–708.

# Index

**D**



**E**



**F**



**G**

**H**



**I**



**K**



**L**



**M**

**N**



**O**



**P**



**Q**



**R**



**S**

## T



## U

**V**



**W**



**Z**